GEOTECHNICAL SPECIAL PUBLICATION NO. 88

ANALYSIS, DESIGN, CONSTRUCTION, AND TESTING OF DEEP FOUNDATIONS

PROCEEDINGS OF THE OTRC '99 CONFERENCE

HONORING
Lymon C. Reese

CO-SPONSORED BY
The Offshore Technology Research Center
Geo Institute of the American Society of Civil Engineers

April 29-30, 1999
Austin, Texas

EDITED BY
Jose M. Roesset

1801 ALEXANDER BELL DRIVE
RESTON, VIRGINIA 20191–4400

Abstract: This proceedings, *Analysis, Design, Construction, and Testing of Deep Foundations*, consists of papers presented at the OTRC '99 Conference held in Austin, Texas, April 29-30, 1999. While deep foundations are used extensively for the support of structures in both the terrestrial and marine environments, significant uncertainties still exist in predicting the field behavior of such foundations. This proceedings presents some of the latest research done in this area. The papers present research on such topics as the mechanics of pile foundation collapse, three-dimensional analysis, pile driving techniques, and suction caisson foundations.

Library of Congress Cataloging-in-Publication Data

OTRC'99 Conference (1999: Austin, Texas)
    Analysis, design, construction, and testing of deep foundations: proceedings of the OTRC'99 Conference: honoring Lymon C. Reese / co-sponsored by The Offshore Technology Research Center, Geo-Institute of the American Society of Civil Engineers.
        p. cm.
    Includes bibliographical references and index.
    ISBN 0-7844-0422-4
    1. Foundations—Design and construction—Congresses. 2. Piling (Civil engineering)—Congresses. 3. Strains and stresses—Congresses.
    I. Reese, Lymon C., 1917- . II. Offshore Technology Research Center. III. American Society of Civil Engineers. Geo-Institute. IV. Title.
    TA775.O858 1999
    624.1'5–dc21   99-12876
                                                                              CIP

Geotechnical Special Publications

Foreword

Deep foundations are extensively used for the support of structures in both the terrestrial and marine environments. In spite of extensive research over a period of many decades, there remain significant uncertainties in predicting the field behavior of such foundations. This symposium was organized to honor Dr. Lymon C. Reese, an international authority in deep foundations and an originator of the p-y method widely used in predicting behavior of deep foundations under lateral loading. This symposium is intended to present some of the latest research into analysis, design, construction, and testing of deep foundations. The proceedings include **8** invited lectures by international authorities as well as 11 submitted papers. Other papers that were received too late for publication will be distributed at the conference.

International Steering Committee

Chandra Desai	University of Arizona
Earl Doyle	Shell Deepwater Development, Inc.
John Focht	Consultant
Michele Jamiolkowski	Polictenico di Torino
Richard Jardine	Imperial College
Kjell Karlsrud	Norwegian Geotechnical Institute
Hudson Matlock	Consultant
J. Don Murff	Exxon Production Research
Larry Olson	Olson Engineering
Mike O'Neill	University of Houston
Harry Poulos	Coffey Partners Int., Pty. Ltd.
Mark Randolph	University of Western Australia
Ian Smith	University of Manchester
W. F. Van Impe	University of Gent
Andrew Whittle	Massachusetts Institute of Technology

Local Organizing Committee

Wayne Dunlap	OTRC, College Station
Robert Gilbert	OTRC, Austin
Roy Olson	OTRC, Austin
Jose M. Roesset	OTRC, College Station
Ken Stokoe	OTRC, Austin
Richard Tucker	CII, Austin
Stephen Wright	OTRC, Austin

Biography

Dr. Lymon Reese is the Nasser I. Al Rashid Chair Emeritus and Professor of Civil Engineering at the University of Texas at Austin. Lymon spent much of his early life in Abilene, Texas. He attended the University of Texas at Austin, receiving his B.S. degree in 1949 and his M.S. degree in 1950. For his M.S. research, he worked with Prof. Raymond Dawson and performed load tests with instrumented piles. He continued this work at the University of California at Berkeley with Prof. H. Bolton Seed and received his Ph.D. degree in 1955. Dr. Reese spent three years on the faculty at Mississippi State University. He joined the faculty of UT in 1955, where he has remained ever since. Lymon was Chairman of the Department of Civil Engineering from 1965 until 1972, and Associate Dean of the College of Engineering from 1972 to 1979. He continues at UT as a chair emeritus and lectures in undergraduate, graduate, and seminar courses.

Dr. Reese has done extensive research in the field of geotechnical engineering, principally concerning the behavior of deep foundations. He has conducted pioneering work in performing field studies with instrumented piles and drilled shafts, and has developed analytical methods now widely used in the design of deep foundations. He has authored over 400 technical papers and reports, and presented numerous invited lectures and talks around the world. Among his best known publications are the two manuals he prepared for the Federal Highway Administration: the "Handbook on Design of Piles and Drilled Shafts Under Lateral Loads," in 1984, and the "Drilled Shafts, Construction Procedures and Design Methods," in 1988. His invited lectures include Keynote lectures at the International Conference on Calcareous Sediments in Perth, Australia, in 1988, the International Geotechnical Seminar on Deep Foundations on Bored and Auger Piles, in Ghent, Belgium, in 1988, and the Second Symposium on Geotechnical Problems in Saudi Arabia at King Saud University in Riyadh in 1989 . He was also selected to give one of the General Reports at the 1991 International Conference on Deep Foundations in Paris, France.

Dr. Reese has long been active in the American Society of Civil Engineers (ASCE) and was elected an Honorary Member in 1984. He has held various offices in the Texas Section of ASCE and was President in 1968-79. For several years he served as member of the Executive Committee of the Geotechnical Engineering Division, ASCE, and was Chairman in 1986-87.

Lymon has received many honors during his distinguished career. He was elected to the National Academy of Engineering in 1974. In 1976, he delivered the ASCE Terzaghi Lecture, and in 1983 he received the ASCE Terzaghi award. From the University of Texas, he received the Joe J. King Professional Achievement Award in 1977 and was selected as a Distinguished Engineering Graduate in 1984. He was the first person to become an Honorary Member of the International Association of Foundation Drillers in 1984 and he received the Offshore Technology Conference Distinguished Achievement Award for Individuals in 1985.

Lymon has also been a devoted husband, father, grandfather and great-grandfather. He has been married to Eva Lee (Jett) for over 50 years. They have two daughters, a son, eleven grandchildren, and three great-grandchildren, and they are very proud of all of them.

Contents

Other Contributions

Presentations that could not be included in the proceedings were:

Some Fundamental Aspects of the Performance of Piles in Clay
    K. Karlsrud

Quantifying the Disturbance Caused by Pile Installation
    A.J. Whittle

Calibration Chamber Tests and Pile Design
    M. Jamiolkowski

DEEP FOUNDATIONS IN THE FUTURE

by Lymon C. Reese,[1] Honorary Member, ASCE
and William M. Isenhower,[2] Member, ASCE

INTRODUCTION

The design and construction of deep foundations have experienced significant improvements in the past several decades and new methods are appearing frequently at present. It is of interest to note, however, that most developments are arising from the work of commercial firms and not from a concerted effort by any funding agency. The contributions of the offshore industry are continuing but at a much slower pace than formerly when their research contributed importantly to a better understanding of the behavior of deep foundations. This paper presents a overall view with respect to deep foundations of the future, applicable to both onshore and offshore structures. Plainly, an exhaustive treatment of any topic is impossible and discussion is limited to only a few significant features.

CHARACTERIZATION AND EVALUATION OF SOIL

Any comprehensive investigation of subsurface conditions must involve some boring and sampling. However, recent trends suggest that a larger effort be directed to in situ testing than for sampling for laboratory testing. In his classic report, Hvorslev (1948) showed sampling disturbance graphically and Emrich (1971) showed that wire-line techniques, common in offshore investigations, cause more disturbance than the best onshore methods. No feasible sampling method can eliminate the change in the state of stress as a sample is removed from the tube for

---

[1] Nasser I. Al Rashid Chair Emeritus, Department of Civil Engineering, The University of Texas at Austin, Austin, Texas 78712, Founding Principal, Ensoft, Inc., Austin, Texas.
[2] Project Manager, Ensoft, Inc./Lymon C. Reese & Associates, P. O. Box 180348, Austin, Texas 78718, ensoft@ensoftinc.com.

laboratory testing. The double-ring system developed by Dames & Moore (Dames, 1940) solved some problems, but sampling disturbance remained.

In situ tests alleviate problems of sample disturbance and are fast gaining prominence, especially in Europe. The tests that are in common use are the cone penetrometer test (CPT), the cone penetrometer test with pore-pressure measurement (CPTU), the pressuremeter test (PMT), and the dilatometer test (DMT).

A sketch of a modern penetrometer is shown in Fig. 1. The force on the cone is measured electronically. Not shown are transducers for measurement of the tilt of the cone and seismic parameters.

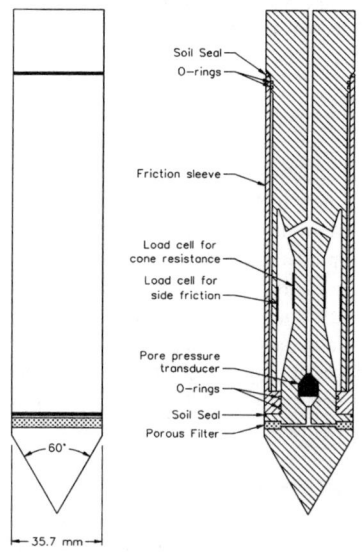

Fig. 1. Cone Penetrometer

Holeyman, et al. (1997) report that the CPT is used extensively in Belgium for "soil profiling, identification of potential end bearing layer, quantification of base and shaft resistance." The French developed extensive rules for design based on the pressuremeter test (Baguelin, et al., 1978). Van Impe and Peiffer (1997) presented data from the use of the plate-dilatometer test on the influence of the installation of screw piles on the state of stress in the soil.

Geophysical methods (Woods and Stokoe, 1985) have been used extensively. Many of these methods have utilized measurements of sonic-wave velocities, propagated in the cross-hole and down-hole directions (Hoar and Stokoe, 1977, 1981). Other methods have utilized surface-wave refraction and spectral analysis of surface waves (Stokoe et al., 1989, 1998). Geophysical methods have also been applied with some success to making tomography measurements of drilled shaft foundations (Olson, et al., 1993). Another interesting in-situ method results from the instrumentation of the drill rig; the measurement of torque and force from crowd gives information that helps to characterize the soil being drilled (De Paoli, et al., 1988).

Many of the in-situ techniques for soils testing can be extended to offshore practice. Advancing the borehole can be done with the drill string as done commonly at present. A package of in situ tools can be lowered on a wire line, the package locked into slots on the bottom of the drill string, the operation of rams controlled by transmission lines, and data can be acquired electronically. Such systems have been tried on a limited basis.

A variety of the methods noted can be expected to yield more accurate properties of soil in the future, but the engineer is cautioned to base the design of piles on properties derived from soil tests at sites where load tests were performed. For example, if in-situ methods at a site yield values of shear strength of the clay 50% higher than that from tube samples, the use of the higher values in design will plainly be unwise if the experimental data base on load transfer is based on results from tube samples. As improved methods are developed for obtaining more accurate values of shear strength, consideration can be given to the use of these methods at sites where load tests of high quality were performed. New and more accurate load-transfer parameters for pile design can then be developed.

TESTING OF FULL-SIZED PILES

Regardless of the complexity and sophistication of models, advances in predicting the response of piles to loading has been dependent on the testing of full-sized piles. Instruments are available for determining the stresses in a test pile as a function of depth; these stresses can be correlated with soil properties in devising methods of prediction.

Preparation of the system for the application of axial load, whether by anchorages or by dead weight, is time-consuming and expensive. The Osterberg cell or O-cell is (Osterberg, 1998) is a down-the-hole device that expands to apply a known load and deflection that has proven to be effective in the testing of bored piles, yielding either ultimate end bearing or ultimate side resistance. Special versions of the Osterberg cell may be used with driven piles. The O-cell, while a welcome contribution to pile testing has not displaced the need to test piles by top loading.

The testing of piles by applying a dynamic load, either with a pile hammer (Hussein et al., 1996) or by a controlled explosion (Berminghammer, 1999) have experienced wide use. To ensure that the behavior of the pile under dynamic load yields information for design under static load, the dynamic test must be done at a time when the soil around the pile has reached an equilibrium, both in terms of total stress and pore-water pressure.

Sufficient evidence is at hand to emphasize the importance of the testing of piles under axial load, using an appropriate testing technique, to validate methods of predicting the behavior of piles under axial loading. However, the use of data bases, compiled from published reports of tests of piles under axial loads, can frustrate efforts to develop correlations between load transfer and soil properties because of misleading or missing information. There is a strong need to classify axial-load tests, A, B, C and so on, depending on the completeness and quality of the data.

The system for testing of full-sized piles under lateral loading can be assembled readily. The loads will be smaller that for axial load and the reaction can be achieved by testing two piles simultaneously or by using an auxiliary anchorage. However, the performance of a proof test as for axial loading is not usually possible because the condition at the pile head, whether fixed or free or in between, cannot be made to simulate that of the production piles. Therefore, an analytical model of the pile and the soil must be used in interpreting the results of lateral-load tests.

As realistic models become known for single piles under axial load, single piles under lateral load, and pile groups under generalized loading, the results of additional high quality load tests will appear to allow better methods of prediction of the response of piles to loading. The methods should yield the loading at collapse of the system and the movements at the working load, both as a function of time.

PILE SYSTEMS AND THEIR SELECTION

Pile systems can be described by the material in the pile, the method of installation, and the method of transmitting load. The selection of one or a combination of the systems is the problem facing the engineer in planning. Considering possibility of litigation, the owner and engineer are inclined to select a "standard" foundation system and to write the construction documents to eliminate responsibility for "change of conditions." However, as costs increase and a variety of foundation systems become available, emphasis will be placed on selecting the best system for a particular application. Factors to be considered are: loading, nature and properties of soil, necessity to perform load tests, local availability of required equipment, cost, and the consequences of a failure.

Two relatively new foundation systems have appeared that justify mention: the soil-cement or GeoJet® pile (Spear, et al., 1994), and the screw pile (Van Impe and Peiffer, 1997). Sketches in Fig. 2 show the method of construction of a soil-cement pile: a drilling machine is used to create a column of soil-cement, and a structural unit is lowered into place. Depending on the kind of soil being drilled, sufficient water is added through the auger to reduce the strength of the soil cement to allow a pipe or H-section to fall into position under self weight, and sufficient cement grout is added to achieve the desired compressive strength.

An example of a screw pile is shown in Fig. 3. A section of a pile is fabricated from steel plate in the shape of a wood screw. The pile is installed by torque and crowd. A relatively large power unit is required to seat the pile to an appropriate depth.

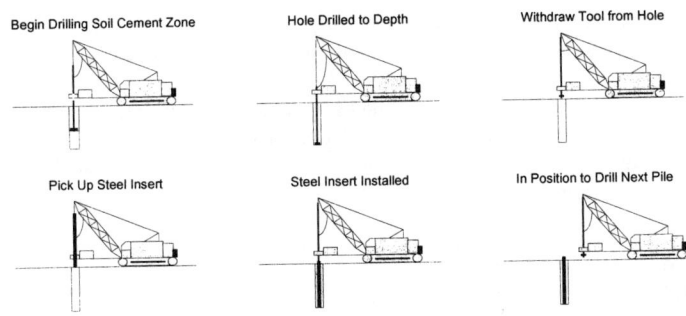

Begin Drilling Soil Cement Zone   Hole Drilled to Depth   Withdraw Tool from Hole

Pick Up Steel Insert   Steel Insert Installed   In Position to Drill Next Pile

Figure 2. Soil-Cement Pile

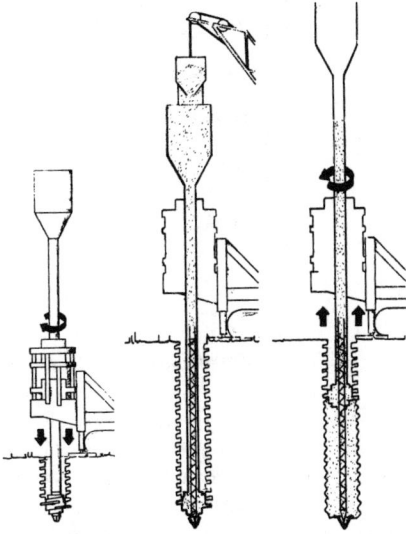

Figure 3. Example of Screw Pile

Either of these methods could be used onshore or offshore. If an open-ended pipe is driven to refusal at an offshore location but still has not achieved the required tensile capacity, the pile could be extended by using a soil-cement pile or possibly a screw-pile.

The selection of a foundation system for a particular structure appears to be an ideal application of artificial intelligence (AI). The preparation of AI codes, taking design, construction, and inspection into account, will require a significant effort. The uses of AI in foundation engineering have been minimal but increased use can be expected. At first, the engineer would probably only use AI to check that significant parameters have been considered.

MODELING

The most effective models for analysis of single piles are shown in Fig. 4 (axial loading) and in Fig. 5 (lateral loading). In these models, the pile is considered to obey the equations of mechanics and the soil is replaced by a set of nonlinear mechanisms. Iteration is required to achieve a solution because the response of the soil, as found from the deflection of the mechanism as a particular point, will depend

6 DEEP FOUNDATIONS

on the deformation of the pile, which depends on the loading from the soil. Convergence of solutions is rapid with modern personal computers.

The two models can be combined if desirable. However, experimental data have shown that the near surface soil is predominant in resisting lateral loading and is usually not very effective in resisting axial loading. Thus, the analyses can be separated without appreciable error.

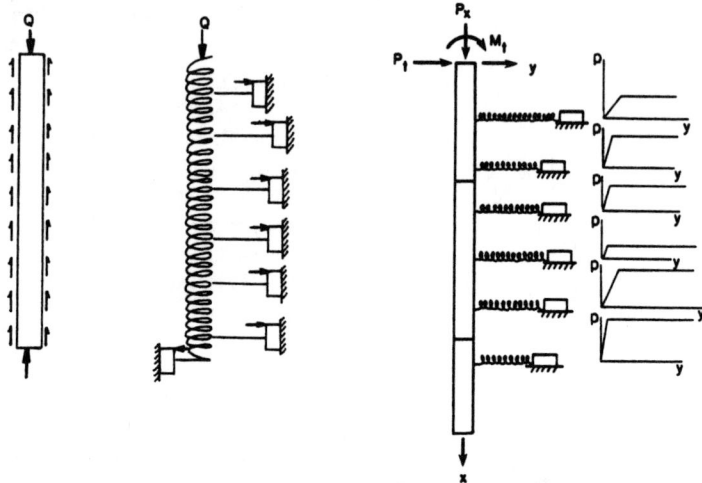

Fig. 4. Model for Axial Loading   Fig. 5. Model for Lateral Loading

Predictions for the load-transfer functions for axial loading, t-z curves for side resistance and q-w curves for end bearing, have been developed from experimental data for several soils, based on pile dimensions and method of construction. Similarly, predictions have been developed for load-transfer functions for lateral load, p-y curves. The two models continue to be improved as nonlinear stiffness is introduced for the piles to reflect the behavior of reinforced concrete and as additional experiments are performed with instrumented piles to obtain experimental load-transfer functions.

The models in Figs. 4 and 5 are based on the Winkler concept of discrete soil reactions, but the models can be made more general if the continuum effect is included explicitly. That is, each of the resistance functions would be modified by accounting for the movement of the pile above and below the point considered. It is of interest to note, however, that the so-called continuum effect is accounted for in the resistance functions derived from an experiment with an instrumented pile. The curves for values of t-z, q-w, and p-y are based strongly on results from experiment in addition to theory. In spite of their simplicity, the models in Figs. 4 and 5 have proved to be quite effective in predicting the behavior of single piles under axial or lateral loading.

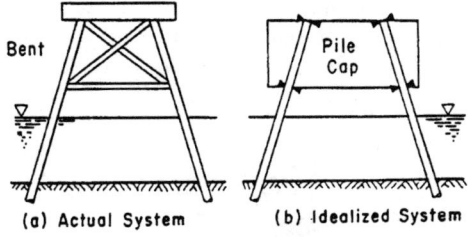

(a) Actual System (b) Idealized System

Fig. 6. Pile Group

The models in Figs. 4 and 5 can be implemented into a model for a group of piles under inclined and eccentric loading, such as that illustrated in Fig. 6. The piles supporting the pile cap are replaced by nonlinear resistances that have appropriate values of axial force, lateral force, and bending moment, depending on the movement of the pile cap. In computing the response of each pile, the load-transfer curves are modified to take soil-structure interaction into account for closely spaced piles. Again, convergence is rapid with available computer codes. The model can be made more general to account for moving soil, as due to liquefaction during an earthquake or due to the settlement of a fill.

A finite-element model for the analysis of a pile during driving is shown in Fig. 7. The model can be used effectively, but calibration is required because the constitutive models used for soil and slip elements may be incomplete. The energy when a pile is impacted is taken by the pile and by the surrounding soil. The mass of the soil must be included in Fig. 7 for the model to be complete. Significant work has been done in improving the model (El-Mabsout, 1991), but a more comprehensive model awaits development.

The model shown in Fig. 8 will be implemented in due course. The pile may be modeled as a beam-column, but the soil will be modeled by finite elements. The model must be three-dimensional and capable of predicting the response of single piles and pile groups under short-term (static) loadings, sustained loadings, cyclic loading, and dynamic loadings, from earthquake or pile driving. The predictions must be time dependent to account for thixotropy and pore-pressure increase and dissipation. Each finite element must include a mass and have nonlinear stress-deformation characteristics. Nonlinear geometry must also be considered. Computing power will be at hand shortly but an enormous amount of research must be done in the laboratory, in the field, and with theory.

COMPUTATIONAL POWER

In the past, evolution of software tools was limited due to two primary factors; limitations of computing hardware and availability of software. Early microcomputers were limited by the amount of available random-access memory. Many engineering programs were main-frame programs, developed either by university researchers or by governmental agencies, that were recompiled for use on the microcomputer. Early on, microcomputers had less memory than was available on university mainframes. This is less true today, when many engineering workstations have more memory available to the user than the university supercomputer of a decade ago.

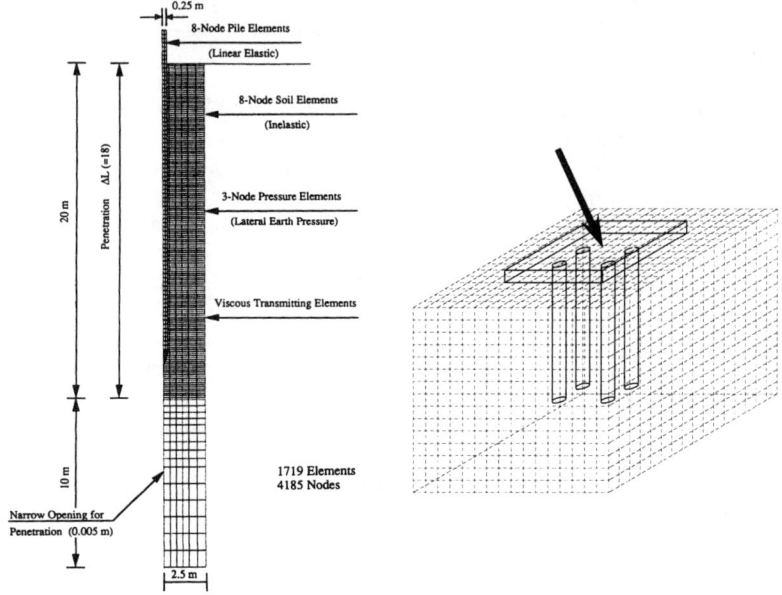

Fig. 7. Finite Element Model of Pile   Fig. 8. Example FEM Model of Pile Group
         During Pile Driving

For the practicing engineer, computational power is derived from the combination of computing hardware and software. In the past, limitations due to computing hardware were significant and did effect the quality of solutions obtained by the designing engineer. However, with the improvements in computing hardware during the late 1990's, limitations due to hardware capabilities have vanished for some types of analyses and are of lesser concern for others.

In spite of the growth of computing power since the early 1960's, some problems related to deep foundations remain intractable due to limitations in available software. For example, solutions to problems involving a complex constitutive relationship used for a nonlinear soil and nonlinear, three-dimensional geometry remains unavailable to many practicing engineers due to unavailability of software.

In the last decade, the following developments in computer software used for engineering applications have been realized:

- use of programming languages other than Fortran for engineering applications,
- development of graphical user-interfaces and visualization tools,
- development of object-oriented programming, and
- improvement of numerical methods applied to engineering problems.

Solutions or partial solutions to the following problems have been developed by the research community:

- nonlinear soil behavior using constitutive models, including changes in hydraulic conductivity due to shearing strain deformations,
- nonlinear structural material models, including fracture,
- behavior during low-frequency cyclic loading with hysteretic and degrading load-transfer relationships,
- behavior during seismic loading including inertia effects and radiation damping,
- behavior of piles during and after installation,
- three-dimensional soil profiling, and
- analysis of reliability and risk assessment.

Successful development of the techniques developed by the research community into viable engineering software utilized by the engineering profession depends on two factors. First the software developer must provide technical assistance and program maintenance for the software over a long period of time. Second, the software sales must generate enough income to justify the effort by the software provider. Thus, the software developer can provide software and support only for widely used applications. These two factors imply a professional relationship between the design professionals and the software-development community. The design professionals must recognize that marketplace for advanced software for the design of deep foundations is small, so the expense per user for developing software must be larger than that for widely-used applications. Thus, the cost of software cannot decrease with time in the same manner the cost of computing hardware.

INSTRUMENTATION

Innovative technologies have developed around instrumentation for geotechnical investigations. A variety of applications of the basic devices; electrical-resistance gauges, vibrating-wire gauges, micro-levels and data loggers have led to the acquisition of useful information on response of piles to loading. Better small, rugged, and versatile instruments can be expected to be developed in the future; for example, gauges applied along a driven pile could send data through the pile wall on changes of pore pressure with time.

As the expense of instrumentation drops, instrumentation can be included on routine construction projects for purposes of quality control and assurance. This requirement is likely to originate on large government-funded projects first and gradually be adopted by contractors for use on many projects. Successful adoption of new instrumentation will necessitate that owners be knowledgeable about the natural variation of both soil properties and construction practice.

INFLUENCE OF INSTALLATION

Complex and little known effects on soil properties arise due to pile driving or even to the installation of bored piles. Appropriate models for response under loads of single piles and groups of piles cannot be devised until such effects can be predicted as a function of time and pile installation practice. Instrumentation, micro-sampling, sonic methods, and observation of performance of the total system can lead to better understanding of the massive changes in soil characteristics that occur. These time-dependent characteristics are currently beyond predictive capability, but can be understood better and better in time as research progresses.

A recent conference on the topic (Andersen and Myers, 1993) had contributions from 26 invited engineers. Many important concepts were presented, but the proceedings of the conference were characterized by a dearth of experimental data. Examples of data that should be collected are shown in Figs. 9 and 10.

The data presented in Fig. 9 are derived from tests with the cone penetrometer in the vicinity of a pipe pile driven into sand (Cox, et al., 1974). The increase in cone-penetration resistance occurred even though the sand was judged to have a initial value of high relative density. Many examples of such data can be expected in the future, leading to methods of predicting changes in the properties of granular soil due to pile driving.

The data in Fig. 10 are derived from load tests of piles in clay soils (Reese, 1990). The increase in the capacity of the pile with time is striking. Further, the time for a pile to reach a given percentage of its capacity increases with the amount of displacement of the pile. Data, not shown here, indicate that most of the increase in capacity occurs with a decrease in excess pore-water pressure and that the initial sharp increase in pore pressure correlates with pile displacement. Some increase in capacity with time may be due to thixotropy. A tentative conclusion can be reached that the ultimate capacity of a group of closely-spaced piles will occur more slowly that that shown in the figure.

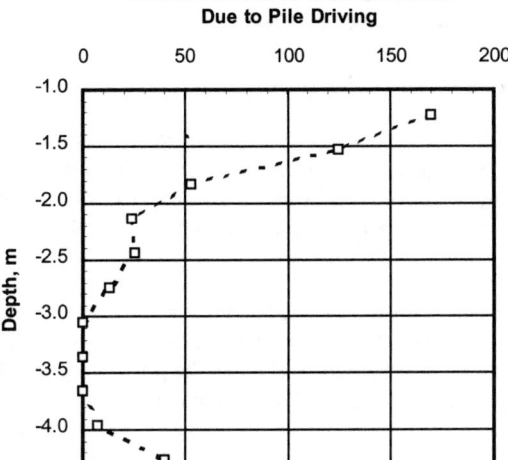

Fig. 9. Changes in Cone Resistance Due to Pile Driving

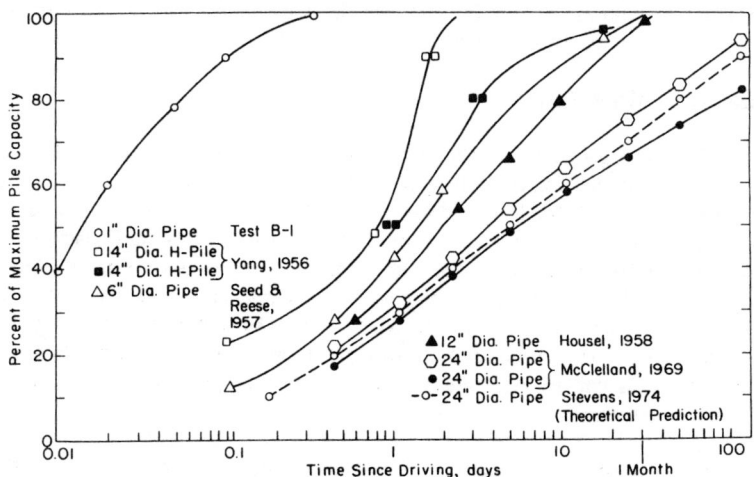

Fig. 10. Pile Capacity vs. Time (after Vesic, 1977)

Data can be expected in the future on many aspects of the influence of the driving of piles into clays. Pore-pressure and total-pressure measurements around single piles and pile groups, coupled with data on soil characteristics, can lead to the formulation of theories on pore-pressure generation and dissipation as a function of time and position with respect to the piles.

PERFORMANCE MONITORING

A variety of instrumentation and non-destructive testing techniques are being developed for use in performance monitoring of pile-supported structures. Evaluation of predictions of performance is necessary for improvement of design practice. These evaluations will be based on the data obtained through performance monitoring.

Considerable benefits to the profession can be derived from observations of the response to loading of completed foundations. Not only can information be gained on the manner in which loads come to the piles in a group, but the settlement of the piles with time can be measured. Techniques are at hand for collecting such data.

Van Impe (1999) described a project where a five-story building was founded on 55 piles. A load cell was placed on the top of each pile before construction of the superstructure. Loads to each pile were measured as the construction progressed to completion. The data, of course, revealed some very interesting phenomena. Numerous examples of similar projects in the future will allow significant improvements in methods of prediction.

DESIGN STANDARDS

Several organizations have progressed in the production and implementation of standards for the design and construction of deep foundations. Some design standards were developed to address the needs for special types of structures. Examples of special applications include offshore oil production platforms and navigation structures on waterways. Other design standards have been developed to bring modern design practices into wide-spread engineering practice, such as for highway-bridge structures. Most current design standards fail to address the interaction between geotechnical and structural engineering. Development of unified design standards can improve the practice of foundation engineering for safe and economical structures.

Efforts have been made to implement load-and-resistance-factor design (LRFD) codes by AASHTO in the US and by other agencies in North American and European nations (DiMaggio, 1998). The approaches taken in the US have varied from those used in Europe.

The US approach is to use an approach where the sum of factored loads equal the characteristic resistances times a reduction factor. Efforts have been made to examine data bases of load tests in order to select values for load factors and reduction factors on the basis of statistical analysis. A logical extension of this approach would be a refinement of load and resistance factors to reflect desired levels of reliability. The development in Europe of partial-safety factors preceded LRFD in the US, but the method awaits general implementation.

One benefit of improved design codes is an improvement in the efficiency of allocation of engineering resources. Not all structures or projects warrant high levels of effort and financial resources during design. The relationship between resistance reduction factors and levels of effort in field and laboratory testing will be refined as methodologies are improved for use of LRFD. This will allow economic studies to identify when money can be saved by utilizing additional engineering resources such as additional field exploration studies, pile load tests, or pile driving retap tests.

CONCLUDING COMMENTS

In the first half of the twentieth century, it was generally recognized that the design of deep foundations was in transition from an art form to a scientifically-based engineering practice. This period marked an era where fundamental research in soil mechanics and foundation engineering was conducted by a few inspired individuals. As their contributions were appreciated, the range of research was expanded to solve problems associated with modern structures, notably offshore-oil production platforms. By the end of the third quarter of the twentieth century, designing engineers were confident to extrapolate the size of deep foundations being designed beyond the range of sizes verified by experiment. Unfortunately, engineering management began to believe that the design of deep foundations did not require any expertise beyond that gained in the college classroom. The economic losses due the wasted capacity in the designs being used was not recognized. In last quarter of the twentieth century, research in deep foundations has languished because there is no coordinated effort to promote and disseminate such research. Instead, research funds were directed to areas that were perceived to be unexplored.

Improvement in the methods used for design of deep foundations will have universal benefits. The rate of developments in the methods used for design and construction of deep foundations is uncertain at present. Currently, there is no central organization in the United States that serves either as a central agency for coordination and funding of research. Instead, research is funded by numerous agencies in the Federal and state governments. Virtually no international cooperative research efforts exist for deep foundations. A coordinating organization can achieve the following:

- promoting continuing education for both contractors and designing professionals
- promoting cooperative research between construction contractors, design professionals, and owners,

- promoting technology transfer between different industries and nations,
- establishing legal principals to define responsibilities and obligations of designing professionals, construction industry, and governing agencies,
- promoting periodic workshops to review the state of current knowledge and construction practices and to develop a consensus for critical research needs, and
- other functions related to professional design, construction, and legal practices.

What does the future hold for deep foundations? The following advances can be predicted with confidence.

- Methods of characterizing soil and evaluating properties with accuracy.
- New types of deep foundations and improved methods of installing current types.
- Clear understanding of the time-dependent influence of installation and loading on soil properties.
- Comprehensive data bases from tests of instrumented foundations on response of pile to complex loadings, including static, sustained, dynamic, and seismic.
- Analytical models of single piles, pile groups, and the supporting soil that enable time-dependent predictions of response of each element in the system as a function of time.
- More knowledgeable owners, engineers, and contractors and method of writing contractual documents to assign risk properly and eliminate legal conflicts.

The tools are at hand or will be in the near future to accomplish each of the advances, but the rate of development will depend on the geotechnical engineering community. Based on historical developments, success will extend will into the next century, but a cooperative effort could accomplish most of the goals by 2025.

REFERENCES

Andersen, J. M. , and Myers, W. M., Editors, "Proceedings of the Workshop on Effects of Piles on Soil Properties," US Army Waterways Experiment Station, 13-15 July 1993, Miscellaneous Paper GL-95-2, August 1995.

Berminghammer Foundation Equipment, *http://www.berminghammer.com/stat.htm*

Baguelin, F., J. F. Jézéquel, & D. H. Shields, *The Pressuremeter and Foundation Engineering*, Trans Tech Publications, Switzerland, 1978, 617 pages

Cox, W. R., Reese, L. C., and Grubbs, B. R., "Field Testing of Laterally Loaded Piles in Sand," *Proceedings*, Offshore Technology Conference, 1974, Vol. 2, pp. 459-472.

Dames, T. R., "Practical Shear Tests for Foundation Design," *Civil Engineering*, December, 1940.

De Paoli, B., G. Viola, & A. Tomiolo, "The use of drilling energy for soil classification," *2nd International Symposium on Field Measurements in Geomechanics,* Sakurai (ed), Balkema, Rotterdam, 1988, pp. 313-321.

DiMaggio, J. A., et al., "Summary Report of the Geotechnical Engineering Study Tour, FHWA, December, 1998, 76 pp.

El-Mabsout, M., "A Finite Element Model for the Analysis of Pile Driving," Ph.D. Dissertation, The University of Texas at Austin, 1991, 145 pp.

Emrich, W. J. "Performance Study of Soil Sampler for Deep-Penetration Marine Borings," *Sampling of Soil and Rock*, ASTM STP 483, 1971, pp. 30-50.

Hoar, R. J., and Stokoe, K. H., II, "Generation and Measurement of Shear Waves In-Situ," *Dynamic Geotechnical Testing*, ASTM STP 654, 1977, pp. 3-29.

Hoar, R. J., and Stokoe, K. H., II, "Crosshole Measurement and Analysis of Shear Waves," *Proceedings*, 10th Intl. Conf. on Soil Mechanics and Foundation Engr., Stockholm, 1981, Vol. 3, pp. 223-226.

Holeyman, A., C. Bauduin, M. Bottiau, P. Debacker, F. De Cock, E. Dupont, J. L. Hilde, C. Legrand, & N. Huybrechts, "Design of axially loaded piles--Belgian practice," *Design of Axially Loaded Piles--European Practice,* De Cock & Legrand (eds), Balkema, Rotterdam, 1997, pp. 57-82.

Hussein, M., Likins, G., Rausche, F., (1996), "Selection of a Hammer for High-Strain Dynamic Testing of Cast-in-Place Shafts," *Proceedings*, 5th International Conference on the Application of Stress-Wave Theory to Piles, Orlando, 1996.

Hvorslev, M. J. "Subsurface Exploration and Sampling of Soils for Civil Engineering Purposes," US Army Waterways Experiment Station, 1948, 465 pp.

Kjellman, W., Kallstenius, T., and Wager, O., "Soil Sampler with Metal Foils," Royal Swedish Geotechnical Institute Proceedings, No. 1, 1950, 75 pp.

Olson, L. D., Jalinoos, F., Aouad, M. F., and Balch, A. H., " Acoutic Tomography and Reflection Imaging for Nondestructive Evaluation of Structural Concrete." NSF Phase I Final Report, SBIR Industrial Innovation Interface Division, Washington, DC.

Osterberg, J. O., "The Osterberg Load Test Method for Drilled Shafts and Driven Piles - The First Ten Years," *Proceedings*, Seventh International Conference on Piles and Deep Foundations, Vienna, Austria, June 1998.

Quiros, G. W., Personal communication, January, 1999.

Reese, L. C. "Load versus Settlement for an Axially Loaded Pile," *Proceedings*, Symposium on Bearing Capacity of Piles, Part 2, Central Building Research Institute, Roorkee, 1964.

Reese, L. C., "The Action of Soft Clay along Friction Piles: Bay Mud Revisited," *Proceedings*, H. Bolton Seed Memorial Symposium, 1990, Vol. 2, pp. 135-153.

Schmertmann, J., Hayes, J., Molnit, T., and Osterberg, J. O., "O-cell Testing Case Histories Demonstrate the Importance of Bored Pile (Drilled Shaft) Construction Technique," *Proceedings*, Fourth International Conference on Case Histories, University of Missouri, Rolla, Missouri, March, 1998.

Spear, D., Reese, L. C., Reavis, G. T., and Wang, S.-T. (1994), "Testing of GeoJet Units Under Lateral Loading," *Proceedings*, International Conference on the Design and Construction of Deep Foundations, Orlando, Florida, Vol. 2, pp. 969-979.

Stokoe, K. H., II, Rix, G. J., and Nazarian, S., "In-Situ Seismic Testing with Surface Waves," *Proceedings*, 12th Intl. Conf. on Soil Mechanics and Foundation Engr., Rio de Janeiro, 1989, Vol. 1, pp. 331-334.

Stokoe, K. H., II, Wright, S. G., Bay, J. A., and Roësset, J. M., "Characterization of Geotechnical Sites by SASW Method," ISSMFE Technical Committee 10 for XIII ICSMFE, Geophysical Characteristics of Sites, pp. 146-186.

Van Impe, W. F., Personal communication, January, 1999.

Van Impe, W. F., and H. Peiffer, "Influence of screw pile installation on the stress state in the soil," *Design of Axially Loaded Piles--European Practice,* De Cock and Legrand (eds), Balkema, Rotterdam, 1997, pp. 3-19.

Vesic, A. S., "Design of Pile Foundations," National Cooperative Highway Research Program, Synthesis of Highway Practice, No. 42, 1977.

Woods, R. D., and Stokoe, K. H., II, "Shallow Seismic Exploration in Soil Dynamics," *Proceedings*, Richart Commemorative Lectures, ASCE, 1985, pp. 120-151.

Lateral loading response of piles in calcareous sediments

Gerard Dyson[1] and Mark Randolph[2]

ABSTRACT

The lateral response of piles embedded in calcareous sediments has received relatively little attention, in spite of the widespread occurrence of such sediments. This paper summarises the results of an extensive experimental investigation of lateral pile response in calcareous soils ranging from fine-grained silts to artificially cemented calcarenite. General expressions are presented for load transfer curves, linking the magnitude of lateral resistance to the local cone resistance. Modification factors for the lateral resistance are introduced to account for effects of different methods of pile installation, partially drained loading conditions, and cyclic loading.

INTRODUCTION

Calcareous sediments are found extensively in offshore regions throughout the world, and can range in type from fine-grained muds to strongly cemented calcarenites (Murff, 1987; Fookes, 1988). In general, calcareous sediments are more compressible than comparable terrigenous clays and sands, which is mainly attributable to the relatively high void ratios of natural deposits (Semple, 1988). Individual soil grains are usually highly angular, being composed of fragments of small organisms such as foraminifera, leading to high intrinsic friction angles.

Design models for the lateral response of piles have been developed over several decades of research for silica sands and clays. However, these models have been found to be inappropriate for calcareous sands (Wesselink et al, 1988), due to the combined high compressibility and high frictional strength of this material.

In order to provide a framework for design of laterally loaded piles in calcareous sediments, an extensive program of centrifuge model tests was conducted

[1]Research student, [2]Director, Special Research Centre for Offshore Foundation Systems, The University of Western Australia, Nedlands, WA 6907, Australia

over the previous three years. Instrumented model piles, at a nominal scale of 1:160, were tested in soil conditions ranging from fine-grained silts to cemented sands, although the majority of tests were conducted in uncemented silty sand. The study included investigation of the effects of the method of pile installation, the rate of loading, and cyclic loading representative of offshore conditions.

The test results have been interpreted using conventional load transfer (P - y) curves, following the basic formulation proposed by Wesselink et al (1988) from their studies of lateral pile response in uncemented calcareous sands from Bass Strait. It has been found that the key load transfer parameter may be linked directly to the cone resistance, for a wide range of soil conditions (including cemented sediments).

MODELLING DETAILS

The experimental work was carried out on the geotechnical centrifuge at the University of Western Australia (Randolph et al, 1991). This centrifuge has a platform radius of 1.8 m, and a maximum acceleration level of 200 g. The present work was carried out at an acceleration of 160 g, and thus a linear scale of 1:160. All results are presented in appropriate non-dimensional form, or in prototype units.

Soil Conditions

Two different soil types were used for the experimental program. These soils were recovered from the seabed from two locations on the North-West Shelf of Australia. The coarser material is a fine calcareous sand from the vicinity of the Goodwyn A platform in 120 m water depth, while the finer material is a silt from the region of the Gorgon gas field, in 200 m water depth. In order to compensate for some loss of fines during sample recovery and drying of the silt, additional 15 % by weight of commercial calcium carbonate fines ('Omyacarb') was added to the silt.

The main properties of the two soil types are summarised in Table 1. Both soil types had calcium carbonate contents in excess of 95 %, with a specific gravity of particles of around 2.75. The tests were conducted using silicon oil as the pore fluid, with a viscosity 100 times greater than water, in order to help simulate prototype drainage conditions. As will be seen, for the range of loading rates explored (covering an appropriate range for offshore conditions) the sand behaved in a partially drained manner, while the silt was effectively undrained.

Strength profiles were obtained during each test by carrying out miniature cone penetrometer tests and, for the silt, T-bar tests (Stewart and Randolph, 1994). The relative scarcity of each soil type necessitated re-use of the same material for several series of pile tests, and this was accompanied by a gradual decrease in the strength profile, which was attributed to slight degradation of the material and decrease in mean particle size. Results from different tests have been compared by normalising using the cone resistance or undrained shear strength.

Table 1 Representative soil properties

Property	Sand	Silt
Mean particle size, d_{50} (mm)	0.1	0.03
Effective unit weight, γ' (kN/m^3)	8.1	7.0
Critical state friction angle, ϕ_{cv} (°)	38	38
Peak friction angle, ϕ_{peak} (°)	43	42
Coefficient of consolidation, c_v (mm^2/s)	10	0.1

Figure 1 shows profiles of cone resistance for the calcareous sand, and also profiles of resulting N_q values (ratio of cone resistance to effective overburden stress). In any given test, the cone resistance increases approximately proportionally with depth, resulting in a uniform value of N_q, which ranged between 20 and 60 for the various tests. The presence of a cemented layer (see later) is shown for one test, with a peak cone resistance of 10.5 MPa for material with an unconfined compression strength of 0.5 MPa. (Note that the cone resistance in the 15 mm thick model cemented layer was measured with a special miniature cone, 3 mm in diameter, compared with the normal 10 mm diameter cone.)

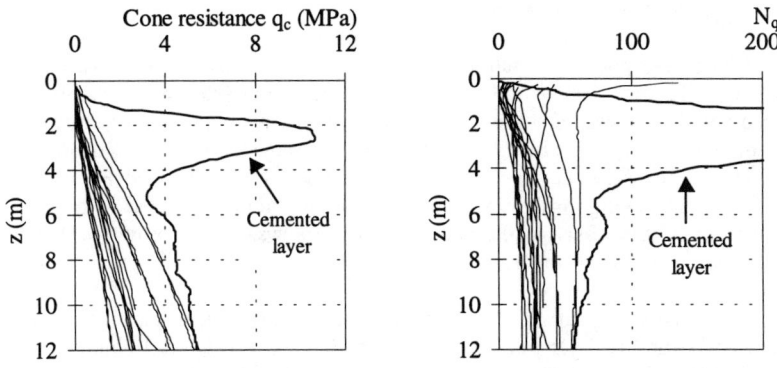

Figure 1. Profiles of cone resistance and N_q in calcareous sand

Figure 2 shows corresponding profiles of net cone resistance and T-bar resistance for the silt. Again, each profile shows an approximately linear increase with depth, corresponding to a shear strength gradient in the range 2 to 3.4 kPa/m. The ratio $s_u/\gamma'z$ (where s_u is the undrained shear strength, and $\gamma'z$ is the effective overburden stress) lies in the range 0.3 to 0.5. For interest, a cone resistance profile is also shown for a cone test that was carried out following a cyclic loading test on a pile, after allowing for re-consolidation of the soil. The cone was a distance of 4 diameters from the pile, but there is clear evidence of increased soil density and strength (by a factor of 2.5 to 3) as a result of the pile test.

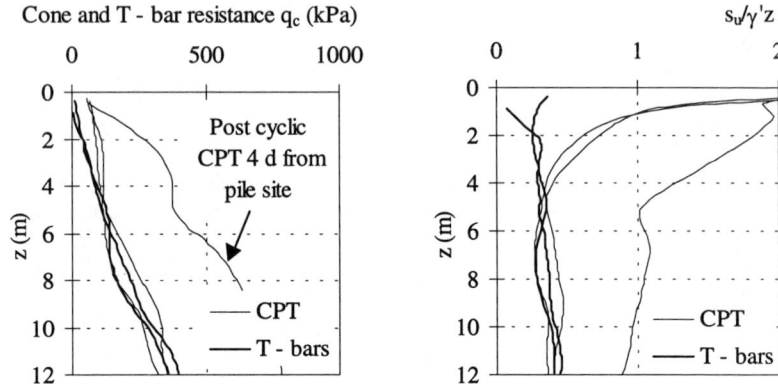

Figure 2. Profiles of cone resistance and $s_u/\gamma'z$ in calcareous silt

Model Piles and Testing Arrangements

The model piles were fabricated out of aluminium tubing, with an external diameter of 12 mm and wall thickness of 1 mm. The piles were instrumented externally with 13 levels of strain gauge bridges to measure bending moments, with a layer of epoxy, 0.5 mm thick, protecting the instrumentation. The overall pile diameter at prototype scale was 2.08 m, and the piles were embedded 20 diameters into the soil. Allowing for the lower modulus of aluminium compared with steel for prototype piles, the effective d/t ratio for the piles was 55 (a prototype bending rigidity of EI = 26,700 MNm²).

The testing layout in the centrifuge strongbox, model pile and loading arrangements for 'free' (or pinned) and 'fixed' (or restrained) head conditions, are shown in Figure 3. The piles were loaded at a distance of between 2 and 4 diameters above the soil surface. This distance was maintained constant in any one series, for example to explore the effects of loading rate, but varied from one series to another. The restrained loading condition did not provide complete fixity, but the rotational stiffness was estimated to lie in the range 3 to 8 GNm/rad (prototype scale).

In the main testing program, the piles were installed (closed-ended) by jacking them at 1 g into the prepared soil bed. A separate set of tests was undertaken in order to quantify the effects of different methods of pile installation, and the effect of the pile tip condition (closed or open) on the lateral response. That series included driving using a miniature pneumatic drop hammer.

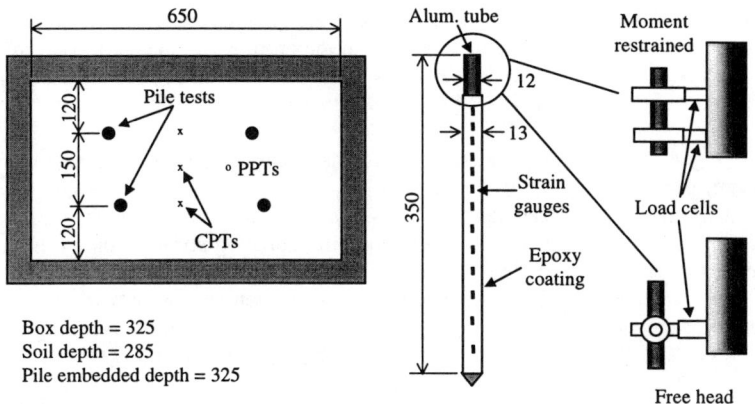

Box depth = 325
Soil depth = 285
Pile embedded depth = 325

Figure 3. Testing layout, model pile and loading arrangements (dimensions in mm)

Interpretation of Load Transfer Curves

Two different approaches were used to obtain load transfer curves from the instrumented pile tests. The direct approach was to use conventional double differentiation of a fitted bending moment profile to obtain local values of force per unit length (P), and double integration to obtain displacements (y). The bending moment data were fitted using two fourth order polynomials over the embedded part of the pile, and a linear fit over the exposed section. Continuity of shear force and force per unit length was ensured between each section. Further details are given by Dyson and Randolph (1997) and Dyson (1999). Integration boundary conditions included measurement of the pile head deflection and rotation (using two laser displacement transducers) and a tip condition of zero displacement, since the pile tip was embedded by 1 diameter into an underlying gravel layer). Redundancy in the boundary conditions allowed assessment of the accuracy of the procedure.

An indirect approach, similar to that described by Wesselink et al (1988), was also used to optimise parameters for a generic form of P – y curve. This procedure involved numerical analysis of the pile response for an assumed set of P – y curves, and then iterative adjustment of the P – y curve parameters until minimum error was achieved between measured and computed bending moments and pile head deflection (for a given applied load).

The generic form of P – y curve is the same as that adopted by Wesselink et al (1988), from their pile tests on Bass Strait calcareous sand, expressed as:

$$P = Rd\left(\frac{z}{z_0}\right)^N \left(\frac{y}{d}\right)^\gamma \qquad (1)$$

where R is a coefficient (in units of stress), d is the pile diameter, z is the depth, z_0 is 1 m, and N and γ are coefficients. There are therefore 3 parameters to optimise in order to obtain a best fit generic load transfer curve. Wesselink et al (1988) derived best fit parameters of $N = 0.7$, $\gamma = 0.58$ and $R = 650$ to 850 kPa.

Figure 4 shows a typical set of load transfer curves, expressed in terms of net pressure, $p = P/d$, normalised by $\gamma'd$ (where γ' is the effective unit weight) plotted against normalised displacement, y/d. Two sets of curves are shown, comparing 'experimental' data (i.e. obtained directly by differentiating and integrating the fitted bending moment profiles) with 'optimised' data based on the generic form of $P - y$ curve. As may be seen, there is good agreement between the two sets of curves, and for most of the work the direct approach has been used, but with the resulting curves then fitted using the form of load transfer curve expressed in equation (1).

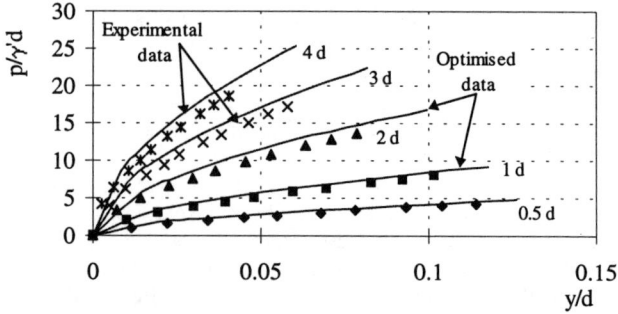

Figure 4. Comparison of experimental and optimised load transfer curves

MONOTONIC RESPONSE IN CALCAREOUS SAND

The program of model tests to investigate the monotonic response in calcareous sand included a range of soil strength profiles (as commented above), two different pile head boundary conditions, and modelling of models (using a model pile of 1.65 times the diameter of the one shown in Figure 3, and a correspondingly reduced acceleration level). These are documented fully in Dyson (1999). The tests showed a high degree of consistency, once normalised by the soil cone resistance.

It was found that the load transfer curves were best expressed in the form:

$$\frac{p}{\gamma'd} = R\left(\frac{q_c}{p_a}\right)^{0.72}\left(\frac{y}{d}\right)^{\gamma} \tag{2}$$

where the cone resistance, q_c has been normalised by atmospheric pressure (p_a =

100 kPa). Since the cone resistance is approximately proportional to depth, the exponent of 0.72 is very similar to the power of $N = 0.7$ in the original formulation of Wesselink et al (1988) (see equation (1)).

The best fit values of R and γ differed slightly for the two different boundary conditions at the pile head, as detailed in Table 2. However, the average fit still gives excellent agreement, and coincides precisely with the average values from Wesselink et al (1988), after reformulating the latter using the reported cone resistance profiles. Figure 5 shows typical results for the pile head load-displacement response, and bending moment profiles, for the two different pile head conditions. The computed response using the average parameters of Table 2 gives very good agreement with the experimental data.

Table 2   Deduced load transfer parameters

Parameters from Equation (2):	R	γ
Free head	10.37	0.64
Restrained head	9.34	0.52
Average values	9.85	0.58
Average Wesselink et al (1988) (Bass Strait)	9.85	0.58

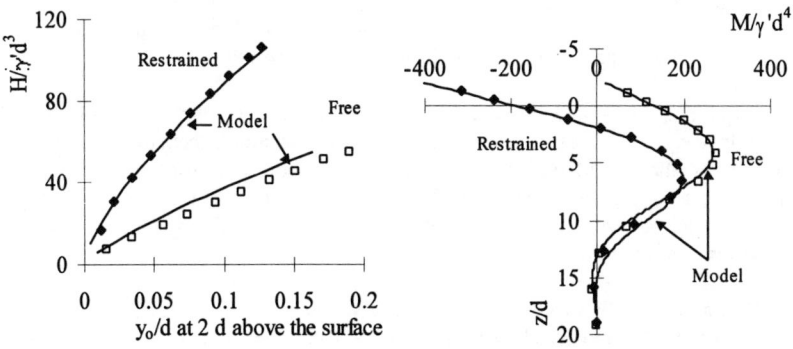

Figure 5. Typical responses for free and restrained pile head conditions

Effect of Installation Method

The above results were obtained for piles that were jacked (closed-ended) into the sand sample at 1 g. The method of installation will affect the stress conditions in the soil around the pile, as discussed by Craig (1980), and different methods of pile installation were explored in order to quantify this effect. The different installation methods included pre-installing the pile (and pouring the soil around the pile), at 160 g on the centrifuge, and driving at 160 g. Both open and closed ended tip conditions were explored (using uninstrumented piles for the former, and comparing the pile head load-displacement response). In addition, tests were undertaken to explore the effect of stopping the centrifuge, and thus relieving the stresses induced by pile installation at 160 g, before returning to 160 g and loading the pile.

Detailed discussion of these results has been presented by Dyson and Randolph (1998), and so the results will just be summarised here. The effect on the pile head load-displacement response is shown in Figure 6(a). The initial pile response is similar, but for displacements greater than 5 % of the pile diameter, there is increasing divergence of the curves. Example load transfer curves (normalised using the cone resistance, following equation (2)) are shown in Figure 6(b) for two different depths, and these show much greater variation, but with a trend of reducing difference as the depth increases. Finally, in Figure 6(c), multiplicative factors are shown (to be applied to the value of R in equation (2)) as functions of depth. The range of values corresponds to the effect of stopping the centrifuge between initial installation and loading, with the higher values in lighter shade being obtained for piles that were loaded directly after installation, without stopping the centrifuge.

Effect of Loading Rate

The preceding results were obtained for a relatively low rate of loading, corresponding to essentially drained conditions. As a prelude to exploring the response under cyclic loading, a series of tests were conducted at different loading rates to span the range between drained and undrained conditions.

For shallow foundations, or cone penetration tests, the extent of drainage during loading can be expressed conveniently by the non-dimensional ratio, vd/c_v, where v is the foundation velocity, d the diameter and c_v the coefficient of consolidation. Finnie and Randolph (1994) found that for shallow foundations, fully drained conditions were obtained for vd/c_v of the order of 0.01, while fully undrained conditions required vd/c_v of about 30.

In a lateral pile test, the local velocity varies down the length of the pile, and so even if the pile velocity close to the seabed is sufficient for undrained conditions, the velocity at a depth of a few diameters may well be partially drained. As such, the effect of the rate of loading may be quite moderate.

Figure 7 shows the pile head load-displacement response, and load transfer curves at a depth of 2 diameters, for a range of loading rates covering vd/c_v values of 0.05 to 5. The pile head stiffness reduces by a factor approaching two, as the rate of loading is increased, but the load transfer curves show a much greater effect.

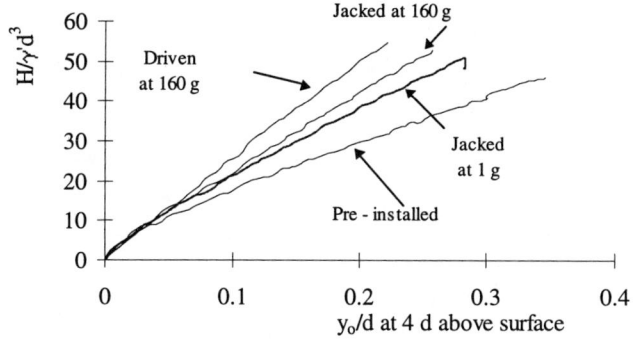

(a) Pile head load-displacement response

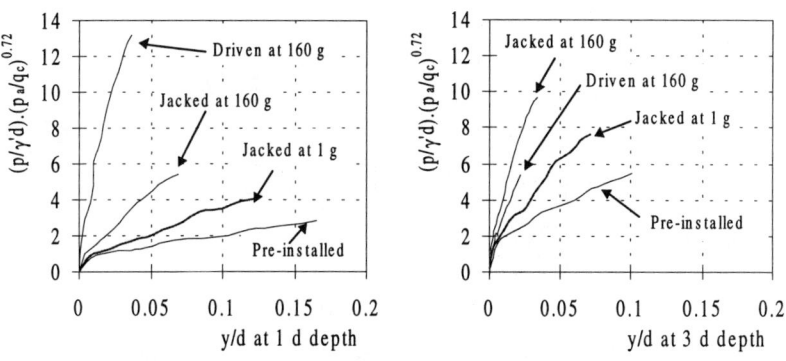

(b) Load transfer curves at depths of 1 and 3 diameters

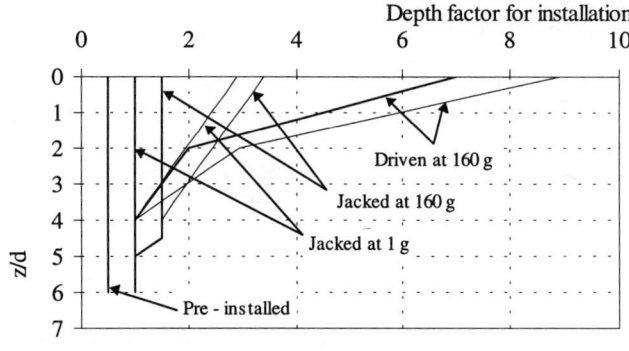

(c) Correction factor for R in equation (2)

Figure 6. Effect of pile installation method on pile response and load transfer curves

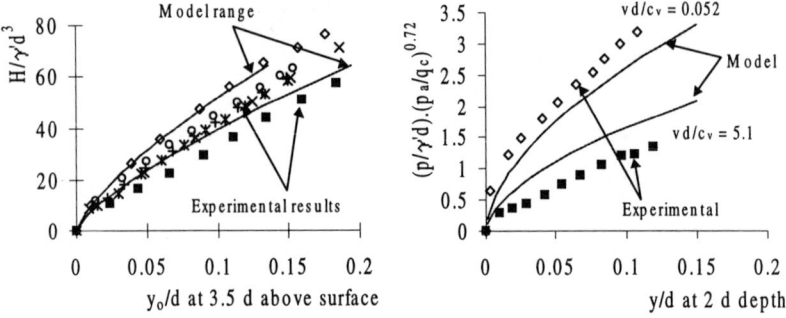

Figure 7. Effect of loading rate on pile response and load transfer curves

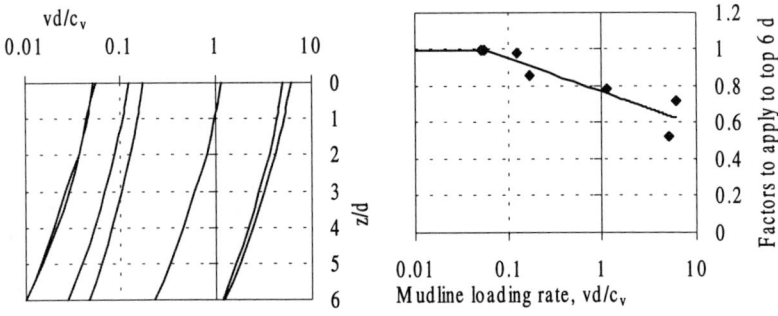

Figure 8. Velocity profiles and adjustment factor for load transfer curves

The model results in Figure 7 have been obtained by factoring the controlling load transfer parameter, R (equation (2)) over the upper 6 pile diameters in the soil, according to the non-dimensional loading rate. This approach is somewhat approximate, because of the variation of velocity down the length of the pile, as discussed above. However, the model results agree reasonably with the experimental data, particularly in respect of the overall load-displacement response. Figure 8 shows velocity profiles down the pile, and also the adjustment factor that has been used in the model (over the upper 6 pile diameters), as a function of the normalised mudline velocity of the pile.

Effect of Cemented Layer

Calcareous sediments frequently contain layers of cemented soil, often lying just below surficial uncemented material. In order to assess the effect of such a layer, a series of tests were undertaken with a cemented layer of just over 1 pile diameter thick, situated 3.5 m below the surface, as shown in Figure 9. The cemented layer was prepared using gypsum as the cementing material, and water was used as the pore fluid throughout the sample (as opposed to the silicon oil used in the remaining tests). The effect on the load-displacement response, and load transfer curves at a depth of 2 diameters from two different tests, are shown in Figure 10. As may be seen, the cemented layer leads to a stiffening of the pile head response by some 20 %.

The results have been modelled using the load transfer curve and parameters detailed in equation (2) and Table 2, based on an average cone resistance in the cemented layer of $q_c = 6.5$ MPa (see Figure 1). The model gives excellent prediction of the pile head response, and also of the load transfer data up to a displacement of about 6 % of the pile diameter. Beyond that point, the experimental data appear to reach a plateau, with a limit pressure of about 670 kPa, compared with the unconfined compression strength of 500 kPa.

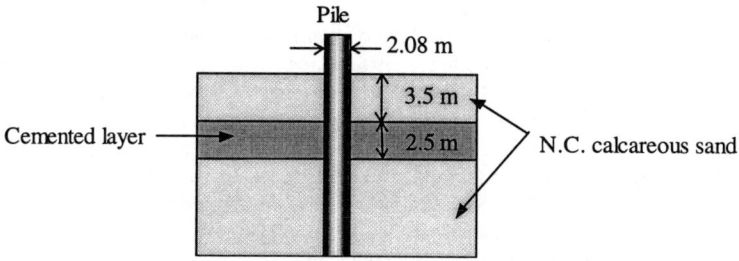

Figure 9. Stratigraphy of sample with cemented layer

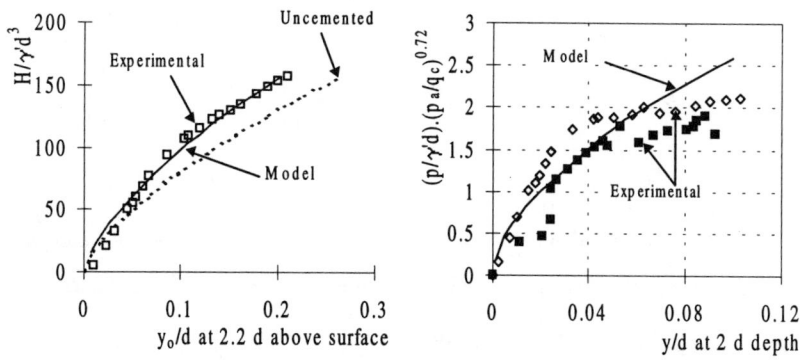

Figure 10. Pile head response and load transfer curves for cemented layer case

MONOTONIC RESPONSE IN CALCAREOUS SILT

The other material type considered in the experimental program was a calcareous silt, with a coefficient of consolidation two orders of magnitude lower than that of the sand.   The normal rate of loading for the calcareous sand tests corresponded to a normalised velocity at the mudline of $vd/c_v \sim 6$, and so was close to fully undrained. As such, it was found more useful to re-cast the load transfer curve, normalising directly by the local undrained shear strength, to give:

$$\frac{p}{s_u} = R\left(\frac{y}{d}\right)^\gamma \tag{3}$$

The shape of the load transfer curves showed a greater tendency to level out for displacements in excess of about 5 % of the pile diameter, and this was captured through a reduced value of the exponent, γ. The best fit set of parameters was found to be $R = 26$, $\gamma = 0.49$.

Figure 11 shows the pile head load-displacement response, and resulting load transfer curves at 2 diameters depth, for tests carried out at different rates with a range of vd/c_v between 0.65 and 4040. The model load transfer curve given by equation (3) is also shown for comparison.   The load-displacement responses are essentially identical at the different rates. Superficially, the load transfer curves at the two higher velocities appear to be steeper initially, and to flatten out at a slightly lower pressure of 7 to 8 times the local shear strength. However, interpretation of the load transfer curves at these higher velocities was prone to errors introduced by inertial effects, and the confidence in these curves is less high than for the slower tests.

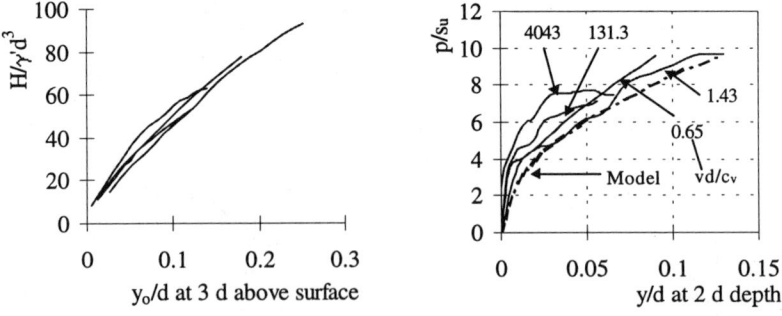

Figure 11. Pile head response and load transfer curves for calcareous silt

COMPARISON WITH CURRENT DESIGN METHODS

Before discussing effects of cyclic loading, which were relatively complex, it is useful to compare the measured performance and deduced load transfer curves with conventional design approaches for silica sands and clays. For the calcareous sand, comparison has been made with the design approach proposed by Reese et al (1974), and also with the current API RP2A method (API, 1993), based on the report to API by O'Neill and Murchinson (1983).

The critical parameters for these two methods are the effective unit weight of the soil and the friction angle (taken here as ϕ_{cv}), and these are given in Table 1. For the Reese et al method, the initial gradient of the load transfer curve has been taken as $E_{si} = kz$, where z is the depth and k has been taken as 33 MPa/m. A similar value was obtained from the charts in API (1993).

Figure 12 shows load transfer curves at a depth of 2 diameters derived from these two methods (and also using the Wesselink et al (1988) method), compared with results obtained with different methods of pile installation. On the right hand side of this Figure, bending moment profiles are compared with the experimental curve for the pile installed by jacking at 1 g. There are three main observations from these comparisons:

(1)   The API and Reese et al approaches give initial gradients of load transfer curve that agree very well (at a depth of 2 diameters), with the experimental results from the driven pile.

(2)   These approaches give limiting pressures that are lower than the extrapolated curve for the driven pile.

(3)   The bending moment profiles predicted using the API and Reese et al approaches give smaller maximum moments, at a shallower depth, than the experimental data for the jacked pile, and that is consistent with the stiffer load transfer curves.

Overall, the load transfer approach of Wesselink et al (1988) gives predicted responses that are close to the modified approach, where the multiplier, R, is adjusted for the cone resistance as in equation (2). It is also interesting to compare the profiles of net pressure (normalised as $p/\gamma'd$), as indicated in Figure 13. Clearly, the pressures obtained from the API and Reese et al approaches are much too high at depths greater than 3 diameters, and correspondingly too low at shallow depths. Indeed, the agreement at 2 diameters depth may be seen to be somewhat fortuitous. The limited depth for significant enhancement of the load transfer curves due to driving the pile is also well illustrated in this Figure.

A similar comparison has been undertaken for calcareous silt, using the original Matlock (1970) method, and also the more recent approach proposed by Dunnavant and O'Neill (1983). These methods require the undrained shear strength profile (taken as $s_u = 3z$ kPa, where z is in m) and also values of the strain to 50 % failure, ε_{50}, in triaxial compression. These values were obtained from site investigation data at the Gorgon site, and result in a y_{50} value of 0.0041 to 0.0042 m over the upper 5 m of soil (2.5 diameters).

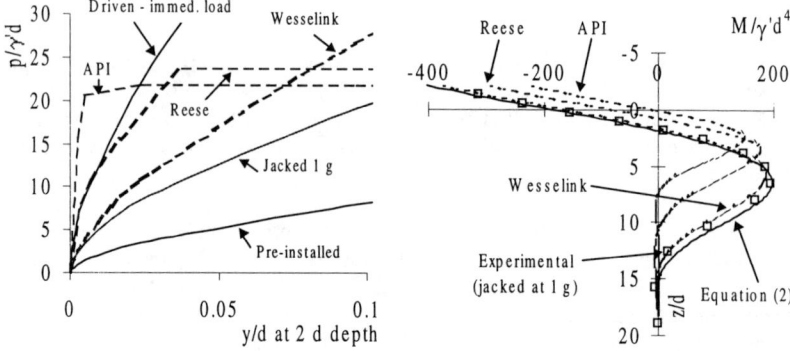

Figure 12. Comparison of load transfer curves and moment profiles for calcareous
sand, with approaches of Reese et al (1974) and API (1993)

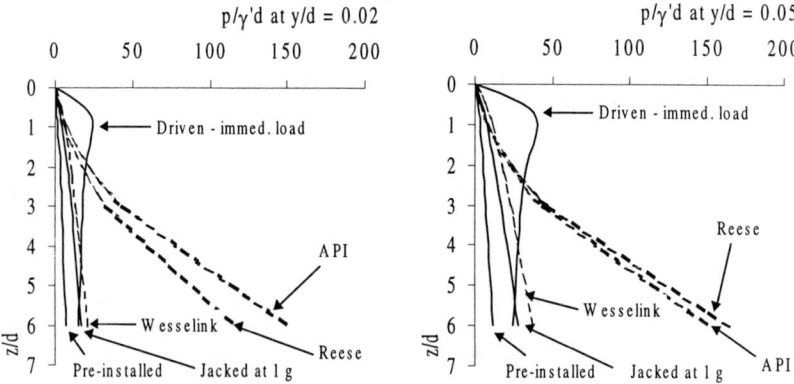

Figure 13. Profiles of normalised net pressure at two different displacements

Figure 14 shows the resulting load transfer curves at a depth of 2 diameters,
and also bending moment profiles. For this soil, the original Wesselink et al (1988)
approach is (not surprisingly) too stiff, since it was derived for relatively coarse
calcareous sand. The Matlock and Dunnavant & O'Neill approaches give too stiff a
load transfer curve response at low displacements, and too low a limiting pressure at
high displacements. These compensating errors lead to reasonable prediction of the
bending moment profiles.

Profiles of normalised pressure, p/s_u, are shown in Figure 15 for two different
displacements. The present approach (equation (3)) is generally softer at a pile
displacement of 2 % of the diameter, but consistent with the Matlock and Dunnavant
& O'Neill approaches at a displacement of 5 % of the pile diameter.

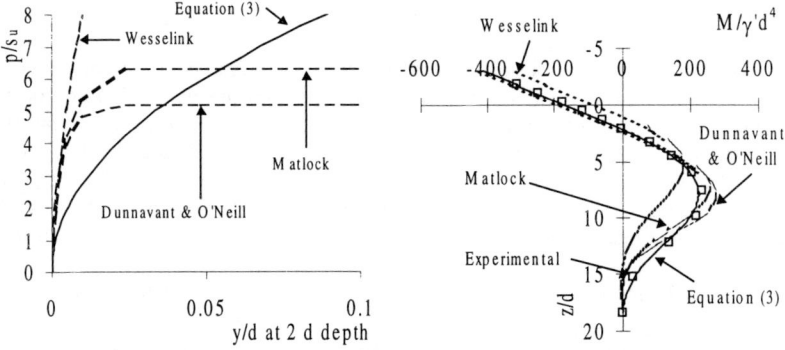

Figure 14. Comparison of load transfer curves and moment profiles for calcareous silt, with approaches of Matlock (1970) and Dunnavant & O'Neill (1993)

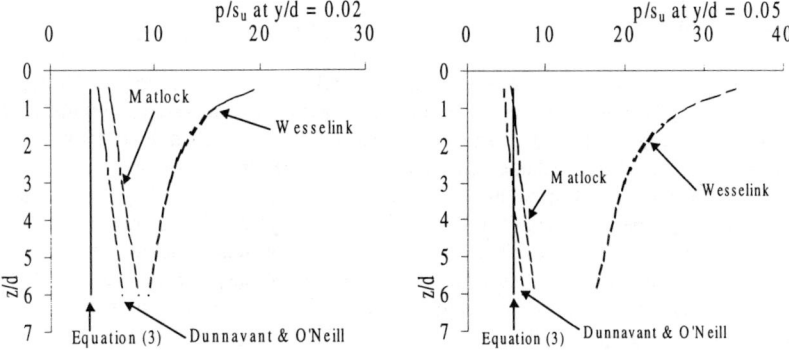

Figure 15. Profiles of normalised net pressure at two different displacements for calcareous silt

CYCLIC LOADING

Cyclic loading tests were carried out in both calcareous sand and silt, at frequencies that were largely controlled by the equipment and the displacement amplitude at the pile head. The type of loading was limited to a strongly biased load range, which led to asymmetric accumulation of displacement.   A system was developed to capture data from all bending moment gauges and other instrumentation at a rate of up to 500 Hz per channel. Typical loading frequencies averaged about 4 Hz, corresponding to a prototype period of 64 seconds (allowing for the 1:100 reduction in the viscosity of the silicon oil, compared with water, but a scaling ratio of 1:160). This is a factor of 4 slower than typical wave periods, and will have resulted in a higher degree of consolidation than for the prototype situation.

Calcareous Sand

The pile response in calcareous sand was partially drained, with consolidation and consequential stiffening overcoming initial pore pressure generation and partial liquefaction. Table 3 gives the loading ranges, expressed as $H/\gamma'd^3$, for the four pile tests conducted, with 4 or 5 loading stages in each test, and 100 load cycles in each stage. This set of tests allowed investigation of the sequence of load packets, with gradually increasing packets in the first test, compared with the final test where the most severe load packet was applied first, before reverting to an increasing sequence.

Table 3 Nominal load ranges ($H/\gamma'd^3$) for cyclic tests in calcareous sand

Test	Stage 1 Min	Stage 1 Max	Stage 2 Min	Stage 2 Max	Stage 3 Min	Stage 3 Max	Stage 4 Min	Stage 4 Max	Stage 5 Min	Stage 5 Max
1	-4	18	-5	27	-7	36	-11	53	-14	71
2	-7	36	-11	53	-14	71	-4	18		
3	-11	53	-14	71	-4	18	-7	36		
4	-14	71	-4	18	-7	36	-11	53		

Limitations of space prevent addressing all the issues that emerged from the cyclic loading. However, a typical pattern of behaviour is illustrated in Figure 16 for the second stage of Test 1. The Figure shows the pile head load-displacement response and the deduced load transfer curves at a depth of 2 diameters. The pile head response shows accumulation of displacement over the 100 cycles, but also a softened 'S' shaped response for intermediate load cycles (10 to 50), followed by stiffening over the final cycles. The corresponding load transfer curves show evidence of almost complete liquefaction (in the immediate vicinity of the pile) with essentially zero resistance at that depth for cycles 10 to 50, followed by stiffening. By the end of the load packet, the load transfer stiffness is actually greater than for the initial cycle, due to consolidation and densification of the soil around the upper part of the pile, but translated to a mean displacement of 10 % of the pile diameter.

The effect of the localised liquefaction is illustrated in the bending moment profiles shown in Figure 17, where the maximum moment increases, and occurs at much greater depth, for cycles 10 to 50, but decreases again by the final cycle. An important effect for the profiles at minimum load is the gradual accumulation of significant bending moments in the lower region of the pile, due to the accumulated lateral displacement under the biased loading regime. This has important consequences for the design of offshore piles, in particular the choice of pile depth over which the wall thickness should be increased.

The effects of cyclic loading are discussed in full detail by Dyson (1999). It was found that the load transfer curves during initial and final cycles could be described by the model of equation (2), but with parameters of R = 4.0, γ = 0.32 initially (differing from the monotonic values in Table 2, due to the higher rate of loading), increasing to R = 26.8, γ = 0.32 by the final cycle. Consolidation within the 100 cycles of a loading packet is consistent with the total (non-dimensionalised) elapsed time of T_v (= $c_v t/d^2$) ~ 1.5. For a higher loading frequency, corresponding to a prototype wave period of 12 seconds, T_v would have been only 0.3, giving reduced consolidation and stiffening in the later stages of each loading packet.

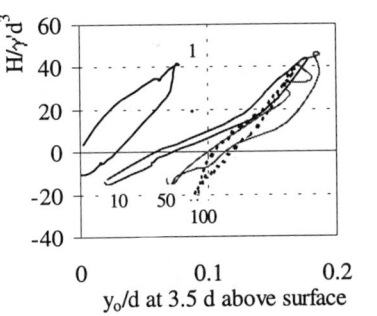
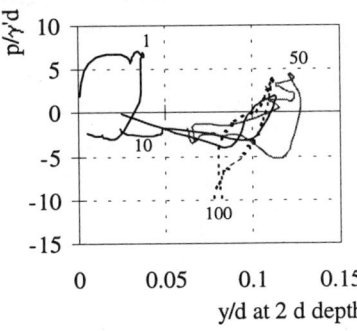

Figure 16. Pile response and load transfer curves for cyclic loading in calcareous sand

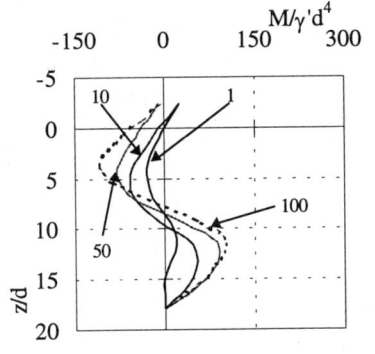
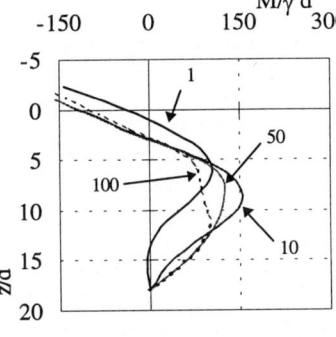

(a) Minimum load   (b) Maximum load

Figure 17. Bending moment profiles at loading extremes during cyclic loading

A final point of interest is the similarity in the final load transfer response of the densified material near the pile head, with that for piles that were initially driven into position at 160 g (as opposed to being jacked at 1 g). Figure 18 shows the load transfer response at a depth of 1 diameter for the two cases, which may be taken to apply to pile response in a fully densified calcareous sand.

Calcareous Silt

By contrast with the sand, the cyclic tests conducted in calcareous silt showed no evident effect of consolidation during the time scale of the loading. A similar style of loading was applied as for the tests in sand, and a typical response is shown in Figure 19, with bending moment profiles at loading extremes shown in Figure 20.

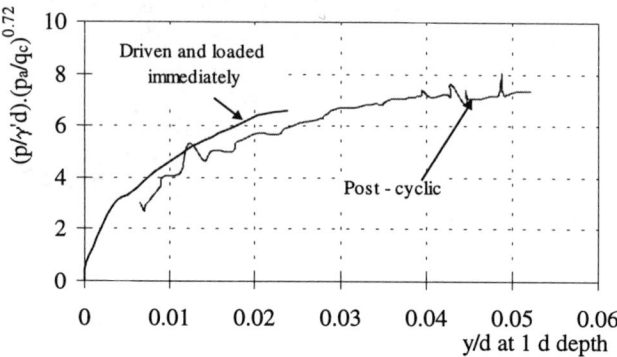

Figure 18. Comparison of load transfer curves for jacked pile after cyclic loading, and initial monotonic response of driven pile in calcareous sand

It may be seen that complete liquefaction occurs at the depth shown, and typically down to at least 3 or 4 pile diameters, accompanied by an increase in pile head displacement by a factor of over 6. There was also evidence of rigid body lateral motion of the pile, which was only embedded by 18 diameters. It is clear that liquefaction, and the consequential increase in pile head displacement and maximum bending moment, would be a critical design issue for piles in calcareous silt.

CONCLUSIONS

This paper has presented a brief summary of an extensive series of centrifuge model tests of laterally loaded piles in calcareous sand and silt. The results have confirmed that, in calcareous sand, the formulation proposed by Wesselink et al (1988) for piles in Bass Strait works extremely well for the finer, softer, materials of the North West Shelf. The original load transfer formulation of Wesselink et al has been generalised by replacing the depth correction term by a term linked to the local cone resistance of the soil. It was found that this normalisation can be extended right up to cemented materials, with cone resistance as high as 10 MPa at shallow depths.

The tests in calcareous silt indicated essentially undrained behaviour, where the load transfer response could be normalised by the undrained strength. A power law relation between net pressure and normalised displacement was still found to provide the best fit to the data.

For both soil types, the deduced load transfer curves were found to be very different from conventional API formulations for silica sands and clays. This difference was also evident from cyclic loading tests, which showed partial liquefaction followed by consolidation and densification for the tests in sand, but almost complete liquefaction and loss of resistance in the upper part of the pile for the tests in silt. The gradual build up of residual bending moments in the pile at large depths, due to accumulated lateral displacement under biased cyclic loading, is an important feature that cannot be captured by design approaches based solely on softened load transfer curves.

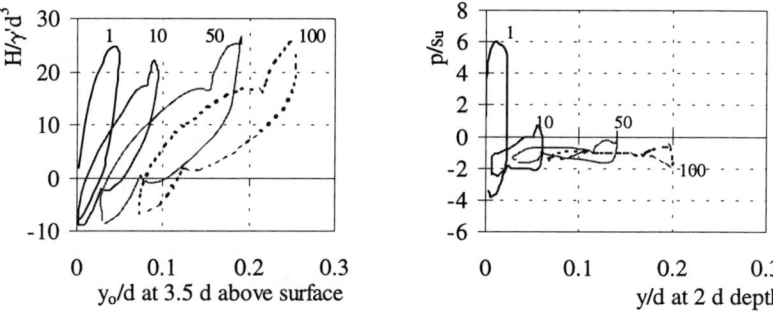

Figure 19. Pile head response and load transfer curves for during cyclic loading

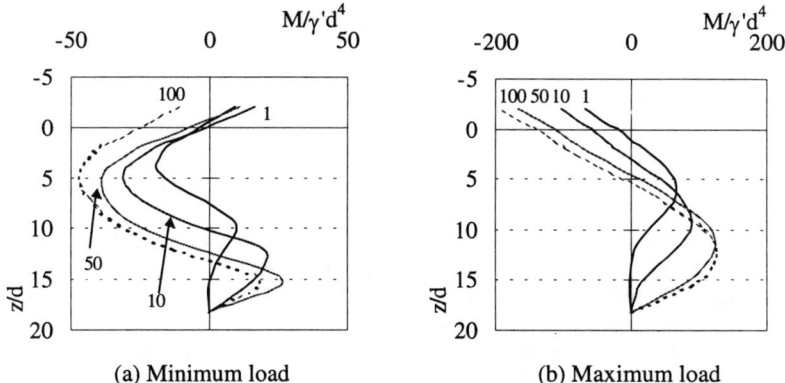

(a) Minimum load   (b) Maximum load

Figure 20. Bending moment profiles at loading extremes during cyclic loading

ACKNOWLEDGEMENTS

The work described here forms part of the research program of the Special Research Centre for Offshore Foundation Systems, established and supported under the Australian Research Council's Research Centres Program. Financial support for the project was provided by the Minerals and Energy Research Institute of Western

Australia, West Australian Petroleum Company and Woodside Energy. The first author was supported by an Australian Postgraduate Award during his doctoral research. This financial support is gratefully acknowledged. Particular thanks are due to Mr Don Herley and Mr Rudi Reimann for assisting with the centrifuge model tests.

REFERENCES

American Petroleum Institute (1993) "Recommended practice for planning, designing and constructing fixed offshore platforms – working stress designs" 20th ed. API-RP2A-WSD, A.P.I., Washington, D. C.

Craig, W. H. (1980) "Installation studies for model piles", Proc. Symp on Application of Centrifuge Modelling to Geotechnical Design, Manchester, 440 – 452.

Dunnavant, T. W. & O'Neill, M. W. (1989) "Experimental p – y model for submerged stiff clay" J. of Geot. Eng. Div., ASCE, 115(GT1), 95 – 114.

Dyson, G. J. & Randolph M. F. (1997) "Load transfer curves for piles in calcareous sand", Proc. 8th Int. Conf. On the Behaviour of Offshore Structures. Elsevier Science, England, 3, 245 – 258.

Dyson, G. J. & Randolph M. F. (1998) "Installation effects on lateral load-transfer curves in calcareous sands", Proc. of Centrifuge 98 Conf., Balkema, Rotterdam, 545-550.

Dyson, G. J (1999) Forthcoming PhD Thesis, The University of Western Australia

Finnie, I.M.S. & Randolph, M.F. (1994), "Punch-through and liquefaction induced failure of shallow foundations on calcareous sediments", Proc. Int. Conf. On Behaviour of Offshore Structures, BOSS '94, Boston, 217-230.

Fookes, P.G. (1988) "The geology of carbonate soils and rocks and their engineering characterisation and description", Engineering for Calcareous Sediments, Balkema, Rotterdam, 2, 787-806.

Matlock, H. (1970) "Correlations for design of laterally loaded piles in soft clay" Proc. of 2nd Annual Offshore Technology Conference, Houston, OTC1204.

Murff, J.D. (1987), "Pile capacity in calcareous sands: state of the art", J. Geot. Eng. Div., ASCE, 113(GT5), 490-507.

O'Neill M. W. & Murchinson, J. M. (1983) "An evaluation of p-y relationships in sand" Report to the American Petroleum Institute.

Randolph, M.F., Jewell, R.J., Stone, K.J.L. & Brown, T.A. (1991) "Establishing a new centrifuge facility", Proc. Centrifuge 91, Boulder, Colorado, 3-9.

Reese, L. C., Cox, W. R. & Koop, F. D. (1974) "Analysis of laterally loaded piles in sand" Proc. of 6th Annual Offshore Technology Conference, Houston, OTC2080.

Semple, R.M. (1988) "The mechanical properties of carbonate soils", Engineering for Calcareous Sediments, Balkema, Rotterdam, 2, 807-836.

Stewart, D.P. & Randolph, M.F. (1994) "T-Bar penetration testing in soft clay", J. Geot. Eng. Div., ASCE, 120(GT12), 2230-2235.

Wesselink, B. D., Murff, J. D., Randolph, M. F., Nunez, I. L. & Hyden, A. M. (1988) "Analysis of centrifuge model test data from laterally loaded piles in calcareous sand" Engineering for Calcareous Sediments, Balkema, Rotterdam, 1, 261 – 270.

SOIL-STRUCTURE ANALYSIS OF FOUNDATIONS WITH UNIFIED CONSTITUTIVE MODELS FOR SOILS AND INTERFACES

Chandrakant Desai[1]

Abstract

This paper presents a review of the analyses of problems in shallow and deep foundations performed by the author and coworkers. The importance of appropriate constitutive models for soils and interfaces, and associated laboratory testing is discussed. A recently developed unified constitutive modelling approach based on the disturbed state concept (DSC) allows for elastic, plastic and creep behavior, and microcracking leading to degradation and softening for both solids and interfaces. It also allows for the identification of liquefaction based on critical disturbance at which microstructural instability occurs. The DSC model is implemented in a nonlinear finite element procedure based on generalized Biot's theory for coupled deformation and pore water pressure behavior. The numerical procedure is used to solve three typical problems: pile in clay, pile in sand, and shake table test. The predictions are compared with field and laboratory observations for stresses, shear transfer, pore water pressures and liquefaction. The DSC model and the computer procedure can provide improved and realistic simulation of soil-structure interaction problems in deep foundations.

INTRODUCTION

Interaction between geologic and structural materials play an important role in the analysis and design of (deep) foundations subjected to static and dynamic loading. Usually, the spatial extents involved require three-dimensional analysis; however, often one- and two-dimensional idealizations are also used for specific configurations. Appropriate characterization of the behavior of materials and interfaces assume vital importance for realistic predictions of the behavior of foundations. Hence, it is necessary to develop, calibrate based on (laboratory) tests,

---

[1] Regents' Professor, Dept. of Civil Eng. and Eng. Mech., The Univ. of Arizona, Tucson, AZ 85721

and utilize appropriate constitutive models for soils and interfaces. The behavior of saturated soils and interfaces require additional considerations compared to those for dry conditions. Under cyclic and dynamic loads, saturated soils and interface may require consideration of instability and liquefaction, in addition to computations of displacements, strains, stresses and pore water pressures.

SCOPE

The subject of deep foundations is very wide in scope. It is not intended here to present a comprehensive review of many works; it is appropriate to mention the significant contributions of Professor Lymon Reese in the area of deep (pile) foundations.

In this paper, the main attention is given to the contributions of the author and his coworkers. Brief statements are first provided on earlier contributions in the areas of one-, two- and three-dimensional analyses of axially and laterally loaded piles, pile groups and anchors in soils. Then, major attention is given to recent contributions including the unified constitutive model for soils and interfaces, definition of the model parameters based on laboratory tests, validation of the model, its implementation in nonlinear computer (finite element) procedures, and solution of practical (field) problems. The latter includes dynamic analysis of piles in a marine clay, in a saturated sand, and simulated soil-structure problem in a shake table test, including liquefaction.

PREVIOUS WORK: A BRIEF REVIEW

The author and coworkers have been involved in analysis of various problems in shallow and deep foundations. Desai and Reese (1970) analyzed footings on cohesive soils using axisymmetric finite element method including validations with respect to laboratory model tests. Here, a nonlinear elastic model simulated by using piecewise linear segment representation of stress-strain behavior was used. Analysis (axisymmetric) of axially loaded piles in sands with a special interface representation, and including comparison with field data, have been reported by Desai (1974) and Armaleh and Desai (1987). Here nonlinear elastic hyperbolic representation for soils was used. Computer predictions for axially and laterally loaded piles using one-dimensional idealization have been presented in Desai and Kuppusamy (1979). Here, nonlinear elastic representation for soils was used.

Special equivalent two-dimensional idealization of a pile-group supported navigation lock using the finite element method with nonlinear hyperbolic model for soils and interfaces were used to study and validate the field behavior (Desai, et al., 1974). Pile groups were analyzed by idealizing piles as one-dimensional and pile cap as plate elements in Desai, et al. (1981). Three-dimensional analysis of piles using nonlinear elastic hyperbolic model was performed by Desai and Appel

(1976). Three-dimensional finite element analysis of soils, interfaces and piles in a pile-pile cap-soil system, tested in the laboratory using nonlinear elastic idealizations for soils and interfaces was reported by Muqutadir and Desai (1986). Three-dimensional analysis of short structures (tillage tools) moving in soils including large strains and deformations, and with a plastic hardening (cap) model, was reported by Desai, et al. (1982). Three-dimensional analysis for anchors in sands tested in the field was reported by Desai, et al. (1986a), by using a version of the hierarchical single surface (HISS) plasticity model (see below). Nonlinear three-dimensional seismic analysis of soil-pile-structure system is presented by Cai, et al. (1998).

A majority of the works above employed simplified stress-strain models such as nonlinear hyperbolic and plasticity to characterize the soil and interface behavior. It has, however, been recognized that such models do not provide a realistic characterization of soils and interfaces. For instance, they do not account for volume changes before peak stress, stress path dependent behavior, effect of both deviatoric and volumetric plastic strains on hardening and microcracking leading to softening, and degradation during cyclic loading. As a result, the author and coworkers have developed improved models that can provide unified characterizations for both soils and interfaces.

The constitutive models described below are based on the recent disturbed state concept (DSC) with the associated hierarchical single surface (HISS) plasticity approach. This is a unified approach which provides for the characterization of "solids" (soils, rocks, concrete, ceramics, metal alloys) and interfaces and joints in the same mathematical framework. It allows, in a hierarchical manner, elastic, plastic and creep strains, microcracking, fracture leading to softening, and healing or stiffening under thermomechanical loading. The most important advantages of the approach is that it provides the user flexibility to choose a version(s), elastic, elastoplastic, viscoplastic, and disturbance (damage and softening) depending upon the behavior of the given material. In other words, only parameters relevant to the given version for a specific problem are required to be determined from laboratory tests and used.

The DSC model and its versions have been applied successfully to characterize behavior of cohesionless soils, cohesive saturated soils, rocks, concrete, ceramic composites, metal alloys like solders in electronic packaging problems, and silicon with dislocation and impurities, and interfaces and joints (Desai, 1995, 1998; Desai and Fishman, 1991; Desai and Ma, 1992; Katti and Desai, 1995; Desai and Toth, 1996; Alanazy, 1996; Desai, et al., 1998a, 1998b; Desai and Rigby, 1997; Park and Desai, 1997; Shao and Desai, 1998). The model has been implemented in static and dynamic finite element procedures for dry and saturated materials and interfaces (Desai, et al. 1997; Shao and Desai, 1998). Brief descriptions of the DSC model and the computer procedures as they relate to the pile and soil-structure problems included in this paper, are given below.

THE DISTURBED STATE CONCEPT

The disturbed state concept (DSC) is based on the simple idea that a deforming material element can be considered to be a mixture of two (or more) reference component states. In the beginning, the material element can be treated as relatively intact (RI) or as continuum. As the load is applied, the RI part transforms, through a process of the natural adjustment of its microstructure, to fully adjusted (FA) state; the element thus involves randomly distributed material parts in the RI and FA states, Fig. 1.

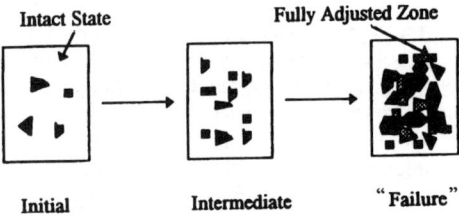

Figure 1. Relative Intact and Fully Adjusted States

The observed behavior of the material involves the coupled response due to the interaction of the responses of the parts in the reference RI and FA states. It is expressed in terms of the two responses by using the disturbance (D), which represents an interpolation and coupling mechanism, and the deviation of the observed behavior from those of the reference states, Fig. 2.

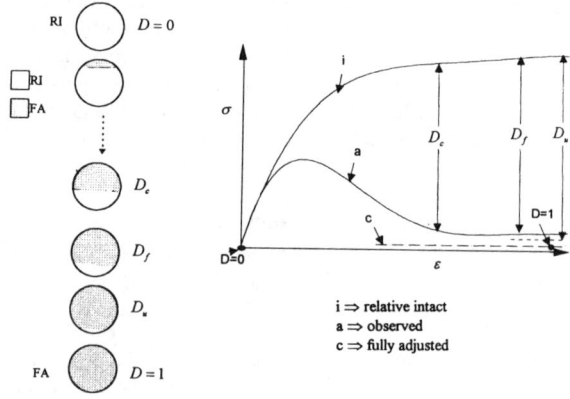

Figure 2. Representations of DSC

Based on the equilibrium of forces in the RI and FA parts and the observed force, the following incremental constitutive equations are obtained:

$$d\sigma^a = (1-D)\underset{\sim}{d\sigma^i} + D\underset{\sim}{d\sigma^c} + dD(\underset{\sim}{\sigma^c} - \underset{\sim}{\sigma^i}) \tag{1a}$$

or $\qquad d\underset{\sim}{\sigma^a} = (1-D)\underset{\sim}{C^i} \, d\underset{\sim}{\varepsilon^i} + D\underset{\sim}{C^c} \, d\underset{\sim}{\varepsilon^c} + dD(\underset{\sim}{\sigma^c} - \underset{\sim}{\sigma^i}) \tag{1b}$

where σ and ε = stress and strain vectors, respectively, a, i and c denote observed, RI and FA states, respectively, d denotes increment or rate, $\underset{\sim}{C}$ = constitutive matrix and D = (scalar) disturbance, which can be expressed as a tensor if appropriate laboratory tests are available (Desai and Toth, 1996). In the DSC, the stresses and strains in the RI and FA states can be different; however, if they are assumed to be compatible (i.e., $d\underset{\sim}{\varepsilon^a} = d\underset{\sim}{\varepsilon^i} = d\underset{\sim}{\varepsilon^c}$), the formulation is simplified and only Eqs. (1) are required. Equations (1) include elastic, elastoplastic (or viscoplastic) models or versions as special cases when D = 0. If D ≠ 0, disturbance (microcracking and damage) is allowed, Desai and Toth (1996), Desai (1995, 1998).

The RI response can be characterized by using such continuum theories as elasticity, elastoplasticity and viscoplasticity, with thermal effects. Here, the elastoplastic model in the HISS approach (Desai, et al., 1986) described later, is used. In the FA state, the material parts can carry no stress at all, as they act as cracks or voids, like in the classical damage model, or it can carry hydrostatic stress but no shear stress like a constrained liquid, or it can carry shear stress reached up to that state for a given mean pressure and deform in shear without volume change as in the critical state concept (Roscoe, et al., 1957; Desai, 1995). The latter two are considered to be realistic and provides for important interaction between the RI and FA parts, and are used often. In this paper, the FA state is characterized by using the critical state concept.

Hierarchical Single Surface (HISS) Plasticity Model

The δ_0-associative, isotropic hardening version of the HISS approach (Desai, et al., 1986b) is used to characterize the behavior of the RI part. The yield function, F, that allows continuous yielding with continuous surface used is given by

$$F = \bar{J}_{2D} - \left(-\alpha\bar{J}_1^n + \gamma\bar{J}_1^2\right)\left(1 - \beta S r\right)^{-0.5} = 0 \tag{2}$$

where J_{2D} = second invariant of the deviatoric stress tensor, S_{ij}, J_1 is the first invariant of the total stress tensor, S_r = stress ratio = $\sqrt{27}/2\,J_{3D} \cdot J_{2D}^{-1.5}$, J_{3D} is the third invariant of S_{ij}, the overbar denotes nondimensionalization with respect to the atmospheric pressure, p_a, γ and β = parameters associated with the ultimate (or failure) surface, n = parameter associated with transition from contraction to dilation, $\bar{J}_1 = (J_1 + J_{1s})/p_a$, J_{1s} is the bonding (tensile or cohesive strength), and α is the hardening or growth function given by

$$\alpha = \frac{a_1}{\xi^{\eta_1}} \tag{3}$$

where a_1 and η_1 = hardening parameters, ξ is the plastic strain trajectory:

$$\xi = \int \left(d\underset{\sim}{\varepsilon}^p \, d\underset{\sim}{\varepsilon}^p \right)^{1/2} \tag{4}$$

and $\underset{\sim}{\varepsilon}^p$ = plastic strain vector. The plots of F in $J_1 - \sqrt{J_{2D}}$ spaces is shown in Fig. 3. As it is shown (Desai, 1995, 1998), the yield function, Eq. (2), can include as special cases various other plasticity models such as classical (von Mises, Drucker-Pragr, Mohr Coulomb), critical state, cap, Vermeer, Lade, etc.

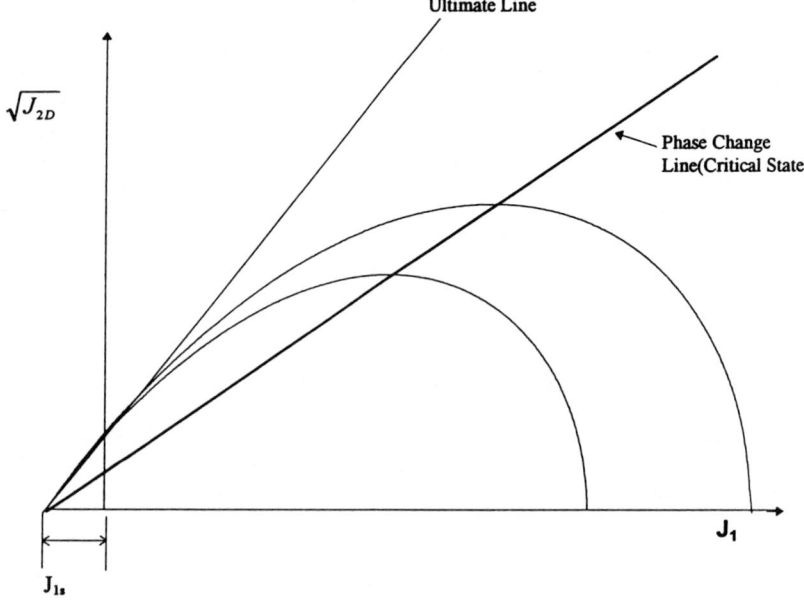

Figure 3. Plots of Yield Surface, F

Fully Adjusted or Critical State

For the material in the FA state characterized by using the critical state concept (Roscoe, et al., 1958; Desai, 1995), the following two equations are involved:

$$\sqrt{J_{2D}^{c}} = \overline{m}\, J_{1}^{c} \tag{5a}$$

$$e^{a} = e_{o}^{c} - \lambda\, \ell n(J_{1}^{c}/3p_{a}) \tag{5b}$$

where c denotes critical, $\overline{m}$ is the slope of the critical state line, λ is the slope of consolidation (hydrostatic) line, and e = void ratio.

Unloading and Reloading

Details of the simplified procedures to simulate unloading and reloading are given elsewhere (Desai, et al., 1997; Shao and Desai, 1998). Their characterization requires two additional parameters, the slope of stress-strain curve at the end of unloading (E^{u}) and the irreversible strains during unloading cycle (ε^{p}).

Disturbance Function

The disturbance function, D, is defined based on phenomenological considerations. It is expressed as

$$D = D_{u}\left(1 - e^{-A\xi_{D}^{Z}}\right) \tag{6}$$

where D_{u} = ultimate disturbance, A and Z are parameters, and ξ_{D} is the trajectory of deviatoric plastic strains. Plot of disturbance vs ξ_{D} is shown in Fig. 4. The disturbance is expressed based on laboratory test data in terms of measured stresses, void ratio (volume), effective stress and nondestructive properties such as ultrasonic velocities (Desai, 1995, 1998). For example, D is expressed as (Fig. 2)

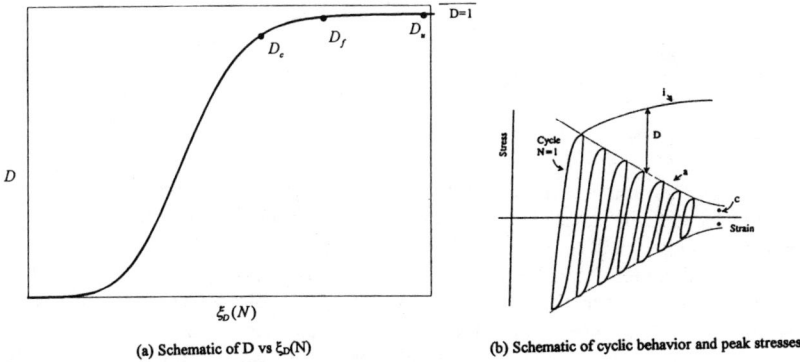

(a) Schematic of D vs ξ_{D}(N)　　　　　　　(b) Schematic of cyclic behavior and peak stresses

Figure 4. Disturbance Function and Cyclic Behavior

$$D = \frac{\sigma^i - \sigma^a}{\sigma^i - \sigma^c} \qquad (7)$$

where σ = measured stress such as axial stress (σ_1), stress difference ($\sigma_1 - \sigma_3$), shear stress (τ) and second invariant, $\sqrt{J_{2D}}$.

Instability and Liquefaction

A deforming material element undergoes microstructural changes that may involve threshold transitions such as transitions from contraction to dilation, peak stress and ultimate conditions. One such transition is related to the microstructural instability that can lead to liquefaction. This state is identified on the basis of the critical disturbance, D_c, at which the rate of increase in D stabilizes towards its "saturation" state, Fig. 4. Details of threshold transitions and microstructural instability leading to cyclic fatigue failure and liquefaction are given in Desai, et al. (1998c). In the dynamic analysis herein, the critical disturbance criterion is used to identify the initiation of liquefaction.

INTERFACES

The foregoing formulation for solids can be specialized to characterize behavior of interfaces (or joints) idealized as thin layers between structural and geologic materials (Desai, et al., 1984). Details are given by Desai and Ma (1996), Desai and Fishman (1991), Desai and Rigby (1997), Park and Desai (1997), and Shao and Desai (1998). For a planar, two-dimensional interface, Fig. 5(a), Eq. (2), specializes as

$$F = \tau^2 - \alpha\, \sigma_n^{\bar{n}} - \gamma\, \sigma_n^2 = 0 \qquad (8)$$

where τ and σ_n = shear and normal stresses, $\bar{n}$ = state or phase change parameter, γ is the ultimate parameter, and α is the growth or hardening function; plots of F in τ - σ_n space are shown in Fig. 5(b).

PARAMETERS

The parameters in the DSC model have physical meanings as they are related to specific states during deformation. Their number is lower than that in other available models of comparable capabilities. For example, for the general disturbance model that allows for elastic and plastic strains including continuous yielding, and microcracking leading to softening, the parameters involved are two elastic (E, ν); four plasticity (γ, β, a_1 and η_1), three critical state (e_o, λ, $\bar{m}$) and three for disturbance (A, Z, and D_u). These parameters can be found from standard triaxial (or multiaxial) tests; the disturbance parameters are found based on the softening or cyclic degradation response, Fig. 4.

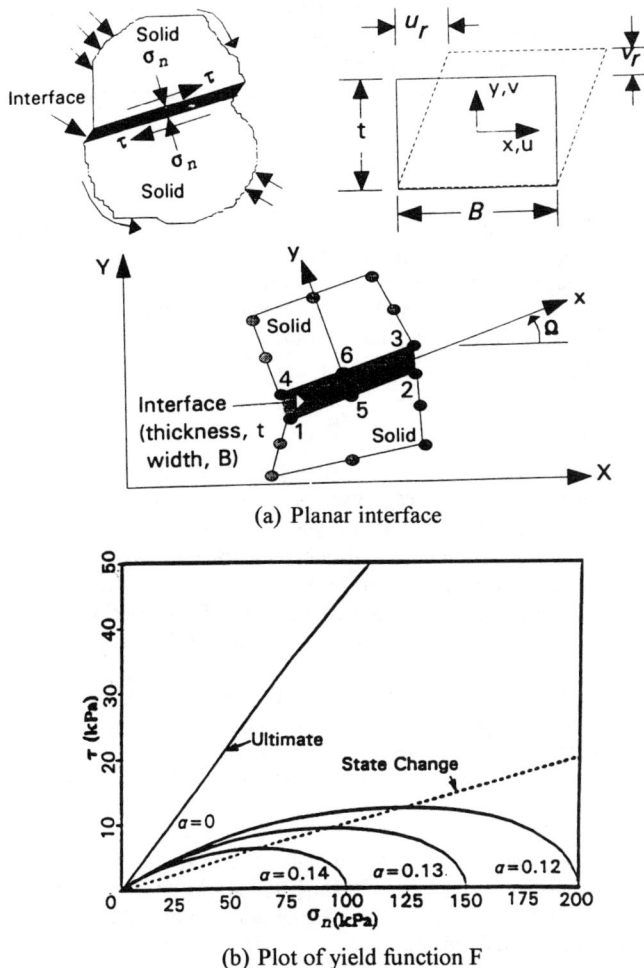

(a) Planar interface

(b) Plot of yield function F

Figure 5. Interface and Yield Surface

FINITE ELEMENT PROCEDURE

The generalized Biot's theory is used to formulate the finite element (FE) procedure for coupled deformation and pore water pressure responses. The procedure allows different versions; e.g., dynamic, consolidation and static. The DSC model for loading, unloading and reloading is implemented in the FE procedure. The results from the code include time dependent displacements, strains, stresses (total and effective), pore water pressures and disturbance. Contours of disturbances with time in the finite element mesh provide identification of zones in which the critical disturbance, D_c, is reached. It denotes initiation of liquefaction, and allows tracing of the growth of disturbance and liquefaction under subsequent loading cycles.

APPLICATIONS

A number of problems involving static, repetitive and dynamic loading with elastoplastic, viscoplastic and DSC models have been solved by using the computer code. In most cases, the computer predictions are compared with observations in the field and simulated laboratory models. Here, three typical examples involving a pile in clays, pile in sand, and dynamic soil-structure interaction in a shake table test are presented.

Pile in Marine Clay

Instrumented pile segments were tested in the field at Sabine, Texas by Earth Technology Corporation (ETC, 1986). The field program included tests for measurement of *in situ* stresses, installation of pile segments with different diameters and cutting shoes, monitoring consolidation, performing axial tension tests at different levels of consolidation and cyclic axial load tests at the end of consolidation. The measurements included total lateral stresses and pore water pressures at the pile wall, and shear transfer versus pile displacement.

Undisturbed specimens of soil were obtained by using Shelby tubes and specially designed rectangular tubes for cyclic triaxial and multiaxial testing, respectively. Interface tests between the marine clay and pile (steel) were conducted by using the cyclic multi degree-of-freedom shear device (Desai and Rigby, 1997).

The results presented here are based on the finite element procedure including the disturbed state model for loading, unloading and reloading (Desai, et al., 1997; Shao and Desai, 1998). Various sequences such as initial conditions, pile driving, consolidation, axial tension tests and cyclic loading were simulated in the finite element analysis. The FE mesh is shown in Fig. 6. The inside nodes on the soil-pile interface were subjected to cyclic (vertical) displacements as applied in the field.

Figures 7 (a) and (b) show comparisons between the predictions with the DSC model and field behavior after consolidation and one-way cyclic loading, respectively. They also show predictions by using anisotropic hardening plasticity (HISS) model without degradation (disturbance) (Wathugala and Desai, 1993; Desai, et al., 1993). Figures 8(a) and (b) show comparisons between predictions and field test data for shear transfer vs time and pore water pressure vs time, respectively, for two-way cyclic load test. It can be seen that the DSC predictions correlate very well with the field data, and the provision for degradation (disturbance) provides improved predictions compared to those from the HISS model.

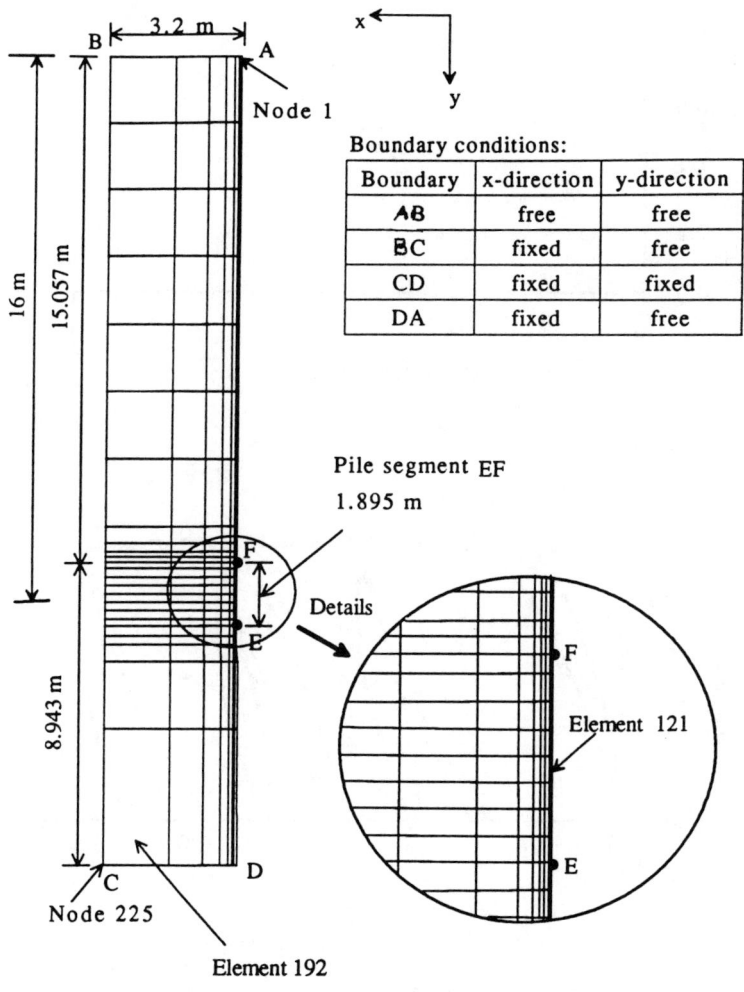

Figure 6. Finite Element Mesh (Shao and Desai, 1998)

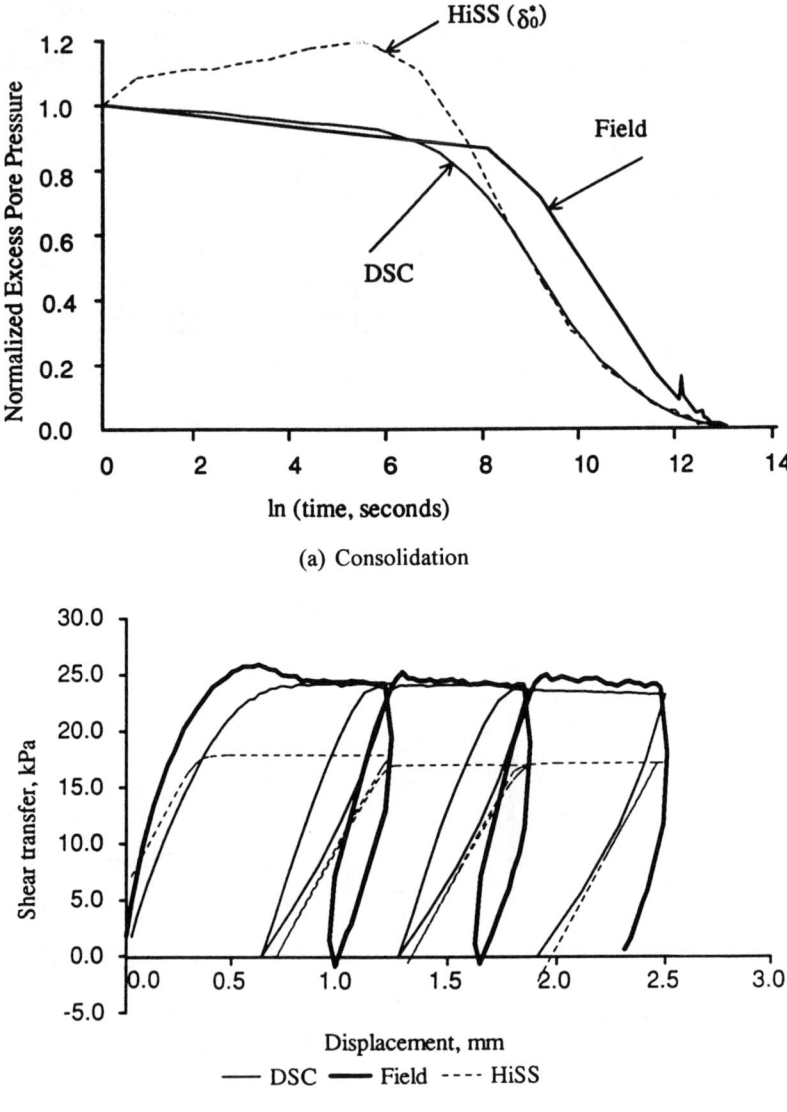

(a) Consolidation

(b) One-way

Figure 7. Comparison between Field Consolidation and One-way Behavior

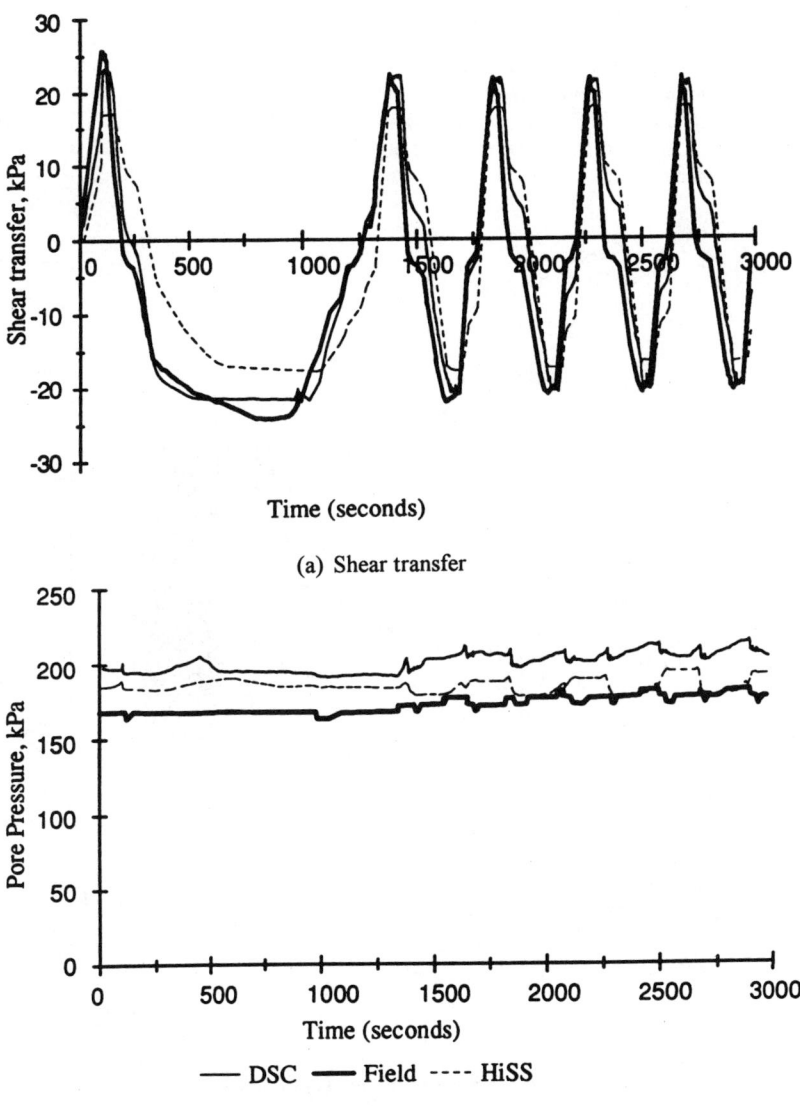

(a) Shear transfer

— DSC —— Field ---- HiSS

(b) Pore water pressure

Figure 8. Comparison between Field Measurement and Predictions from
DSC and HISS Models

Pile in Saturated Sand

In order to study the interaction and liquefaction behavior, a steel pile in saturated Ottawa sand was simulated by using the DSC model and the finite element procedure. Figure 9(a) shows the pile-sand details, and Fig. 9(b) shows mesh details near the interface in which sinusoidal displacement loading was applied at the nodes.

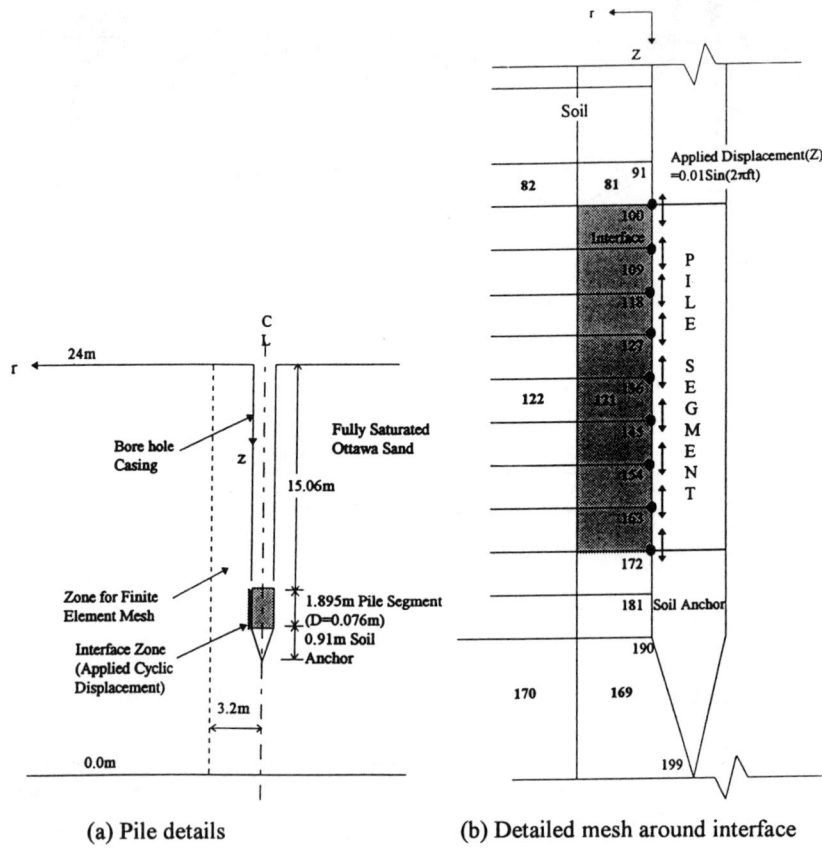

(a) Pile details   (b) Detailed mesh around interface

Figure 9. Pile in Saturated Sand (Park and Desai, 1997)

The cyclic DSC behavior of the soil was characterized from a series of tests using the multiaxial test device with 10 x 10 x 10 cm saturated specimens (Gyi, 1996). The interface DSC behavior was characterized based on a series of interface tests between steel and sand using the CYMDOF-P device (Alanazy, 1996).

Computer analyses were performed for two conditions: (1) no interface, i.e., pile and soil are compatible, and (2) with interface, i.e., relative motions are allowed between the soil and pile. Figures 10(a) and (b) show computed (vertical) displacements with time at typical notes 136 and 137; the former is on the pile and

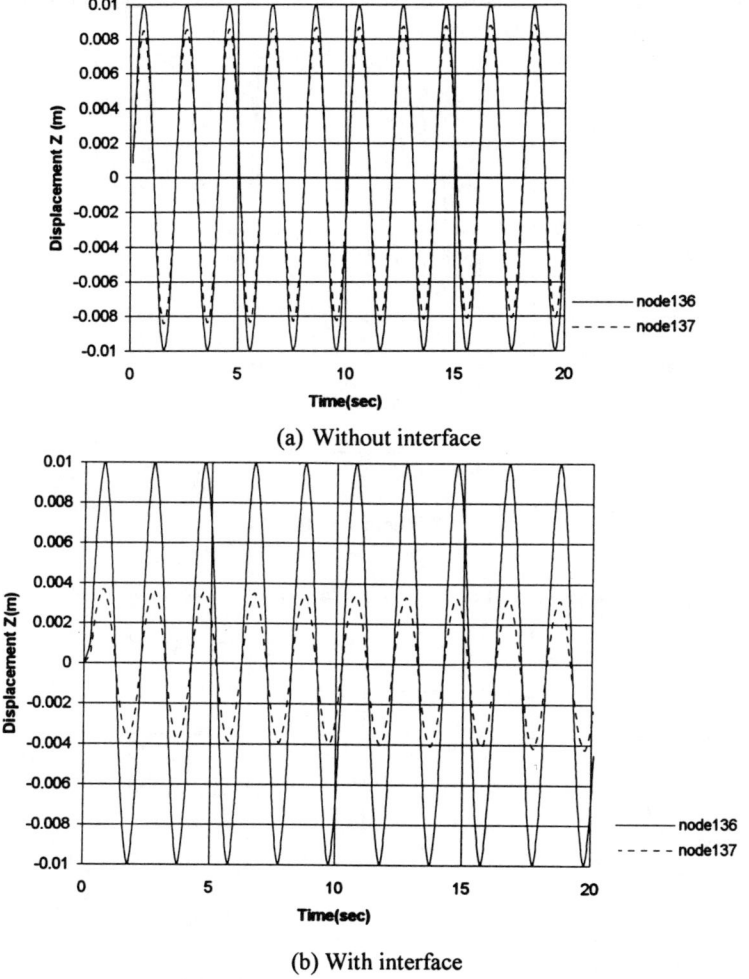

(a) Without interface

(b) With interface

Figure 10. Displacement at Typical Nodes

the latter in the soil. Figures 11(a) and (b) show computed pore water pressures in typical (soil) element 121 with time. Figures 12(a) and (b) show disturbance in soil elements 121 and 122, Fig. 9(b), with time. It can be seen that provision of the interface, i.e., relative motions, modify the computed results significantly. There are significant relative displacements between pile and soil with the interface provision. The variation in the magnitudes of cyclic pore water pressures are much lower with the interface. The disturbances in the vicinity of the interface are generally smaller with the interface.

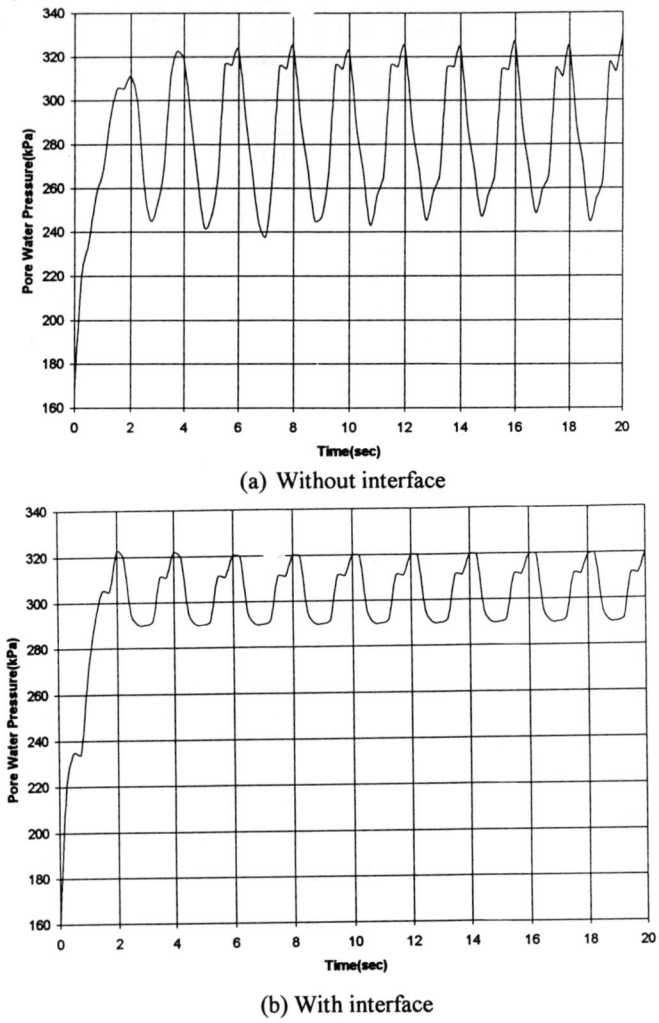

(a) Without interface

(b) With interface

Figure 11. Pore Water Pressure in Element 121

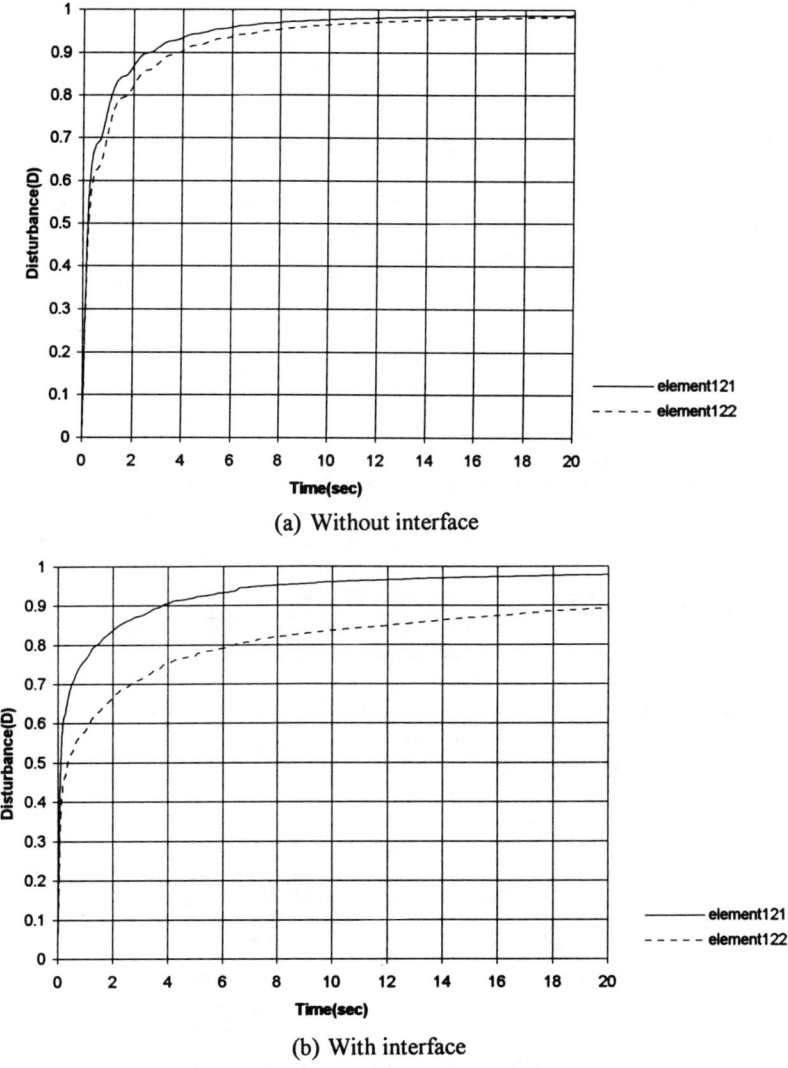

(a) Without interface

(b) With interface

Figure 12. Disturbances in Typical Elements

Based on laboratory tests (Gyi, 1996; Park and Desai, 1997; Desai, et al., 1998c), it was found that liquefaction initiated at the average critical disturbance of about 0.84. Figure 12(a) without interface shows that the value of D_c is reached after about two seconds (i.e., four cycles of loading, frequency = 0.5 Hz). On the other hand, Fig. 12(b) shows that D_c is reached in the vicinity of the interface at average cycles of about 12 cycles. Thus, provision of interface indicate liquefaction potential at higher cycles.

Shake Table Test: Soil-Structure Interaction

Akiyoshi, et al. (1996) reported shake table tests with a saturated sand. Figure 13(a) shows the details of the model, which was subjected to cyclic sinusoidal displacement (x) at the bottom of the test box, given by

$$X = \bar{x} \sin\left(2\pi\, ft\right) \tag{9}$$

where $\bar{x}$ = 0.0013 m, frequency f = 5.0 HZ and t = time.

The finite element mesh is shown in Fig. 13(b); the idea of repeating side boundaries was used. The material parameters for the Ottawa sand used were obtained from multiaxial cyclic tests (Gyi, 1996).

Figure 14 shows comparisons between computed and observed pore water pressures at node (137) at the depth of 300 mm. Figure 15 shows the growth of disturbance near node 137 with time. The critical disturbance, D_c = 0.84, at which liquefaction initiated. This compared well with the observations in which it was found that liquefaction occurred at time ≈ 2.00 secs, Fig. 14. Plot of contours of disturbance at different time (t = 0.50, 1.0, 2.0, 10.0 secs) showed that liquefaction initiated in the zones at and below the depth = 300 mm when D_c ≈ 0.84 was reached (Park and Desai, 1997). Then, as the disturbance increased with cycles beyond D_c = 0.84, liquefaction expanded essentially in the entire zone at t = 10 secs, as was observed in the laboratory test.

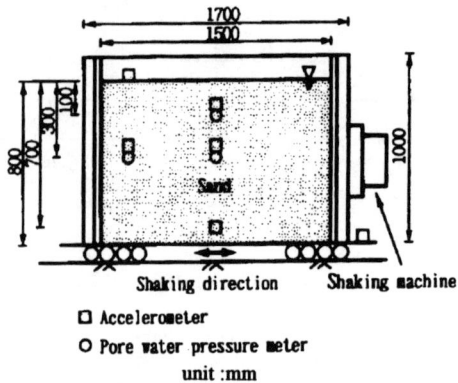

Figure 13. (a) Shake Table (Akiyushi, et al., 1996)

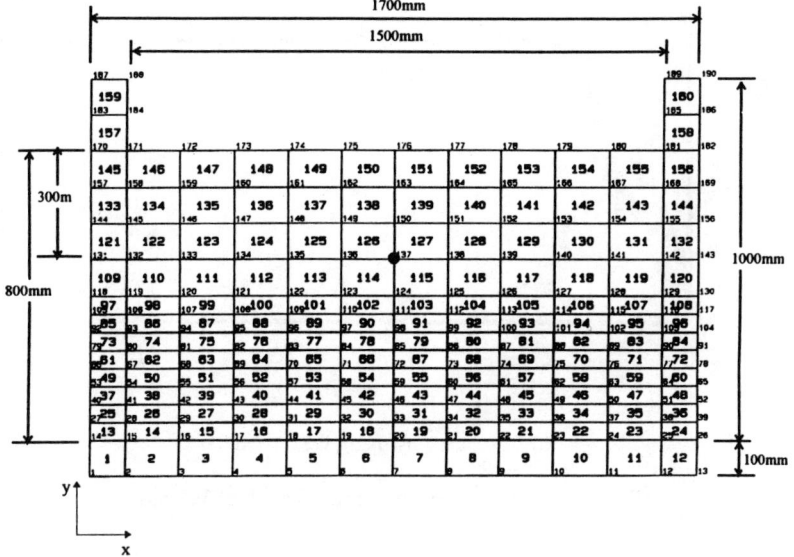

(a) Finite Element Mesh

Figure 13. Set Up Model Test and Mesh

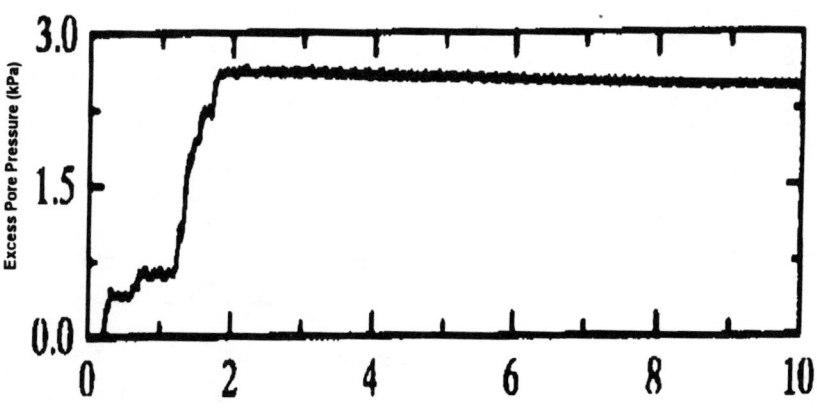

Figure 14. (a) Measured (Akiyoshi, et al., 1996)

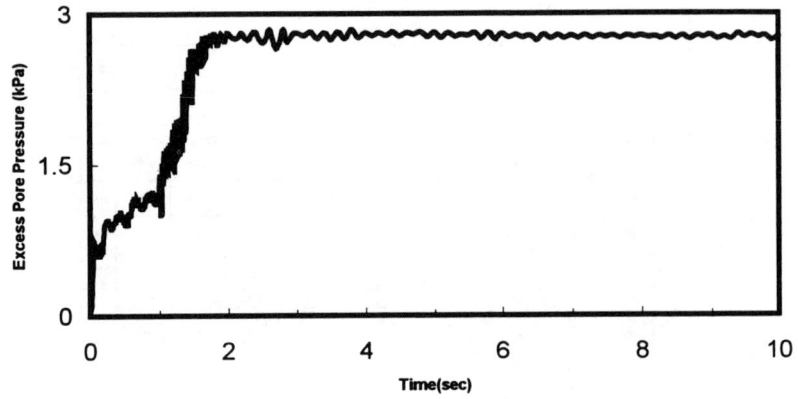

(b) Computed by using DSC model

Figure 14. Excess Pore Pressure at Depth of 300 mm

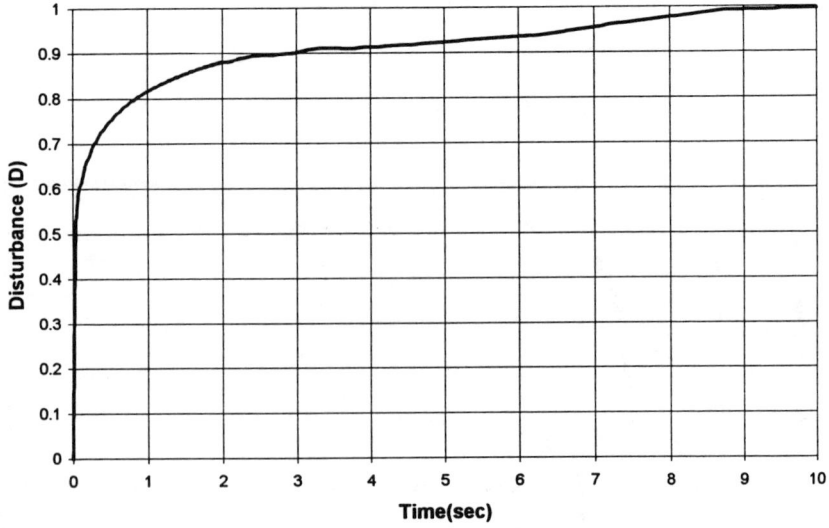

Figure 15. Disturbance at depth of 300 mm

SUMMARY AND CONCLUSIONS

One-, two- and three-dimensional analyses of foundations including soil-structure interaction effects performed by the author and coworkers are discussed. A brief review of previous work involving footings, axially and laterally loaded piles, pile-pile cap-foundation systems and anchors is provided. The need for improved and realistic constitutive models for soils and interfaces is identified, and the recently developed unified and hierarchical model based on the disturbed state concept (DSC) is described. The required material parameters, their determination from laboratory tests and validations for a wide range of materials and interfaces, are discussed. The DSC model is implemented in a general nonlinear finite element procedure. Typical practical problems involving piles in clays and sands and shake table test are solved by using the numerical procedure. The numerical predictions are compared with observed data in the field and laboratory tests. The subject of liquefaction during dynamic loading is considered and the idea of the critical disturbance in the DSC is used to identify initiation of liquefaction and its growth.

It is believed that the proposed unified DSC constitutive model for soils and interfaces and the associated computer procedure can provide improved analysis and design of dynamic and static soil-structure problems in foundation engineering.

ACKNOWLEDGMENTS

Parts of the research results were obtained under various grants (Nos. CSM 9115316, CES 8711764, MSM 8618914) from the National Science Foundation, Washington, DC. A number of students and coworkers contributed (see references cited) to the results presented herein. For the analysis and problems presented in the paper, contributions of Dr. C. Shao and Dr. I.J. Park are acknowledged.

REFERENCES

Akiyoshi, T., Fang, H.L., Fuchida, K., and Matsumoto, H. (1996). "A nonlinear seismic response analysis method for saturated soil-structure system with absorbing boundary." *Int. J. Num. and Analyt. Meth. Geomech.*, 20(5), 307-329.

Alanazy, A.S. (1996). "Testing and modeling of sand-steel interfaces under static and cyclic loading." *Ph.D. Dissertation*, Dept. of Civil Eng. and Eng. Mechs., The Univ. of Arizona, Tucson, Arizona.

Armaleh, S.H., and Desai, C.S. (1987). "Load deformation response of axially loaded piles." *J. Geotech. Eng. Div.*, ASCE, 113(12), 1483-1500.

Cai, X., Gould, P.L., and Desai, C.S. (1998). "Nonlinear analysis of 3-D seismic interaction of soil-pile structure systems and application." *J. of Eng. Structures*, in press.

Desai, C.S. (1974). "Numerical design-analysis of piles in sands." *J. Geotech. Eng. Div.*, ASCE, 100(6), 613-635.

Desai, C.S. (1995). "Constitutive modelling using the disturbed state as microstructure self-adjustment concept." Chapter 8 in *Continuum models for materials with microstructure*. H.B. Mühlhaus (editor), John Wiley, Chichester, United Kingdom.

Desai, C.S. (1998). *Mechanics of materials and interfaces: the disturbed state concept*. To be published.

Desai, C.S., and Appel, G.C. (1976). "Three-dimensional finite element analysis of laterally loaded structures." *Proc., 2nd Int. Conf. on Num. Meth. Geomech.*, ASCE, Vol. 1, 405-415.

Desai, C.S., Basaran, C., Dishongh, T., and Prince, J. (1998a). "Thermomechanical analysis in electronic packaging with unified constitutive model for materials and joints." *Components, Packaging and Manuf. Tech., Part B: Advanced Packaging*, IEEE Trans., 21(1), 87-97.

Desai, C.S., Dishongh, T., and Deneke, P. (1998b). "Disturbed state constitutive model for thermomechanical behavior of dislocated silicon with impurities." *J. of Appl. Physics*, 84, 59-77.

Desai, C.S., and Fishman, K.L. (1991). "Plasticity based constitutive model with associated testing for joints." *Int. J. Rock Mech. and Min. Sc.*, 28(1), 15-26.

Desai, C.S., Johnson, L.D., and Hargett, C.M. (1974). "Analysis of pile supported rocks." *J. Geotech. Eng. Div.*, ASCE, 100(9), 1009-1029.

Desai, C.S., and Kuppusamy, T. (1979). "Application of a numerical procedure for offshore piling," *Proc., Int. Conf. Num. Meth. in Offshore Piling*, Inst. of Civil Engrs., London.

Desai, C.S., Kuppusamy, T., and Alameddine, A.R. (1981). "Pile cap-pile group-soil interaction." *J. Struct. Eng. Div.*, ASCE, 107(5), 817-834.

Desai, C.S., and Ma, Y. (1992). "Modelling of joints and interfaces using disturbed state concept." *Int. J. Num. Analyt. Meth. Geomech.*, 16(9), 623-653.

Desai, C.S., Muqtadir, A., and Scheele, F. (1986a). "Interaction analysis of anchor-soil systems." *J. Geotech. Eng. Div.*, ASCE, 112(5), 537-553.

Desai, C.S., Park, I.J., and Shao, C. (1998c). "Fundamental yet simplified model for liquefaction instability." *Int. J. Num. Analy. Meth. Geomech.*, 22, 721-748.

Desai, C.S., Phan, H.V., and Perumpral, J.V. (1982). "Mechanics of three-dimensional soil-structure interaction." *J. of Eng. Mech. Div.*, ASCE, 108(5), 731-747.

Desai, C.S., and Reese, L.C. (1970). "Analysis of circular footings on layered soils." *J. of Soil Mech. and Found. Eng. Div.*, ASCE, 96(4), 1289-1310.

Desai, C.S., and Rigby, D.B. (1997). "Cyclic interface and joint shear device including pore pressure effects." *J. of Geotech. And Geoenv. Eng.*, ASCE, 123(6), 568-579.

Desai, C.S., Shao, C., and Park, I.J. (1997). "Disturbed state modelling of cyclic behavior of soils and interfaces in dynamic soil-structure interaction." *Proc., 9th Int. Conf. on Computer Meth. and Advances in Geomech.*, Wuhan, China.

Desai, C.S., Somasundaram, S., and Frantziskonis, G. (1986b). "A hierarchical approach for constitutive modeling of geologic materials." *Int. J. Num. Analyt. Meth. Geomech.*, 10(3), 225-257.

Desai, C.S., and Toth, J. (1996). "Disturbed state constitutive modelling based on stress-strain and nondestructive behavior." *Int. J. Solids and Struct.*, 33(11), 1619-1650.

Desai, C.S., Wathugala, G.W., and Matlock, H. (1993). "Constitutive model for cyclic behavior of cohesive soils II: applications." *J. Geotech. Eng.*, ASCE, 119(4), 730-748.

Desai, C.S., Zaman, M.M., Lightner, J.G., and Siriwardane, H.J. (1984). "Thin-layer element for interfaces and joints." *Int. J. Num. Analyt. Meth. Geomech.*, 8, 19-43.

Earth Technology Corporation (1986). "Pile segment tests – Sabine Pass: some aspects of the fundamental behavior of axially loaded piles in clay soils." *ETC Report No. 85-007*, Houston, Texas.

Gyi, M.M. (1996). "Multiaxial cyclic testing of saturated Ottawa sand," *M.S. Thesis*, Dept. of Civil Eng. and Eng. Mechs., The Univ. of Arizona, Tucson, Arizona.

Katti, D.R., and Desai, C.S. (1995). "Modelling and testing of cohesive soil using disturbed state concept." *J. of Eng. Mech.*, ASCE, 121(5), 648-658.

Muqtadir, A., and Desai, C.S. (1986). "Three-dimensional analysis of a pile-group foundation." *Int. J. Num. Analyt. Meth. Geomech.*, 10, 41-58.

Park, I.J., and Desai, C.S. (1997). "Disturbed state modeling for dynamic and liquefaction analysis." *Report to NSF*, Dept. of Civil Eng. and Eng. Mech., The Univ. of Arizona, Tucson, Arizona.

Roscoe, K.H., Schofield, A., and Wroth, C.P. (1958). "On the yielding of soils." *Geotechnique*, 8, 22-53.

Shao, C., and Desai, C.S. (1998). "Implementation of DSC model for dynamic analysis of soil-structure interaction problems." *Report to NSF*, Dept. of Civil Eng. and Eng. Mechs., The Univ. of Arizona, Tucson, Arizona.

Wathugala, G.W., and Desai. C.S. (1993). "Constitutive model for cyclic behavior of cohesive soils I: theory." *J. Geotech. Eng.*, ASCE, 119(4), 714-729.

Validity of the Reese Model for Pile Foundations Using Variational Principles

C.V. Girija Vallabhan[1], F.ASCE

Abstract

The forces applied at the top of a pile are transferred to the soil by means of friction and stresses in the soil on the sides as well as at the bottom of the pile. Design engineers like to settlement of piles under applied load conditions for which often field tests are performed. Also settlement characteristics of a pile are necessary to make a pile group-interaction analysis. For axially loaded piles, Reese introduced an empirical concept of a load transfer mechanism by assuming that the pile is compressible and the magnitude of load transferred into the soil through the sides depends on the movement of the pile at that level. Similar assumptions are made for laterally loaded piles also. Using variational principles, new models are developed and it is shown that the derived equations support the original empirical assumptions made by Reese even if the soil is layered with different material properties.

Introduction

The Reese model as it is known today is used to determine the vertical and horizontal displacements of piles as they are loaded at the top. The technique can be used to characterize piers, caissons etc., however, the word 'pile' is used here to represent all of them. Reese made two types of models; one for analysis of axially loaded piles and the other for laterally loaded piles with Matlock (1960). Many different types of models have been developed for analyzing pile foundations, but to the knowledge of the author, nobody has tried to compare the Reese model with other models. Here, using variational principles, new models are developed for both problems and the assumptions used by Reese are shown to be valid from a practical point of view. The axially loaded pile problem is discussed before the laterally loaded pile problem.

---

[1] Prof. of Civil Engineering, Texas Tech University, Lubbock, Texas, 79409

The Axially Loaded Pile Problem

Many formulas have been proposed by engineers for computing the ultimate axial capacity of piles, however, the determination of the settlement of these piles under working loads is rather quite complex especially when the pile is resting in a layered soil medium. Assuming that the pile is compressible, Seed and Reese (1957) introduced a concept that the magnitude of the load transfer from the pile to the surrounding soil at a particular elevation depends on the displacement of the pile at that level. The coefficient of the load transfer at each level was determined empirically from experimental data. Alternative methods were developed by other engineers (Poulos and Davis, 1980) using Mindlin equations for the load transfer mechanism. They have attempted to extent their theory to layered soils only with some approximations. Randolph and Wroth (1978) developed a semi-analytical model by representing the soil displacement at the surface by a logarithmic function which incidentally becomes infinite as the radial distance becomes infinite. Still other techniques were developed using the so-called finite element or boundary element methods. As mentioned before, no one has tried to compare these models with the Reese model. In this paper, using variational principles, a new simple model is developed for computing the settlement of axially loaded circular piles in a layered soil medium assuming that the soil has linear elastic properties. Perfect compatibility of displacements at the interface of the pile of soil is assumed. The developed equations in the linear range support the original empirical assumptions made by Reese even if the soil is layered with different material properties.

Brief Review of the Reese Model for Axially Loaded Piles

The load transfer mechanism for axially loaded piles in soils is represented by means of empirical nonlinear relations called 't-z' curves, t representing the surface shear stress transferred to the soil corresponding to the z- displacement of the pile. See Fig. 1. Researchers have developed these curves using actual field experimental data and soil material properties. Theoretically, the Reese model uses the classical Winkler concept where the shear transfer depends on the displacement of the pile and the field equations representing this phenomenon can be written as

$$-E_p A_p \frac{d^2 w}{dz^2} + \bar{k}w = 0 \quad \text{for } 0 < z < l \tag{1}$$

where E_p, A_p, l and w are the Young's modulus of elasticity, area of cross section, length and displacement of the pile in the z-direction as shown in Fig. 1 respectively. The parameter $\bar{k}$ varies with the soil and for each depth and is calculated from a series of empirical 't-z' curves as defined by Reese. The bottom load transfer is modeled by another spring with a spring constant K, that has to be determined empirically from the properties of the soil at the bottom of the pile. All these springs are illustrated in Fig. 1. The complex interaction that occurs between the pile and the soil is thus implemented in the model empirically. The model can provide a nonlinear load-displacement behavior of the pile up to its ultimate capacity.

Theory of a New Model for Axially Loaded Piles Using Variational Principles

For simplicity, the theory of the new model is presented first for a two layered soil strata as shown in Fig. 2, and later the model is extended for multi-layered soil. A two layered soil medium carrying a cylindrical pile or pier is shown in Fig. 2. The problem is analyzed using cylindrical coordinates (r, θ, z). It is assumed that each soil layer is linearly elastic, homogeneous with known material properties such as Young's modulus of elasticity and Poisson's ratio. E_p, A_p, R and l are the modulus of elasticity, area of cross section, radius and length of the pile respectively. The displacements at any point (r, z) in the soil medium surrounding the pile are assumed as:

$$\bar{u}(r,z) = 0, \quad \bar{v}(r,z) = 0, \quad \text{and} \quad \bar{w}(r,z) = w(z).\phi(r) \tag{2}$$

where $w(z)$ is the pile displacement and $\phi(r)$ is a mode function such that $\phi(R) = 1$, and $\phi(\infty) = 0$. From practical considerations, for the case of axial lading, it can be assumed that the radial displacement $\bar{u}(r,z)$ in the soil is negligible compared to the vertical displacement $\bar{w}(r,z)$. The total potential energy function for the pile and the soil system in the two regions as shown in the Fig. 2 is given as

$$
\begin{aligned}
\Phi = {}& U_{pier} + U_{soil} - \tilde{P}\, w(0) \\
= {}& \frac{1}{2}\int_0^l E_p A_p \varepsilon_z^2 \, dz + \frac{1}{2}\int_{vol}\left(\sigma_z \varepsilon_z + \tau_{rz}\gamma_{rz}\right) r\, dr\, d\theta\, dz - \tilde{P}w(0) \\
= {}& \frac{1}{2} E_p A_p \int_0^l \left(\frac{dw}{dz}\right)^2 dz + \frac{1}{2}E_2 \pi R^2 \int_l^\infty \left(\frac{dw}{dz}\right)^2 dz + \\
& \sum_{i=1}^2 \pi \int_R^\infty \int_0^l \left\{\overline{E_i}\left(\frac{dw}{dz}\phi\right)^2 + G_i\left(w\frac{d\phi}{dr}\right)^2\right\} r\, dr\, dz - \tilde{P}w(0)
\end{aligned}
\tag{3}
$$

Applying variations in Φ due to variations in w and ϕ and using principles of variational calculus, the following field equations and corresponding boundary conditions are obtained. The first equation describes the behavior of the pile in region 1 and is:

$$-(E_p A_p + 2t_1)\frac{d^2 w}{dz^2} + k_1 w = 0 \quad \text{for} \quad 0 < z < l \tag{4}$$

and for region 2, the equation becomes

$$-(E_2 \pi R^2 + 2t_2)\frac{d^2 w}{dz^2} + k_2 w = 0 \quad \text{for} \quad l < z < \infty \tag{5}$$

with boundary conditions, @ $z = 0$, $\quad -(E_p A_p + 2t_1)\dfrac{dw}{dz} = \tilde{P}_0$, and $\tag{6}$

at $z = l$, $\quad (E_p A_p + 2t_1)\dfrac{dw}{dz}\bigg|_{z=l-} = (E_2 \pi R^2 + 2t_2)\dfrac{dw}{dz}\bigg|_{z=l+}$. $\tag{7}$

and at $z = \infty$, $w = 0$, and

$$k_i = 2\pi G_i \int_R^\infty r\left(\frac{d\phi}{dr}\right)^2 dr, \quad \text{and} \quad 2t_i = 2\pi \overline{E}_i \int_R^\infty r\phi^2 dr,$$ (8)

for the subscript $i = 1, 2$ representing the two regions.

For the soil,

$$-r\frac{d^2\phi}{dr^2} - \frac{d\phi}{dr} + \gamma^2 r\phi = 0 \quad \text{for} \quad R < r < \infty$$ (9)

and $$\left(\frac{\gamma}{R}\right)^2 = \frac{\overline{E}_1 \int_0^l \left(\frac{dw}{dz}\right)^2 dz + \overline{E}_2 \int_l^\infty \left(\frac{dw}{dz}\right)^2 dz}{G_1 \int_0^l w^2 dz + G_2 \int_l^\infty w^2 dz}.$$ (10)

Solving Eq.(5) in closed form for $l < z < \infty$, and applying the boundary conditions,

$$w(z) = w_l e^{\alpha(l-z)} \quad \text{where} \quad \alpha = \sqrt{\frac{k_2}{(E_2 \pi R^2 + 2t_2)}}$$

Substituting $w(z)$ in Eq.(7), one gets @ $z = l$,

$$-(E_p A_p + 2t_1)\frac{dw}{dz} = Kw_l, \text{where} \quad K = \sqrt{k_2(E_2 \pi R^2 + 2t_2)}$$ (11)

This boundary condition is equivalent to a linear spring placed at the bottom of the pile with a spring constant equal to K, as illustrated in Fig. 1. The quantity $(\gamma/R)^2$ in Eq.(10), is further simplified as

$$\left(\frac{\gamma}{R}\right)^2 = \frac{\overline{E}_1 \int_0^l \left(\frac{dw}{dz}\right)^2 dz + \frac{\alpha}{2}E_2 w_l^2}{G_1 \int_0^l w^2 dz + \frac{1}{2\alpha}G_2 w_l^2}$$ (12)

and the quantities with integral signs are computed numerically.

Numerical Model I (Finite Difference Model)

It can be seen that in the new model, Eqs. (4) and (9) are coupled. These second order differential equations are solved by the classical finite difference method. The values of $k, 2t$ and γ are not known initially and therefore some approximate values of k and $2t$ are assumed as equal to E of the soil in the beginning. An iterative procedure developed by Vallabhan and Das (1991a) and Vallabhan et al. (1991b, 1999) for solving beams and plates on elastic foundations is also used here. The parameter used for convergence characteristics is γ/R. The iterative procedure is stopped when two consecutive values of γ/R are within 0.0001, but even a value equal to 0.001 yields satisfactory results.

Numerical Model II (A Simple Finite Element Model)

For piles in multi-layered soil, a finite element model is more convenient than the model I. For the i-th soil layer as shown in Fig. 3, the displacement $w(z)$ is interpolated between the two end nodes of the i-th layer as

$$w(z) = w_1(1 - \frac{z}{h_i}) + w_2\frac{z}{h_i} \tag{13}$$

where w_1 and w_2 are the local nodal displacements of the i-th layer of the pile. The strain energies in the pile, U_{pi} and the soil, U_{si} of the i-th layer are

$$U_{pi} + U_{si} = \frac{1}{2}\int_0^{h_i} E_p A_p \varepsilon_z^2 dz + \frac{1}{2}\int_{vol}(\sigma_z\varepsilon_z + \tau_{rz}\gamma_{rz})\, r\, dr\, d\theta\, dz$$

$$= \frac{1}{2}E_p A_p \int_0^{h_i}\left(\frac{dw}{dz}\right)^2 dz + \pi\int_R^\infty\int_0^{h_i}\left\{\overline{E}_i\left(\frac{dw}{dz}\phi\right)^2 + G_i\left(w\frac{d\phi}{dr}\right)^2\right\} r\, dr\, dz$$

$$= \frac{1}{2}\begin{Bmatrix} w_1 \\ w_2 \end{Bmatrix}'\begin{bmatrix} \alpha_i & \beta_i \\ \beta_i & \alpha_i \end{bmatrix}\begin{Bmatrix} w_1 \\ w_2 \end{Bmatrix} \tag{14}$$

where $\alpha_i = \dfrac{(E_p A_p)_i + 2t_i}{h_i} + \dfrac{k_i h_i}{3}$ and $\beta_i = -\dfrac{(E_p A_p)_i + 2t_i}{h_i} + \dfrac{k_i h_i}{6}$

The values of k_i and $2t_i$ are the same as in Eq.(8). Combining all the soil layers to obtain global behavior, we get the global matrix as

$$[K]\{W\} = \{F\} \tag{15}$$

where the global stiffness matrix is given as

$$[K] = \begin{bmatrix} \alpha_1 & \beta_1 & & & & & \\ \beta_1 & \alpha_1 + \alpha_2 & \beta_2 & & & & \\ & \beta_2 & \alpha_2 + \alpha_3 & \beta_3 & & & \\ & & \beta_3 & \alpha_3 + \alpha_4 & \beta_4 & & \\ & & & \cdot & \cdot & \cdot & \\ & & & & \cdot & \cdot & \cdot \\ & & & & \beta_{n-1} & \alpha_{n-1} + \alpha_n & \beta_n \\ & & & & & \beta_n & \alpha_n + K \end{bmatrix} \tag{16}$$

where n is the total number of degrees of freedom.

$$\{W\}^t = \{W_1 \quad W_2 \quad W_3 \quad W_4 \quad . \quad . \quad W_{n-1} \quad W_n\}^t.$$

The bottom spring constant K can be replaced by including a soil column and the stiffness of the soil layers underneath the pile. Sufficient soil layers are to be included such that the vertical displacements can be assumed to be negligible at the bottom most level.

Example Problem I

This problem consists of a concrete pile resting on a two region soil medium. The dimensions of the pile and the properties of the soil are given as: R = 0.2 m, l = 10.0 m, E_p = 14.0 GPa, E_1 for the soil layer = 40 MPa, μ_1 of the soil = 0.3 and E_2 of the soil at the bottom of the pile = 100 MPa. Applied Force on the top of the pile, $\widetilde{P}_0$ = 400 kN. Good agreement is obtained for the distribution of the surface displacement of the soil $w(r,0)$ with that obtained from the finite element solution. This is shown in Fig. 4. The axial displacement and force distribution data obtained are presented in Figs. 5 and 6. To verify the results, the same problem is solved by a more sophisticated finite element model consisting of 1500 axisymmetric elements and 2000 degrees of freedom. The high degrees of freedom is employed here to model the interface characteristics of the soil accurately. The displacements obtained from the new model were less than the FEM displacements. The new model is a stiffer model because of the simple assumptions used in its development. The force distributions are also similar, even though the axial force in the pile is less than the applied force at the top. This is because the in this model, the pile and the soil is acting in a composite manner as defined by the boundary condition in Eq.(6). In other words, the new and simple model represents the overall pile-soil problem appropriately.

Example Problem II

In this example, a stratified soil with different elastic properties is considered while keeping the same properties for the pile. The variation of the modulii of elasticity of the soil with depth is as follows: For depth 0 – 0.75m, 7 Mpa, 0.75 – 2.5m, 21 Mpa, 2.5 – 5.0m, 42 Mpa, 5.0 - 6.0m, 63 Mpa, 6.0 – 8.5m, 28 Mpa, 8.5 – 10.0m, 35 Mpa, and below 10.0m, 100 Mpa. The results are illustrated in Figs. 7 and 8. Both finite element and the new model solutions are shown in the figures.

Validity of the Reese Model for Axially Loaded Piles

If one modifies the Eq. (6) in the following manner,

$$-E_p A_p \frac{d^2 w}{dz^2} + \left(\frac{E_p A_p}{E_p A_p + 2t} \right) k w = 0,$$ (17)

one gets $\bar{k} = \left(\dfrac{E_p A_p}{E_p A_p + 2t} \right) k$, which is the   same equation as Eq.(1) of the Reese

model, with a spring constant K as given in Eq. (11) placed at the bottom of the pile model. Thus the assumptions made by Reese in his model are valid from a practical point of view. Examining the function $2t$, the new model reveals that in addition to the shear stresses in the surrounding soil, there exist some axial compressive strains in the soil as the load is transferred to the soil. The strain energy due to compressive strains in the soil is about 7% of the total work done by the external load for the problem shown here. Thus, if we ignore the function $2t$ in the above equation, the equation is exactly equal to the one in the Reese model.

The Laterally Loaded Pile Problem

Three different types of models were developed in the literature; one type is the Matlock and Reese model (1960) where they assumed that the lateral resistance offered by the soil depends on the magnitude of the lateral displacement of the pile at that level. The behavior of the soil is modeled by providing horizontal soil springs at each depth. The spring constants vary with depth and the model is extended to consider the nonlinear lateral soil resistance using empirical relations.

The other type of models uses the concept of a semi-infinite soil continuum with linearly elastic properties. Researchers such as Spillers and Stoll (1964), Poulos (1971), Poulos and Davis (1980), Verrujit, Kooijman (1989) and Sun (1993) to name a few, have used a semi-analytical method for the analysis of this complex interaction problem. Extention of their models to consider layered soil medium is approximate. Other researchers have used finite element and boundary element methods and also tried to incorporate some nonlinear characteristics of the soil into their model empirically. In all these types of modeling, the primary properties required are the values of the modulus of elasticity E_s and Poisson ratio v of the soil and if the soil is stratified, similar properties for each layer are required. As mentioned before, so far no one has attempted to compare continuum model with the Winkler model.

Brief Review of the Matlock-Reese Model for Laterally Loaded Piles

Both Matlock and Reese (1960) essentially employed the classical Winkler model for the beam on elastic foundation, where the pile is assumed as a beam and the behavior of the surrounding soil medium is modeled as a series of discrete springs. The value of the horizontal soil spring constant, denoted by 'k', has to be determined from experiments or empirically. The nonlinear load deformation characteristics of the soil are represented by what are known as 'p-y' curves. Each soil layer can have its own p-y curves, and therefore even the stratification and nonlinearity in the soil deformation characteristics can be incorporated empirically. Their mathematical model can be written as:

$$\frac{d^2}{dz^2}\left(E_p I_p \frac{d^2 u}{dz^2} \right) + ku = 0 \text{ for } 0 < x < l \tag{18}$$

with boundary conditions:

$$\text{at } z = 0, \ EI\frac{d^3 u}{dz^3} = \widetilde{H}_0 \text{ and } -EI\frac{d^2 u}{dz^2} = \widetilde{M}_0$$

$$\text{at } z = l, \ EI\frac{d^3 u}{dz^3} = 0 \text{ and } EI\frac{d^2 u}{dz^2} = 0$$

Here $E_p I_p$ is the bending stiffness of the pile cross section, u is the lateral displacement and z is the vertical axis of the pile. $\widetilde{H}_0$ and $\widetilde{M}_0$ are the prescribed force and moment at the top of the pile.

Theory of a New Model for Laterally Loaded Piles Using Variational Principles

This model was developed as an extension of the model for beam on elastic foundation developed by Vallabhan and Das (1991a). For simplicity, the theory is presented first, for a two layered soil and will be extended to multi-layered soil medium. A similar approach was developed by Sun (1993), but the one presented here has two distinct parameters that are used to represent the radial and tangential displacements of the soil. Also Sun's model cannot handle stratified soil medium. A two layered soil medium with a cylindrical pile is the same as in Fig.2 except that the applied forces are $\tilde{H}_0$ and $\tilde{M}_0$. It is assumed that each soil layer is linearly elastic, homogeneous with known material properties such as Young's modulus of elasticity and Poisson's ratio. $E_p I_p$ is the bending stiffness of the pile cross section R and l are the radius and length of the pile respectively. From practical considerations, it can be assumed that the vertical displacement $\overline{w}(r,\theta,z)$ in the soil is negligible compared to the horizontal displacements $\overline{u}(r,\theta,z)$ and $\overline{v}(r,\theta,z)$ thus, it is assumed that the displacements at any point (r,θ,z) in the soil medium surrounding the pile are:

$$\overline{u}(r,\theta,z)=u(z)\varphi(r)\cos\theta, \quad \overline{v}(r,\theta,z)=-u(z)\psi(r)\cos\theta, \text{ and } \overline{w}(r,\theta,z)=0. \text{ (19)}$$

$\phi(r)$ and $\psi(r)$ are non-dimensional mode shapes of the displacements along the radial direction and tangential direction, respectively. Each of these functions has a value of 1 at the pile soil interface and is equal to zero at infinity. Using strain-displacement relations, Hooke's law and strain energy functions, the minimum potential energy function is written as functions of $u(z), \phi(r)$ and $\psi(r)$. Taking variations in $u(z)$, the following differential equation and boundary conditions are obtained for the domain of the pile:

$$\frac{d^2}{dz^2}\left(E_p I_p \frac{d^2 u}{dz^2}\right)-2t\frac{d^2 u}{dz^2}+ku=0 \text{ for } 0<x<l \tag{20}$$

where:

$$k=\pi\left(\lambda+2G\right)\int_0^\infty\left[\left\{\left(\frac{d\psi}{dr}\right)^2+\left(\frac{\phi-\psi}{r}\right)^2\right\}+2\lambda\frac{d\phi}{dr}\frac{\phi-\psi}{r}+G\left(\frac{\phi-\psi}{r}+\frac{d\psi}{dr}\right)^2\right]r\,dr$$

and $2t=\pi G\int_0^\infty\left(\phi^2+\psi^2\right)r\,dr$

with boundary conditions

$$@z=0, \ EI\frac{d^3 u}{dz^3}-2t\frac{du}{dz}=\tilde{H}_0 \text{ and } -EI\frac{d^2 u}{dz^2}=\tilde{M}_0$$

$$@z=l, \ EI\frac{d^3 u}{dz^3}-2t\frac{du}{dz}=0 \text{ and } EI\frac{d^2 u}{dz^2}=0$$

The values of k and $2t$ need to be calculated for each layer and their values vary depending on the values of λ and $2G$ (the Lame parameters) for each soil stratum.

Taking the variations of the total energy function with respect to ϕ and ψ, the differential equations for ϕ and ψ are obtained as:

$$\frac{d^2\phi}{dz^2} + \frac{1}{r}\frac{d\phi}{dr} - \frac{(\lambda+3G)}{(\lambda+2G)}\left(\frac{\gamma}{r}\right)^2 \phi = -\frac{(\lambda+3G)}{(\lambda+2G)}\frac{\psi}{r^2} + \frac{(\lambda+G)}{(\lambda+2G)}\frac{1}{r}\frac{d\psi}{dr} \tag{21}$$

and

$$\frac{d^2\psi}{dz^2} + \frac{1}{r}\frac{d\psi}{dr} - \frac{(\lambda+3G)}{G}\left(\frac{\gamma}{r}\right)^2 \psi = -\frac{(\lambda+3G)}{G}\frac{\phi}{r^2} - \frac{(\lambda+G)}{G}\frac{1}{r}\frac{d\phi}{dr} \tag{22}$$

where the γ parameter for a single layer is given by

$$\gamma^2 = \frac{\left\{\int_0^l u^2 dz + G_z \int_0^l \left(\frac{du}{dz}\right)^2 dz + \sqrt{\frac{k}{2t}}u_l^2\right\}}{\left\{\int_0^l u^2 dz + \sqrt{\frac{2t}{k}}u_l^2\right\}}$$

Numerical Models I and II

In model I, the three governing equations are solved using finite difference technique combined with an iterative procedure. Here the soil is taken as a two layered soil medium. The model II consists of simple beam finite element for a segment of the pile surrounded by soil medium of same depth. Using Hermitian polynomials, the corresponding stiffness matrices of the pile and the soil are separately determined (Vallabhan and Kondur, 1999). The assumed displacement equations are:

$$\begin{aligned}\bar{u} &= \lfloor N(z)\rfloor\{u\}\,\phi(r)\cos\theta \\ \bar{v} &= \lfloor N(z)\rfloor\{u\}\,\psi(r)\sin\theta\end{aligned} \tag{23}$$

where $\{u\}$ are the magnitudes of the displacements and the rotations of the soil along the z - axis and In the above equation, $\lfloor N(z)\rfloor = \lfloor N_1 \quad N_2 \quad N_3 \quad N_4\rfloor$ where N_1, N_2, N_3 and N_4 are Hermitian polynomials of the first order. The nodal displacement vector is $\{u\} = \left\{u_1 \quad \theta_1 \quad u_2 \quad \theta_2\right\}^t$. The element stiffness equations which includes the element stiffness matrix of the pile element K_p appear as:

$$\left[\left[K_p\right] + k\left[K_k\right] + 2t\left[K_{2t}\right]\right]\{u\} = \{P\} \tag{24}$$

The stiffness matrices K_k and K_t representing the soil strip, can be easily obtained by integrating the shape functions and their derivatives. The details of these equations and solutions are given by Kondur (1998). Using these new values, the displacement of the pile and the internal moments and shear forces are computed.

Results obtained for lateral loads as well as moments are compared with solutions obtained from a more sophisticated finite element model. Overall, excellent comparisons are obtained.

Validity of the Matlock-Reese Model for Laterally Loaded Piles

If one examines the role of the term $2t$ in Eq. (20) , it represents the shear strains in the soil surrounding the pile and hence it provides connectivity of the horizontal soil springs which is ignored in the Winkler concept. If one keeps $2t = 0$ in Eq. (20), the Reese model matches with the variational model.

With models for pile-soil interaction using Winkler and continuum and concepts, exact matching is impossible from a theoretical point of view. But, if one could develop a fair relationship between them at least for the case of a uniform soil, then the assumptions of Reese model is justified.

To accomplish this, the following non-dimensional terms were selected by Vallabhan and Kondur 1998). They are the nondimensional depth x_n, length-diameter ratio L/D, nondimensional soil stiffness K_r of the continuum model using the modulus of elasticity of the soil E_s and nondimensional soil stiffness K_w for the corresponding Winkler model using the value of the spring constant k. Nondimensional displacement u_n, nondimensional moments M_n and shear forces V_n are used for comparison. These terms are:

$$x_n = \frac{x}{L}, \ \frac{L}{D}, \ K_r = \frac{E_s L^4}{E_p I_p} \text{ and } K_w = \frac{kL^4}{E_p I_p} \tag{25}$$

$$u_n = \frac{u}{L} \ , \ V_n = \frac{VL^2}{E_p I_p} \text{ and } M_n = \frac{ML}{E_p I_p} \tag{26}$$

Several problems are solved with length/diameter ratios, L/D from 5,10,15, 25, 40, 50 and 100, with values of K_r such as 10,33,100,333,1000,3333 and 10000. The value of Poisson's ratio of the soil is kept as 0.3 in these analyses. Fairly good comparisons were obtained between the two models when the displacements at the top of the pile were matched. One selected comparison is given for different values of L/D ratio = 50 and for a value of $K_r = 1000$ in Fig. 9. Various values of K_r and K_w are plotted on log-log scale in Fig. 10 and it is found that the relationship is almost a straight line in the plot for a particular value of the L/D ratio. These graphs have almost the same slope but they are displaced slightly for different values for L/D. A tentative suggested equation for the relationship is

$$K_w = A (K_r)^\alpha \quad \text{ or } \quad k = AE_s^\alpha \left(\frac{L^4}{E_p I_p}\right)^{\alpha-1} \tag{27}$$

where the values of A and α are empirically computed as:

$$A = 0.58571 + 6.57143\left(\frac{L}{D}\right) + .000084\left(\frac{L}{D}\right)^2$$

$$\alpha = 1.111748 - .001078\left(\frac{L}{D}\right) + .00000337\left(\frac{L}{D}\right)^2 \tag{28}$$

In other words, if the average material properties of the soil are known for a depth of soil, where critical deformation characteristics of the pile are occurring, an equivalent k value can be computed for the linear Winkler model from the equation or from the graphs. These quantities will be helpful in the initial stages of the 'p-y' curves using Reese model.

Conclusions

1. New models are developed for analysis of axially and laterally loaded piles placed in a layered soil medium. Each layer has to be further divided into horizontal strips to increase accuracy with respective material properties. About 50 to 100 strips are sufficient for good accuracy.
2. It is shown that the assumptions used by Reese and others in modeling axially and laterally loaded pile problems yield equations that match with those obtained from the new models developed using variational principles.
3. For the laterally loaded pile problem, an empirical correlation between the nondimensional parameters for the Winkler and continuum models are shown here. In other words, if the average material properties of the soil are known for a depth of soil where critical deformation characteristics of the pile are occurring, an equivalent k value for the Winkler analysis can be computed from the equation.

Limitations

For axially loaded piles, the model assumes linear elastic properties of the soil and no slipping between the pile and the soil is considered. Complete displacement compatibility is assumed at the pile-soil interface surface. No pile-soil separation is considered on the tension side of the pile for the case of laterally loaded piles.

References

1. Kondur, Devanand, "A New Model for the Analysis of Laterally Loaded Piles", Ph.D. Dissertation submitted to Dept. of Civil Engr., Texas Tech University, Lubbock, TX, 1998.
2. Matlock, H., and Reese, L.C., (1971) "Generalized solution for laterally loaded piles", J.,Soil Mech. and Found. Engr. Div., ASCE, 97(5), 711-731.
3. Poulos, H.G., and Davis, E.H., (1980), "*Pile foundation analysis and design*", John Wiley & Sons, New York.
4. Randolf, M.F.,and Wroth, C.P., (1978), "Analysis of Deformation of Vertically Loaded Piles", J. Geotech. Engr. Div., ASCE, Vol. 104, GT12, pp.1465-1488.
5. Spillers, W.R., and Stoll, R.D., (1964), "Lateral response of piles", J. Soil Mech. and Found. Engr. Div., ASCE, 90(6), 1-9.
6. Seed, H.B., and Reese, L.C., (1957), "Action of Soft Clay along Friction Piles", Trans. ASCE, Vol. 122, pp. 731-754.

7. Sun, K.,(1993), "Static analysis of laterally loaded piles", Proc. 11th Southeast Asian Geotech Conf., Singapore, 589-594.

8. Vallabhan, C.V.G. and Das Y.C., (1991a),"Modified Vlasov Model for Beams on Elastic Foundations", J. Geotech. Engr., ASCE, 117(6), pp. 956-966.

9. Vallabhan, C.V.G., Straughan, W.T. and Das, Y.C., (1991b), "Refined Model for Analysis of Plates on Elastic Foundations", J. Engr. Mech., Vol. 117,12, pp. 2830-2844.

10. Vallabhan, C.V.G., and Daloglu, A., (1999), "Consistent FEM-Vlasov Model for Plates on Layered Soil", J. of Struc. Engrg., ASCE, Vol.125, No.1, pp. 108-113.

11. Vallabhan, C.V.G., and Kondur, D., (1998), "k-values for Analysis of Laterally Loaded Pile Foundations from Elastic Continuum Soultions", Proc. of the 12th Engr. Mech, Conf, ASCE, LaHolla, CA, May 17-20, pp. 542-545.

12. Vallabhan, C.V.G., and Kondur, D., (1999), "A New Model for the Analysis of Laterally Loaded Circular Piles in a Layered Soil Medium", Proc. PACAM-VI, App. Mech. of the Americas, Rio-de-Janeiro, Brasil, Vol. 7, pp. 863-866.

13. Verrujit, A., and Kooijjman, A.P., (1989), "Laterally loaded piles in a layered elastic medium", Geotechnique, London, England, 39(1), 39-46.

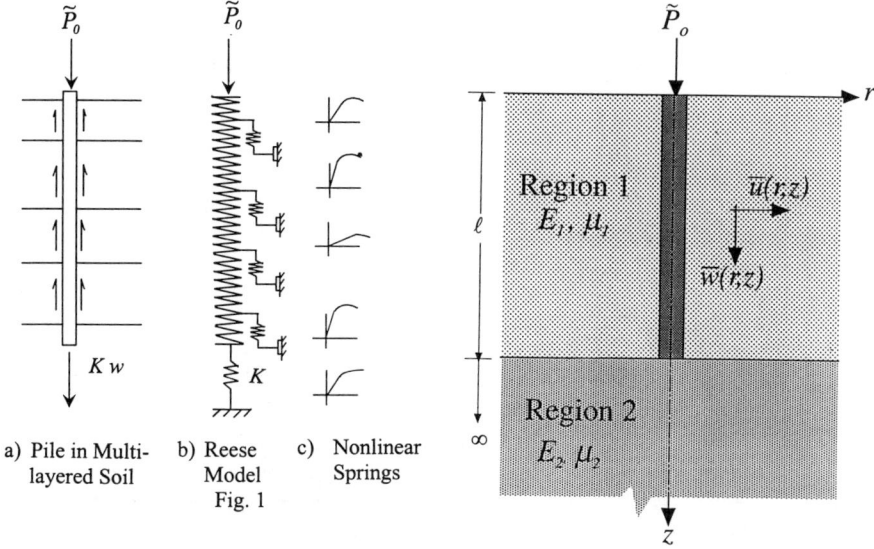

Fig. 2. Pile in Two Layered Soil Medium

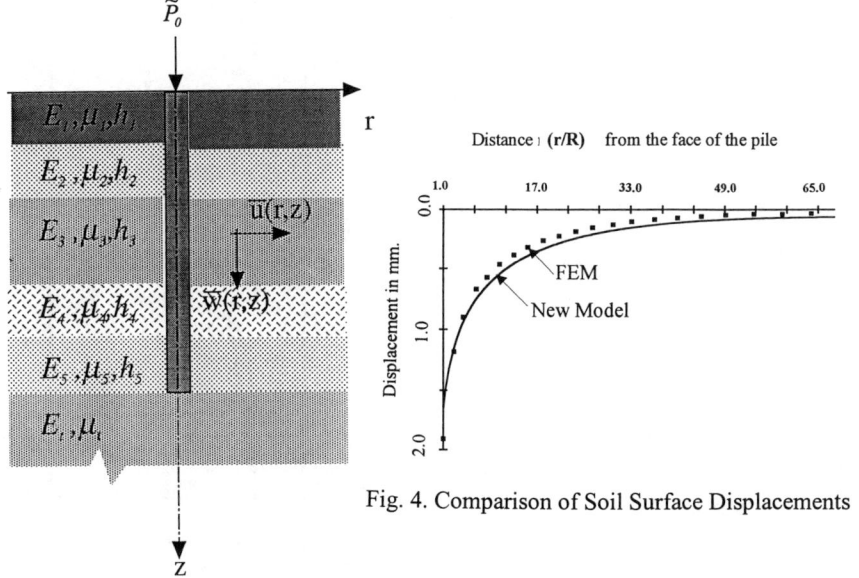

Fig. 4. Comparison of Soil Surface Displacements

Fig. 3. Pile in Multi-layered Soil Medium

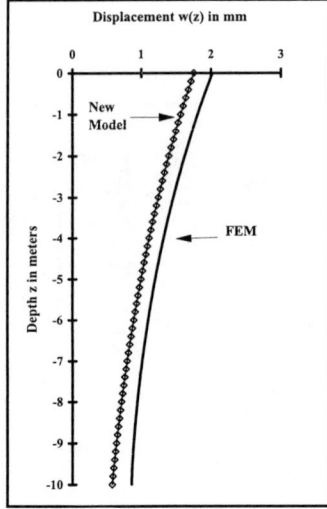

Fig. 5. Axial Displacement vs. Depth

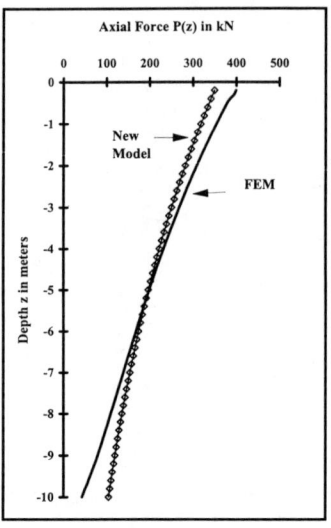

Fig. 6. Axial Force vs. Depth

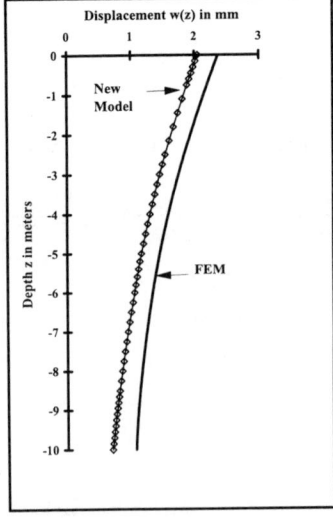

Fig. 7. Axial Displacement vs. Depth

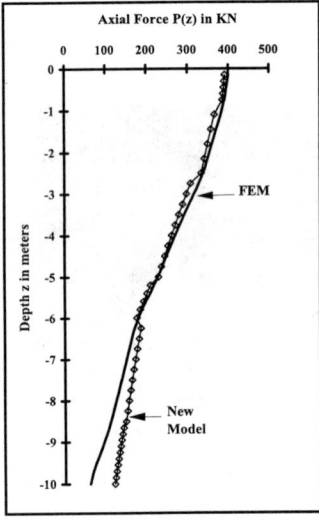

Fig. 8. Axial Force vs. Depth

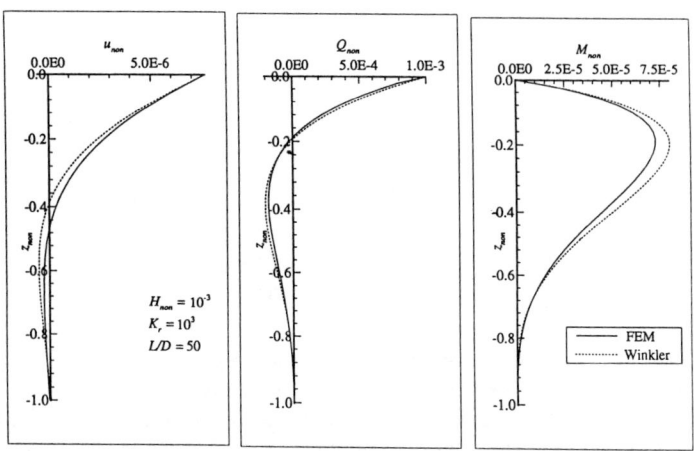

Fig. 9. Comparison of u_n, V_n and M_n from Winkler and New Model

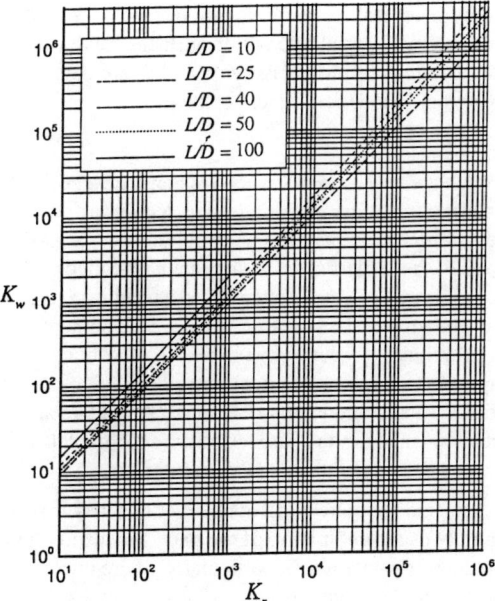

Fig. 10. Relationship between K_W and K_r

The Mechanics of Pile Foundation Collapse
James D. Murff[1], Member ASCE

Abstract

In this paper the mechanics of offshore pile foundation systems collapse is investigated using the general theory of plastic limit analysis. The theory is described and its implementation is demonstrated using the behavior of a single pile under generalized loading. In this method, forces and moments are characterized as generalized stresses and displacements and rotations are characterized as generalized strains. The pile response is represented as a yield surface in terms of axial, shear, and moment loading at the pile head. The approach is then generalized further by analyzing a simple assemblage of pile foundation elements using the upper and lower bound theorems of plastic limit analysis. The upper bound method is then used to analyze a somewhat more general 12 pile foundation system. Various foundation failure modes are explored and useful interaction surfaces are developed in terms of the overall structural loading. Various sensitivity studies are then performed to provide insight into typical foundation system behavior.

Introduction

Most offshore structures are space-frame structures constructed of steel tubular members, and founded on steel pipe piles. These structures are commonly referred to as steel pile jackets or SPJ's. It has become commonplace, as part of the design or re-qualification phases of these platforms, to conduct a so-called push-over analysis to estimate the collapse strength of the system. These results are then used to estimate the 'reserve strength' of the structure against environmental overload which in turn may be used for reliability based procedures. This nonlinear analysis may involve a large complex structural model including a detailed foundation model with nonlinear soil supports for axial and lateral resistance (Arnold, et al. 1977). In spite of the many advancements made in the past few years such models are still time consuming to generate and require substantial computer time.

Possible failure mechanisms for typical offshore structures may involve the structure only, the foundation only or a combination of the two. Frequently there is a need to understand the foundation capacity alone, particularly where uncertainties in the two subsystems vary substantially (Aggarwal, et al. 1996). To assess foundation

---

[1] Sr. Research Advisor, Exxon Production Research Co., PO Box 2189, Houston TX 77001

behavior, the full structural model may be used with a linear superstructure so that the failure mechanism is forced into the foundation. This paper describes a complementary approach, based on the generalized theory of plastic limit analysis, to assessing the foundation capacity. This method is much simpler and much less time consuming than the 'brute force' method described above. It is particularly useful when calibrated with the full nonlinear, structural model results. Because of its ease of use and rapid calculation method, it is practical to conduct literally hundreds of push-over analyses to expand and enhance the fully nonlinear results and to gain improved insight. It is helpful to first consider the basic principles of the proposed method.

Generalized Plastic Limit Analysis

If a system of forces characterizes the stress state in a perfectly plastic structure, the forces can be considered generalized stresses and the corresponding displacements can be considered generalized strains. The treatment used herein (Prager 1959), uses the characterization 'generalized stresses and strains' to emphasize the fact that the variables which characterize the stress and strain state of a plastic structure do not necessarily have dimensions of stress and strain. For a given set of generalized stresses $Q_1 \cdots Q_n$ the generalized strain rates $\dot{d}_1 \cdots \dot{d}_n$ are the work rate conjugates of the stresses, that is

$$\dot{W} = Q_1 \dot{q}_1 + \cdots + Q_n \dot{q}_n \qquad 1$$

where $\dot{W}$ is the work rate of the generalized stresses on the plastic strain rates.

As an example, Figure 1a shows the case of a simple, perfectly plastic structure loaded by two single loads, Q_1 and Q_2 (generalized stresses) with plastic displacements at the point of load applications at yield being d_1 and d_2 (generalized strains). In general the yield condition for the system is a function of the two loads, that is,

$$f(Q_1, Q_2) = 0 \qquad 2$$

Note that where the generalized stresses consist of loads and moments, it is convenient to define the generalized stress and strain due to moment loading as

$$Q_i = M_i / B \qquad 3$$

and

$$d_i = \theta_i B \qquad 4$$

where M_i is a moment, θ_i is the corresponding rotation and B is an associated characteristic length such as a member diameter or width. These terms are work conjugates and also maintain dimensional consistency with loads and displacements.

Figure 1b shows a generalized yield surface for the system. According to theorems of plastic limit analysis (Chen, 1972) this is a convex surface which encloses the coordinate origin in generalized stress space and the generalized strain rates are normal to it, i.e. the surface is a plastic potential. The plastic strain rates are then given by the equation,

$$\dot{d}_i = \lambda \frac{\partial f}{\partial Q_i} \qquad 5$$

where λ is a positive scalar. Thus, the normal to the yield surface dictates the relative magnitudes of the plastic strain rates. The absolute magnitude of the rate is governed by the boundary conditions as will be discussed in subsequent sections.

The yield function is defined such that, for stress states within the surface, the function is negative and for those outside, it is positive. As such, it is then possible to apply the upper and lower bound theorems of limit analysis using the generalized definitions of stress and strain and the yield condition in the form of Eq. 2. The lower bound theorem states,

> "If any stress distribution throughout the structure can be found which is everywhere in equilibrium internally and balances certain external loads and at the same time does not violate the yield condition, these loads will be carried safely by the structure." (Calladine 1969)

To apply the method requires that a stress state (generalized) be postulated that does not violate the yield condition and is in equilibrium with the external loads. It follows that the maximum lower bound is the best solution, i.e. the one closest to the exact solution.

The upper bound theorem states

> "If an estimate of the plastic collapse load of a body is made by equating internal rate of dissipation of energy to the rate at which external forces do work in any postulated (kinematically admissible) mechanism of deformation of the body, the estimate will be either high or correct." (Calladine 1969).

Application of the method requires that a collapse mechanism be postulated and that a method for calculating energy dissipation rate within the deforming material be available. It follows then that the least upper bound solution found is the best solution, that is, closest to the exact solution. Finally note that the theorems of plastic limit analysis do not strictly apply where members have strain softening behavior (load shedding) or where capacity is affected by large geometry changes.

Single Pile Behavior

First we will consider the behavior of a single pile subjected to lateral and axial load and moment at the pile head as shown in Fig. 2. Note that the pile head boundary condition may be fixed or free with specified moment. For many design and analysis situations axial and lateral pile loads are considered independently. For axial loads, designs are typically based on a specified factor of safety against axial soil failure. The possible effect of lateral loads on axial capacity is usually ignored. For lateral loading the typical design is based on working stress methods. In this approach the soil is modeled as a continuous 'bed' of non-linear springs and the pile is modeled as a beam-column. These methods have been developed over the past forty years by Reese, Matlock and their co-workers (e.g. Reese, et al. 1970 and Matlock, et al. 1970) and are almost universally used by offshore engineers. This approach may or may not include the effects of axial load on lateral response. Simplified analyses are conducted to insure the pile stresses remain below allowable values. Thus, it is not standard practice to estimate the pile head lateral load or bending moment capacity of the pile element. However, while such methods are not in common usage for offshore design, the problem of pile capacity under combined loading has been addressed by a number of authors (Broms 1964; Meyerhof 1995; and Murff 1987).

There are two general approaches that can be used to represent the multi-axial yield capacity of single piles. In the first method, a closed form solution or an empirical fit (to numerical model pile results) of a yield surface in terms of lateral load, axial load and moment at the pile head can be developed. This yield surface can then be employed in an analysis of the foundation system capacity using either upper or lower bound approaches. This method is demonstrated in the next section. The second approach is to use an explicit model of the piles, including lateral and axial resistance in an upper or lower bound model. This method is discussed in subsequent sections.

The Yield Surface Approach

Consider a fixed-head, steel pipe pile with diameter, D, wall thickness, t, and yield stress, σ_y, subjected to combined axial and lateral loading as shown schematically in Fig. 2. For simplicity consider the case where the pile is sufficiently long that the failure mechanism is the development of a plastic hinge at some depth, l, rather than pile rotation with the 'kicking out' of the pile tip. For illustration purposes it is assumed that the soil skin friction resistance (axial) in the upper part of the pile (down to the plastic hinge) is small such that the axial load at the plastic hinge is approximately equal to the axial load at the pile top, V. For this example, the lateral resistance of the soil per unit length of the pile, p, is assumed to be constant with depth. The moment at the pile hinge is assumed to be a local maximum thus the shear at the plastic hinge is zero. Moment and horizontal load equilibrium provide the following equations (Murff, 1987):

$$\Sigma M = 0: \ Hl + M = M_p + \frac{pl^2}{2}$$

$$\Sigma H = 0: \ H = pl$$

6

where

$$M_p = M_y \cos(\frac{\pi}{2}\frac{V}{V_y})$$

$$M_y = \frac{\sigma_y}{6}\left[D^3 - (D-2t)^3\right]$$

7

and

$$V_y = \pi(D-t)t\,\sigma_y$$

Note that M_y is the plastic moment capacity of a tubular member of wall thickness, t, with no axial load and V_y is the axial yield capacity of a tubular member with no moment. The cosine term in Eqs. 7 shows the effect of axial load on moment capacity, M_p, for a circular pile cross section with a uniform wall thickness. Simplifying these equations and solving for the depth to the plastic hinge, l, gives,

$$l = 2\sqrt{\frac{M_p}{p}}$$

8

and substituting in the second of Eq. 6 gives

$$H = 2\sqrt{M_p p}$$

9

Eq. 9 provides the equation for the pile head load-interaction surface. Eq. 9 can be written in a normalized form as follows,

$$H' = \sqrt{2\cos(\frac{\pi V}{2V_y})}$$

10

where

$$H' = \frac{H}{\sqrt{2M_y p}}$$

$$V' = \frac{V}{V_y}$$

$$M' = \frac{M}{M_y}$$

11

which is shown plotted in Fig. 3. This result could also be obtained by applying the upper or lower bound method.

Other Approaches: As mentioned above there are other approaches for defining interaction surfaces. For example, numerical analyses such as the finite element method (Desai and Able 1972), using advanced soil constitutive equations, could be used to determine a discrete set of solutions (failure load combinations) for a specific non-dimensional set of parameters. A simpler but more approximate approach is the use of detailed limit analysis methods (Murff and Hamilton 1993) where an explicit soil failure mechanism is postulated. A third option is to conduct physical model or field tests to assess discrete failure load combinations (Meyerhof 1995). Load combinations determined in this manner will normally have to be fit with an analytical function, splines, or some other piecewise continuous surface.

With this rather simple background, we now have the basic tools to apply the method of generalized plastic limit analysis to a system of foundation elements. The following section describes the theoretical basis for this approach.

Simple System Behavior

Now consider the simple planar structure shown in Fig 4. The foundation support elements shown have negligible size and are characterized by the yield surface derived above. In this case a fixed-head pile is assumed so that plastic moments develop at the pile tops as well as at some depth below the soil surface. Thus Eq. 9 can be written as the yield function,

$$H - 2\sqrt{M_p p} = 0 \qquad\qquad 12$$

Now consider using this yield surface in both an upper and lower bound solution.

Lower Bound For the rigid planar structure shown, there are three global equations of equilibrium and six unknown reactions. Assume the global vertical load is known and it is desired to find the lateral load that causes collapse of the structure. According to the lower bound theorem any equilibrium set for which the support reactions are at or below yield is admissible. The problem is formulated as follows:

Governing equations:

$$\sum M = 0 \quad F_H Y_H + F_V X_V - \sum_i M_i - \sum_i V_i x_i = 0$$
$$\sum H = 0 \quad F_H - \sum_i H_i = 0 \qquad\qquad 13$$
$$\sum V = 0 \quad F_V - \sum_i V_i = 0$$

Constraints:

$$H_i - 2\sqrt{M_{yi} Cos(\frac{\pi V_i}{2V_{yi}}) p} \le 0 \qquad\qquad 14$$

where the subscript, i, indicates the support number and the reactions at the support are the optimization variables. Eqs. 13 can be combined to form the objective function with the global lateral load, F_H, as the unknown. This problem can be set up using optimization add-ons that are available for most spread sheets. As additional supports are added or out-of-plane directions are considered the problem generally becomes more sensitive and can require considerable user intervention.

Upper Bound To apply the upper bound method, we assume a failure mechanism. In this case we assume a virtual rotation rate, $\dot{\beta}$, of the rigid structure about a generic point below the foundation with coordinates, x_o, and y_o, then search for the specific point that minimizes the unknown external force F_H. The approach taken here is as follows. The assumed virtual rotation rate of the structure about a specific point specifies the virtual velocities and rotation rates at each pile support in terms of the location of the axis of rotation and the pile head coordinates. These relative velocities and the associated flow rule (Eq. 5) provide two equations in terms of the two unknown forces and the moment at each pile head. Note that Eq. 5 gives three equations but only two are independent. A third equation is provided by the yield surface for each pile. In summary, we then have, for example,

$$\frac{\dot{v}}{\dot{m}} = \frac{\dfrac{\partial f}{\partial V}}{\dfrac{\partial f}{\partial M}}$$

$$\frac{\dot{h}}{\dot{m}} = \frac{\dfrac{\partial f}{\partial H}}{\dfrac{\partial f}{\partial M}} \qquad\qquad 15$$

$$f(V, H, M) = 0$$

where $\dot{h}, \dot{v}, and\ \dot{m}$ are the virtual horizontal and vertical velocities and rotation rate at each pile head respectively. Note that the ratios of the velocities (left hand side) are known from the problem geometry and the failure mechanism and the right hand side can be determined by carrying out the indicated differentiation (the scale factor, λ, cancels in the ratios). The three equations can then be solved (at least in principle) for the unknown pile head forces, V and H, and the moment, M. The total rate of internal energy dissipation, $\dot{I}$, is then,

$$\dot{I} = \sum_i (V_i \dot{v}_i + H_i \dot{h}_i + M_i \dot{m}_i) \qquad\qquad 16$$

From the problem geometry it is also possible to calculate the work rate of the external forces, $\dot{E}$, as

$$\dot{E} = F_V (X_V - x_0)\dot{\beta} + F_H (Y_H - y_0)\dot{\beta} \qquad\qquad 17$$

where F_V is a known load and F_H is the unknown load, respectively, and subscripts V and H refer to vertical and horizontal loads and their coordinates, respectively. Note that it is straightforward to include other known loads in the formulation. However only one unknown load can be determined. It is also possible to apply an envelope of loads which are scaled up to failure, the scale factor being the unknown. For the above formulation, the external work rates are equated to the internal energy dissipation rates and after simplifying gives,

$$F_H = \frac{\sum_{i=1}^{N}\left\{V_i\,(x_i - x_0) + H_i\,(y_i - y_0) + M_i\right\} - F_V\,(X_V - x_0)}{(Y_H - y_0)} \qquad 18$$

where the subscript i refers to the ith pile head. For a given location of the axis of rotation (x_0, y_0), the unknown force, F_H, can then be determined. It follows from the upper bound theorem, that the best solution is the one that minimizes F_H. Since the rotation of the structure about some instantaneous center of rotation includes all possible planar motions then the true minimum upper bound should be the exact solution for the planar mechanism assumed. Thus the system so determined should be in equilibrium. Since we can compute the forces and moment on each foundation element, this can be checked.

Example Consider a specific case of the simple two-pile, planar system as detailed in Table 1. For illustration we assume that the axial capacity below the second plastic moment is such that the full yield strength of the steel can be developed (i.e. axial failure is not limited by the axial soil capacity). It is instructive to calculate the maximum lateral load, F_H, for a range of load application heights. For each height of application we can determine a base shear and base overturning moment at collapse. This result can be plotted on a shear-moment interaction diagram which provides insight as to the nature of the foundation capacity.

In this case, both lower and upper bound methods were formulated using a spreadsheet and optimized as discussed above. Fig. 5 shows the shear-moment interaction diagrams resulting from these analyses. Note first that the lower and upper bound results give virtually identical results indicating that this is an exact solution. At zero moment, indicating the horizontal load is applied right at the mudline, the failure mechanism is essentially pure lateral translation with a peak shear of approximately 23MN. As the load height and hence moment is increased the shear capacity is reduced until the mechanism reaches essentially pure rotation where the overturning moment is the limitation.

Explicit Pile Model

As discussed above, an alternative way to characterize single pile capacity is to use an explicit pile model (Murff and Wesselink 1986) rather than reducing the pile head response to a yield surface. This approach is more straight forward, especially for complex conditions, but it does add some level of difficulty to a foundation system analysis. Fortunately, simplifications can be made to the formulation that maintain the tractability of this method.

For the purposes here the upper bound approach is used to represent the single pile. Figure 6 shows a schematic of the explicit pile model, both rotationally constrained and subject to an inclined load, F, at the pile head at an angle, ψ, to the horizontal. The assumed failure mechanism includes the formation of plastic sections at the pile head (A) and at some depth below the pile head (B). Both axial and rotational deformations can occur at either plastic section according to the axial load-moment interaction surface. As shown, differences in axial deformation between points A and B are resisted by axial shaft resistance and lateral displacement due to rotation of the upper pile section is resisted by the lateral soil capacity. The axial pile capacity at point B can be limited by the axial capacity of the pile material or total soil capacity (shaft resistance plus end bearing), whichever is less. A detailed discussion of this approach (Murff 1987) provides derivations.

For this formulation the optimization variables are (1) the inclination, ω, of the resultant virtual velocity vector, v_o at the pile head (2) the proportion (ξ) of axial deformation at the pile head (δ_t) that occurs plastically at A ($\delta_a = \xi \delta_t$) and B ($\delta_b = (1-\xi)\delta_t$) respectively and (3) the depth to the second plastic section B (l_{ab}). The external work rate is then,

$$\dot{W} = Fv_0(\cos\psi\cos\omega + \sin\psi\sin\omega) = Fv_0\cos(\psi - \omega) \qquad 19$$

where v_0 is the resultant virtual velocity at the pile head. Internal dissipation of energy terms then arise from the yielding of the soil and pile under the imposed deformation pattern. The terms include soil-pile shear due to axial movement of the top pile section,

$$\dot{D}_{aby} = (1 - \xi)v_0 \sin\omega \int_0^{l_{ab}} T(y)dy \qquad 20$$

where $T(y)$ is the axial soil resistance; soil-pile bearing due to lateral movement of the top pile section,

$$\dot{D}_{abx} = v_0 \cos\omega \int_0^{l_{ab}}(1 - \frac{y}{l_{ab}})P(y)dy \qquad 21$$

where $P(y)$ is the lateral soil resistance; and pile deformations D_a and D_b at plastic sections A and B respectively. Derivations of the latter terms are provided elsewhere (Murff 1987). Equating external work rate to internal work rate we obtain

$$F = \frac{D_a + D_b + D_{abx} + D_{aby}}{\cos(\psi - \omega)} \qquad 22$$

Note that the dots are dropped to indicate virtual velocities have been cancelled. Since F is an upper bound it is minimized with respect to ω, l_{ab}, and ξ to obtain the best answer. Independent studies comparing the results of this formulation with more rigorous approaches have shown that these results are usually very accurate for typical configurations.

Example This approach will be demonstrated here for typical offshore piles for a small jacket structure. The pile and soil characteristics are summarized in Table 2. For these purposes we consider a range of load inclinations varying from zero to ninety degrees. Results of this analysis are shown in Fig. 7. Note that the results are presented in the form of an axial load-lateral load interaction surface. As indicated, for situations where the soil capacity of the lower pile section limits the axial capacity, the interaction surface is truncated so that over the lower lateral loads the yield surface is simply a horizontal line.

Analysis of More Complex Systems

The above approach can be used in foundation system analysis in two ways. As suggested in previous sections, interaction surfaces can be generated for complex pile and soil conditions e.g. varying pile wall thickness with depth, varying soil resistances with depth, etc. These results can then be empirically fit with a convenient function which can in turn be used as demonstrated above in the section on simple system behavior. A more straightforward approach is to use the pile model explicitly in an upper bound model of the foundation system. The single pile model in this formulation varies slightly from that described in the previous section since, for system behavior, the structure is assumed to rotate about a specified axis and the specific coordinates of the axis are optimization variables. Thus the velocities at each pile head are prescribed rather than being part of the solution as in the single pile case. Further, the loads on the pile head are implicit as compared with the explicit loads used for the single pile case.

Consider the foundation shown schematically in Fig. 8. Details of the model and the pile characteristics are provided in Table 3. For the purposes here, the resultant platform environmental load can be applied at any height above the sea bed and at any orientation about the platform's vertical axis. It is also assumed that a known vertical load with a known eccentricity is applied at the platform deck. For simplification, the platform/foundation system is reduced to a planar model by projecting it normally onto the plane which contains the vertical axis and the resultant environmental lateral load. The planar structure and foundation are then assumed to translate and rotate in this plane, a constraint that will be discussed shortly (therefore only loads or moments with components in that plane do work, i.e. contribute to the collapse). The upper bound formulation is similar to that described above for Eqs. 15-17. However, rather than calculating energy dissipation for each pile using the yield surface, the various components of dissipation for each pile are explicitly included, i.e at the plastic sections of each pile and within the soil where lateral and axial resistance is developed. Equating external work rate to internal work rate we can then develop a governing equation similar to Eq. 18,

$$F_H = \frac{\sum_{i=1}^{n}(D_{ai} + D_{bi} + D_{abxi} + D_{abyi}) - F_V(X_V - x_0)}{(Y_H - y_0)} \qquad 23$$

where the D's represent the dissipation of various components for each of the n piles, similar to those given in Eq. 22. It is straightforward to include additional known, external loads in the formulation. The best estimate of collapse load is then found by minimizing the right hand side with respect to the coordinates of the center of rotation, x_0 and y_0, the depths to the second plastic hinges of all piles, l_{abi}, and the relative amounts of plastic axial deformation, ξ's, at the two plastic sections for each pile. For a large number of piles the number of optimization variables can be large and require considerable time and effort to achieve a proper solution. However it has been found that the location of the second plastic section is very insensitive to the specific collapse mechanism. Thus, for most cases, the location of the second plastic section can be estimated with a high degree of accuracy at the outset by conducting single pile analysis for each pile under lateral load. These plastic section locations can then be used for all subsequent analyses. The parameters, ξ_i, converges rapidly and can be minimized for each specific mechanism (set of x_0 and y_0 coordinates) considered. These simplifications reduce the global optimization parameters to x_0 and y_0, the coordinates of the center of rotation. These calculations are very efficient allowing many evaluations of relatively complex foundations to be conducted in a matter of seconds. Thus, once a problem is set up, one can explore the effects of the problem parameters of interest, e.g. load locations and orientations and pile and soil parameters, very efficiently. This approach can be used in the design process, fitness evaluation, or for decisions during construction, e.g. pile underdrive, and is especially effective when used in conjunction with a fully nonlinear structural analysis.

Examples For the purposes of illustration, consider the structure/foundation system shown in Fig. 8 and detailed in Table 3. Analyses will be conducted for both broadside (0°) and end-on (90°) loading. The resultant load will be varied from the mudline to several hundred feet above the mudline so that a base shear vs base moment interaction surface can be generated for each direction. Fig. 9 shows the results of these analyses. These results are qualitatively similar to those provided in Fig. 5 for the simple structure. For loads applied near the mudline, the critical failure mechanism is basically translation with the piles failing in shear. For very high loads, the failure is overturning dominated by axial pile behavior. At intermediate heights (near the knee in the curve), the failure involves axial load/ moment interaction. As one would expect the foundation capacity for the end-on loading is the same as broadside loading for shear mechanisms but significantly stronger in overturning. Also shown is a broadside case with battered piles. This shows an increased resistance as the load application becomes higher followed by a drop as interaction/overturning starts to dominate. The pile batter causes the structure to undergo reverse rotation as it translates. This rotation can lift the vertical load causing the increased resistance. These interaction surfaces are helpful in identifying the regimes of critical failure modes and giving an overall sense of understanding of the foundation strength.

A second characteristic of interest is the influence of loading direction on capacity. As mentioned above there is basically no effect of load orientation at low load heights because the shearing capacity of a tubular pile is independent of loading direction. On the other hand the collapse load can be strongly dependent on loading direction for overturning modes and somewhat less so for interaction modes. For illustration, consider conditions in which the load is applied at different horizontal orientations at a location high enough to ensure an overturning mechanism for all directions.

Fig. 10a is an interaction surface for horizontal (x and y) load components. Because the structure is symmetric about both x and y axes, the interaction surface is symmetric in these directions. The interaction surface is a peanut shape with the long axis (strong direction) parallel to the long axis of the structure's cross section as would be expected.

It is important to recall here that each point on the surface represents the capacity of the structure foundation which is constrained to collapse in the plane of the loading. It is possible however that a structure loaded in one direction could fail out of plane. Referring to the interaction surface, this would occur when any out of plane load component reached the interaction surface before the resultant load did. Fig. 10b shows an example of such a situation. Consider a load in the direction O-B. At point B' the resultant has a component (O-A) that will cause failure at A before B is reached. Thus B cannot be reached and the interaction surface is effectively truncated along A-C. The shaded region, A-B-C-A, then becomes inaccessible. Carrying out this construction on each load path results in the improved approximation to the interaction surface shown in Fig. 11 inscribed in the original surface. This surface is also symmetric.

To further illustrate the importance of this analysis, consider situations in which the top, right corner pile strength is changed dramatically during installation e.g. weakening due to underdrive or strengthening due to unexpected dense sand. Interaction surfaces are generated in which the corner pile in the top right quadrant is increased by factor of two in one case and reduced by a factor of two in the other case. The results shown in Fig. 12 give a very clear picture of these effects on all loading directions. Ancillary information in this form has proved to be very useful for making decisions regarding remedial action. Options can be quickly evaluated by the engineer and conveyed to the installation team in a timely manner.

Summary

This paper has outlined a simple, accurate approach for the evaluation of foundation system capacity for conventional pile founded structures. Individual piles may be represented by a detailed model or a yield surface for the loading system imposed. This method is highly efficient when compared with the use of nonlinear structural programs. It is most useful when calibrated with other data or analysis results and used in conjunction with more detailed analytical models.

References

Aggarwal, R. K., Litton, R.W., Cornell, C. A., Tang, W. H., Chen, J. H., and Murff, J. D., 1996. "Development of Pile Foundation Bias Factors Using Observed Behavior of Platforms During Hurricane Andrew", *Proc. Offshore Technology Conference*, Houston

Arnold, P., Bea, R. G., Idriss, I. M., Reimer,, R. B., Beebe, K. E., and Marshall, P. W., 1977. " A Study of Soil-Pile-Structure Systems in Severe Earthquakes", *Proc. Offshore Technology Conference*, Houston

Broms, B. B., 1964. "Lateral Resistance of Piles in Cohesive Soil", *J. Soil Mech. And Found. Div.*, ASCE, 90(2).

Chen, W. F., 1975. *Limit Analysis and Soil Plasticity*, Elsevier Publishing Co., Amsterdam, The Netherlands.

Calladine, C. R., 1969. *Engineering Plasticity*, Pergamon Press.

Desai, C. and Able, J. F., 1972. *Introduction to the Finite Element Method*, Van Nostrand Reinhold, New York.

Matlock, H., 1970. "Correlations for Design of Laterally Loaded Piles in Soft Clay", *Proc. Offshore Technology Conference*, Houston.

Meyerhof, G. G., 1995. "Behavior of pile foundations under special loading conditions", *Canadian Geotechnical Journal*, 32(2), 204-222.

Murff, J. D., 1987. "Plastic collapse of long piles under inclined loading", *Int. J. Numerical and Analytical Methods in Geomechanics*, 11, 185-192.

Murff, J. D., and Hamilton, J. M. 1993. "P-Ultimate for Undrained Analysis of Laterally Loaded Piles", *Journal of Geotechnical Engineering*, 119(1), 91-107.

Murff, J. D. and Wesselink, B. D., 1986. "Collapse Analysis of Pile Foundations", *Proc. Third Int. Conf. On Numerical Methods in Offshore Piling*, Nantes, 445-459.

Prager, W., 1959. *An Introduction To The Theory Of Plasticity*, Addison Wesley, Reading, Mass.

Randolph, M. F. and Houlsby, G. T., 1984. "The limiting pressure on a circular pile loaded laterally in cohesive soil", *Geotechnique*, 34(4), 613-623.

Reese, L. C., Cox, W.R., and Koop, F. D., 1970. "Field Testing and Analysis of Laterally Loaded Piles in Stiff Clay", *Proc. Offshore Technology Conference*, Houston.

Platform Characteristics

Total Vert Load,Kn	75000
Vert Load Eccn, m	0
Distance of H above ML,m	30

Pile Characteristics	Pile 1	Pile 2
Diameter,m	2	2
Wall Thickness,cm	10	10
Yield Strength, Mpa	275	275
Distance from origin,m	-20	20
Batter, Degrees from Vert	0	0
Soil Pult, KN/m	360	360

Table 1   Characteristics of Simple Two-Pile Structure

Description	Value
Pile Diameter, m	0.9146
PileWall Thickness, cm	3.81
Pile Length, m	Varies
Pile Yield Strength, MPa	297.0
Axial Soil Resistance,	
Mudline Shaft Resistance, KN/m	194.0
Strength Gradient, KN/m^2	Varies
Lateral Soil Resistance	
Mudline Bearing Capacity,KN/m	19.7
3 m BML, KN/m	90.5
60 m BML, KN/m	679.2

Table 2   Parameters for Single Pile Analysis

Description	Value
Pile Coordinates	See Figure 8
Pile Diameter, m	0.9146
PileWall Thickness, cm	3.81
Pile Length, m	Varies
Pile Yield Strength, MPa	297.0
Centric Vertical Load, KN	11100.0
Resultant Load Height, m	61.0
Soil Profile	
Mudline Strength, KPa	7.2
Strength Gradient, KPa/m	1.26
Soil Unit Wt., KN/m^3	6.0

Table 3   Parameters for a 12-Pile Jacket Analysis

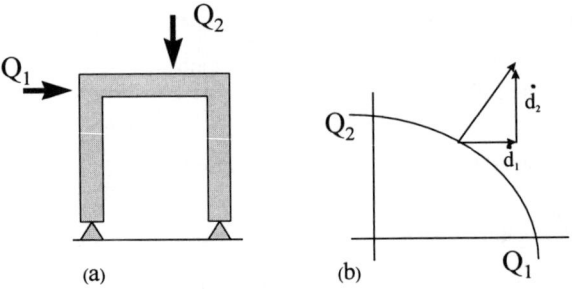

Figure 1   Examples of a Simple Structure (a) with a Corresponding Yield Surface (b)

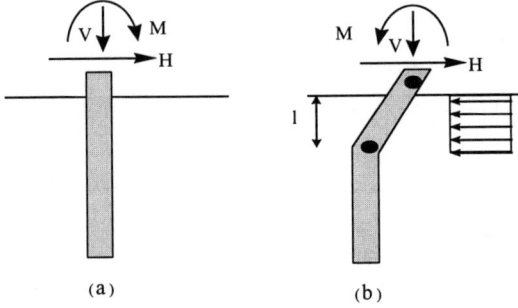

Figure 2   Single Pile Under Combined Load (a) and with Two Plastic Sections (b)

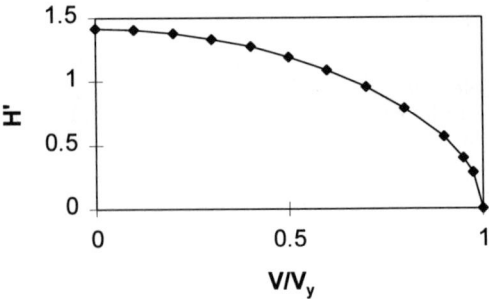

Figure 3   Failure Interaction Surface for a Fixed-Head Pile

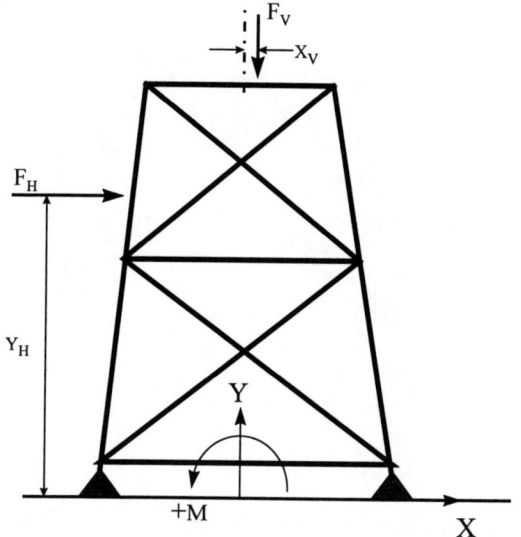

Figure 4 Schematic of Planar Structure on 'Yield Surface' Supports

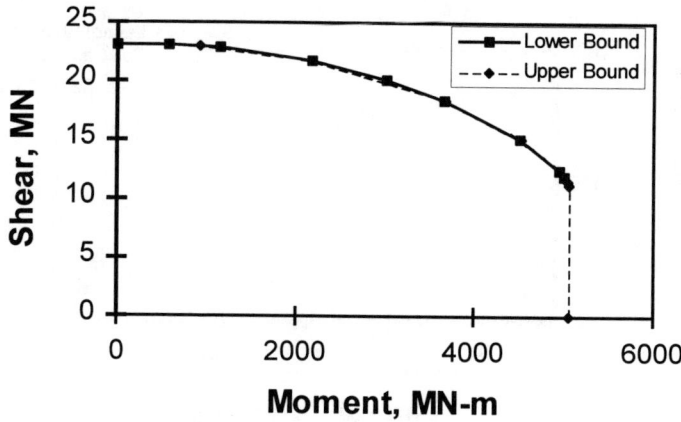

Figure 5 Calculated Upper and Lower Bound Yield Surfaces for Simple Structure

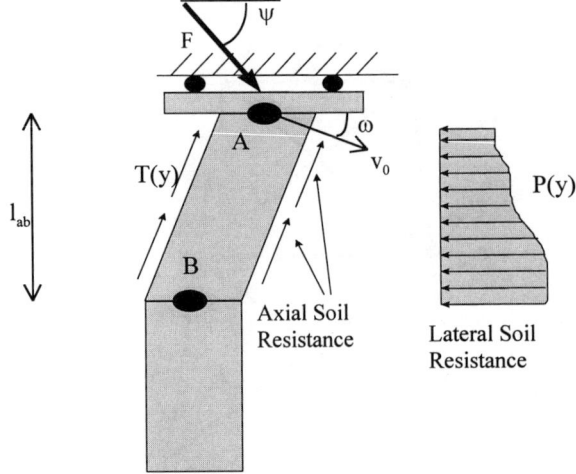

Figure 6   Deformation Mechanism of a Single Pile

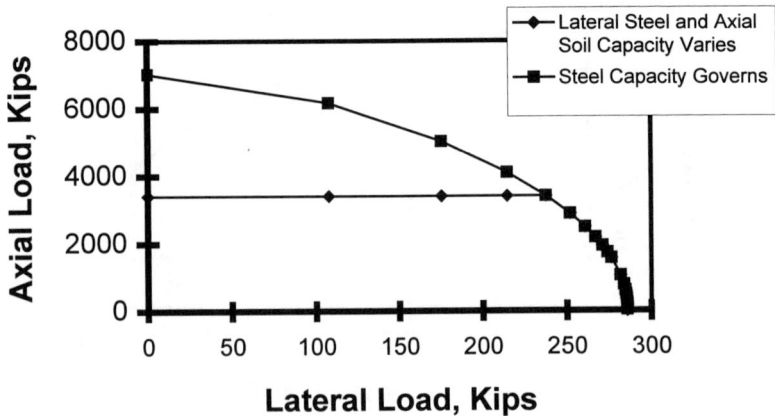

Figure 7   Interaction Surface for a Single Pile

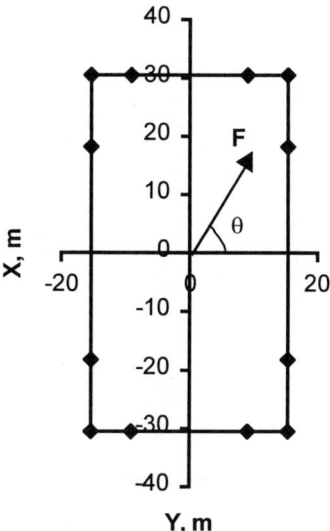

Figure 8   Plan View of a Twelve-Pile Foundation with Loading Direction Indicated

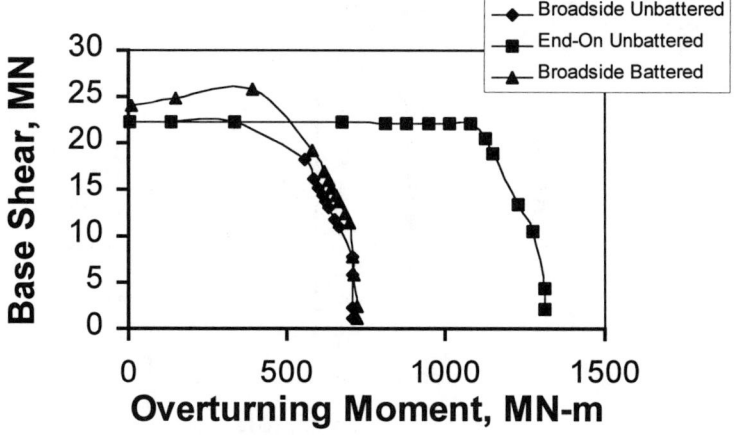

Figure 9   Shear Moment Interaction Diagrams

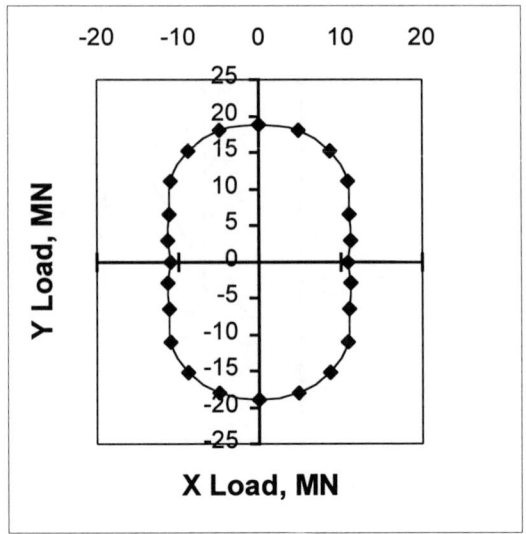

Figure 10a   Interaction Surface Constrained to In-Plane Failure

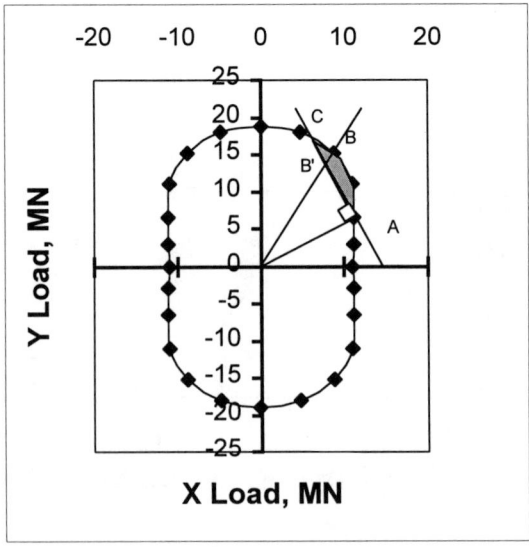

Figure 10b   Graphical Construction to Account for Out-of-Plane Failure

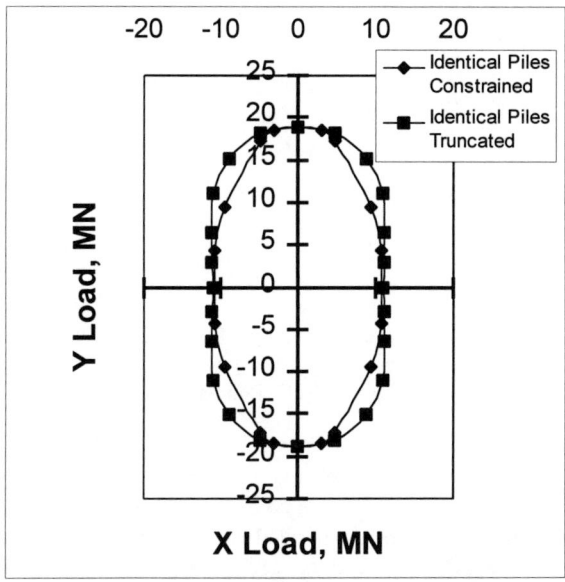

Figure 11 Constrained vs Truncated Yield Surface-- Identical Piles

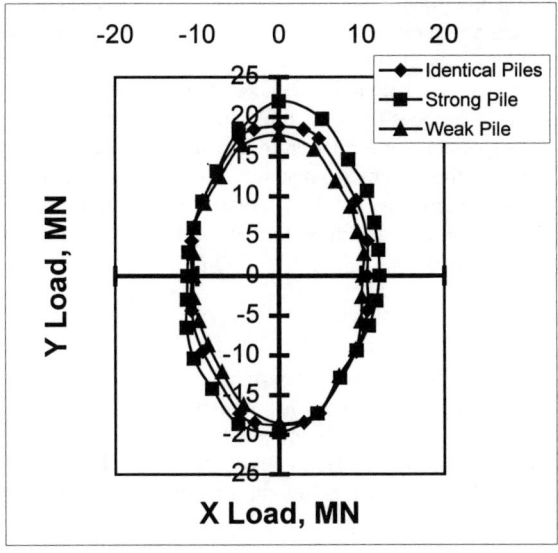

Figure 12   Truncated Yield Surfaces for Varying Strength Corner Pile

Some Aspects of the Design of Piled Raft Foundations

H.G. Poulos[1]

Abstract

This paper outlines the use of piled raft foundations for situations in which the performance of the raft alone does not satisfy the design requirements. Under these circumstances, the addition of a limited number of piles may improve both the ultimate load capacity and the settlement and differential settlement performance of the raft. The philosophy of using piles as 'settlement reducers' is discussed and it is shown that the design of pile-enhanced rafts can be considered to be a two-stage process. The first stage establishes the effects of the number of piles on load capacity and settlement via simplified analyses. The second is a detailed design phase, and involves the use of more refined analysis techniques to assess the optimum number and location of the piles, as well as providing essential information for the structural design of both the raft and the piles. Available methods of detailed analysis are outlined and compared. An example is given to illustrate the potential for achieving an economical foundation system with very little sacrifice in foundation performance. Finally, a typical application of pile-enhanced rafts is described.

Introduction

It is common in foundation design to consider first the use of a raft foundation to support a structure, and then if this is not adequate, to design a fully piled foundation in which the entire design loads are resisted by the piles. Despite such design assumptions, it is common for a raft to be part of the foundation system (e.g. because of the need to provide a basement below the structure). In the past few years, there has

---

[1]Professor of Civil Engineering, Dept. of Civil Engineering, University of Sydney, Australia and Senior Principal, Coffey Geosciences, Sydney. Australia.

been an increasing recognition that the use of piles to reduce raft settlements and differential settlements can lead to considerable economy without compromising the safety and performance of the foundation. Such a foundation makes use of both the raft and the piles, and is referred to here as a pile-enhanced raft or a piled raft. One of the Technical Committees of the International Society for Soil Mechanics and Foundation Engineering (ISSMFE) focussed its efforts in the period 1994-7 towards piled raft foundations, and collected considerable information on case histories and methods of analysis and design, and produced comprehensive reports on these activities (O'Neill et al. 1996; van Impe and Lungu, 1996). In addition, an independent treatise on numerical modelling of piled rafts has been presented by El-Mossallamy and Franke, 1997). Despite this recent activity, the concept of piled raft foundations is by no means new, and has been described by several authors, including Zeevaert (1957), Davis and Poulos (1972), Hooper (1973), Burland et al. (1977), Sommer et al. (1985), Price and Wardle (1986) and Franke (1991), among many others.

This paper describes the philosophy of design of pile-enhanced rafts and outlines circumstances which are favourable for such a foundation. A two-stage design process is proposed, the first being an approximate preliminary stage to assess feasibility, and the second to obtain detailed design information. Methods of analysis are described and compared. An example is outlined to illustrate the potential for cost savings via the use of the piled raft design concept. Finally, a number of applications of piled raft foundation are presented.

Design Concepts

Design Issues

As with any foundation system, a design of a piled raft foundation requires the consideration of a number of issues, including:
1.   Ultimate load capacity for vertical, lateral and moment loadings
2.   Maximum settlement
3.   Differential settlement
4.   Raft moments and shears for the structural design of the raft
5.   Pile loads and moments, for the structural design of the piles.

In much of the available literature, emphasis has been placed on the bearing capacity and settlement under vertical loads. While this is a critical aspect and is considered in detail herein, the other issues must also be addressed. In some cases, the pile requirements may be governed by the overturning moments applied by wind loading, rather than the vertical dead and live loads.

Alternative Design Philosophies

Randolph ('994) has defined clearly three different design philosophies with respect to piled rafts:

- The 'conventional approach', in which the piles are designed as a group to carry the major part of the load, while making some allowance for the contribution of the raft, primarily to ultimate load capacity.
- 'Creep piling', in which the piles are designed to operate at a working load at which significant creep starts to occur, typically 70-80% of the ultimate load capacity. Sufficient piles are included to reduce the net contact pressure between the raft and the soil to below the preconsolidation pressure of the soil.
- Differential settlement control, in which the piles are located strategically in order to reduce the differential settlements, rather than to substantially reduce the overall average settlement.

In addition, there is a more extreme version of the creep piling, in which the full load capacity of the piles is utilized, i.e. some or all of the piles operate at 100% of their ultimate load capacity. This gives rise to the concept of using piles primarily as settlement reducers, while recognizing that they also contribute to increasing the ultimate load capacity of the entire foundation system.

Clearly, the latter approaches are most conducive to economical foundation design, and will be given special attention herein. However, it should be emphasized that the design methods to be discussed allow any of the above design philosophies to be implemented.

Figure 1 illustrates, conceptually, the load-settlement behaviour of piled rafts designed according to the first two strategies. Curve 0 shows the behaviour of the raft alone, which in this case settles excessively at the design load. Curve 1 represents the conventional design philosophy, for which the behaviour of the pile-raft system is governed by the pile group behaviour, and which may be largely linear at the design load. In this case, the piles take the great majority of the load. Curve 2 represents the case of creep piling where the piles operate at a lower factor of safety, but because there are fewer piles, the raft carries more load than for Curve 1. Curve 3 illustrates the strategy of using the piles as settlement reducers, and utilizing the full capacity of the piles at the design load. Consequently, the load-settlement may be nonlinear at the design load, but nevertheless, the overall foundation system has an adequate margin of safety, and the settlement criterion is satisfied. Therefore, the design depicted by Curve 3 is acceptable and is likely to be considerably more economical than the designs depicted by Curves 1 and 2.

Favourable Circumstances for Piled Rafts

The most effective application of piled rafts occurs when the raft can provide

adequate load capacity, but the settlement and/or differential settlements of the raft alone exceed the allowable values. Poulos (1991) has examined a number of idealized soil profiles, and found that the following situations may be favourable:
*   soil profiles consisting of relatively stiff clays
*   soil profiles consisting of relatively dense sands.

In both circumstances, the raft can provide a significant proportion of the required load capacity and stiffness, with the piles acting to 'boost' the performance of the foundation, rather than providing the major means of support.

While the use of piled rafts is not restricted to such circumstances, considerable caution may need to be exercised in some cases, especially where there is a possibility of ground movements occurring after construction (Poulos, 1993; Sinha, 1997).

The Design Process

It is suggested that a rational design process for piled rafts involves two main stages:
*   A preliminary stage to assess the feasibility of using a piled raft, and the required number of piles to satisfy design requirements.
*   A detailed design stage to obtain the optimum number, location and configuration of the piles, and to compute the detailed distributions of settlement, bending moment and shear in the raft, and the pile loads and moments.
The preliminary stage involves relatively simple calculations which can usually be performed without a computer. The detailed stage will generally demand the use of a suitable computer program which accounts in a rational manner for the interaction among the soil, raft and piles, The effect of the superstructure may also need to be considered.

Preliminary Design Stage

In the preliminary stage, it is necessary first to assess the performance of a raft foundation without piles. Estimates of vertical and lateral bearing capacity, settlement and differential settlement may be made via conventional techniques. If the raft alone provides only a small proportion of the required load capacity, then it is likely that the foundation will need to be designed with the conventional philosophy, so that the function of the raft is merely to perhaps reduce slightly the piling requirements. If however the raft alone has adequate or nearly adequate load capacity, but does not satisfy the settlement or differential settlement criteria, then it may be feasible to consider the use of piles as settlement reducers, or to adopt the 'creep piling' approach.

For assessing vertical bearing capacity, the ultimate load capacity can generally

be taken as the lesser of the following two values:
- The sum of the ultimate capacities of the raft plus all the piles
- The ultimate capacity of a block containing the piles and the raft, plus that of the portion of the raft outside the periphery of the piles.

For estimating the load-settlement behaviour, an approach similar to that described by Poulos and Davis (1980) can be adopted. However, a useful extension to this method can be made by using the simple method of estimating the load sharing between the raft and the piles, as outlined by Randolph (1994). This modified method can be computed with the aid of a computer spreadsheet or a mathematical program such as MATHCAD. In this way, it is simple to compute the relationship between the number of piles and the average settlement of the foundation. Such calculations provide a rapid means of assessing whether the design philosophies for creep piling or full pile capacity utilization are likely to be feasible.

Burland's Simplified Design Approach

When the piles are designed to act as settlement reducers and to develop their full geotechnical capacity at the design load, Burland (1995) has developed the following useful simplified process of design:
1.   Estimate the total long-term load-settlement relationship for the raft without piles (see Figure 2). The design load P_o gives a total settlement S_o.
2.   Assess an acceptable settlement S_d, which should include a margin of safety.
3. P_1 is the load carried by the raft corresponding to S_d.
4.   The load excess P_o-P_1 is assumed to be carried by settlement-reducing piles. The shaft resistance of these piles will be fully mobilized and therefore no factor of safety is applied. However, Burland suggests that a 'mobilization factor' of about 0.9 be applied to the 'conservative best estimate' of ultimate shaft resistance, P_{su}.

On this basis, the number of piles required can be estimated as follows:
Furthermore, if the piles are located below columns which carry a load in excess

$$n = \frac{P_0 - P_1}{0.9 P_{su}} \qquad (1)$$

of the ultimate pile load capacity P_{su}, the piled raft may be analyzed as a raft on which reduced column loads act. At such a column the reduced load is then:

$$Q = Q - 0.9 P_{su} \qquad (2)$$

where Q   =   total column load

P_{su} = ultimate shaft capacity of pile

Detailed Design Stage

Once the preliminary stage has indicated that a piled raft foundation is feasible, it is necessary to carry out a more detailed design in order to assess the detailed distribution of settlement and decide upon the optimum locations and arrangement of the piles. The raft bending moments and shears, and the pile loads, should also be obtained for the structural design of the foundation.

Several methods of analyzing piled rafts have been developed, and some of these have been summarized by Poulos et al. (1997). The less simplified methods of numerical analysis tend to fall into the following categories:

- Methods employing a 'strip on springs' approach, in which the raft is represented by a series of strip footings, and the piles are represented by springs of appropriate stiffness (e.g. Poulos, 1991)
- Methods employing a 'plate on springs' approach, in which the raft is represented by a plate and the piles as springs (e.g. Clancy and Randolph, 1993; Poulos, 1994). In the latter paper, the development of a computer program called GARP (Geotechnical Analysis of Raft and Piles) is described. This program has been employed for some of the results presented later in the present paper
- Boundary element methods, in which both the raft and the piles within the system are discretized, and use is made of elastic theory (e.g. Butterfield and Banerjee, 1971; Kuwabara, 1989; Sinha, 1997)
- Methods combining boundary element for the piles and finite element analysis for the raft (e.g. Hain and Lee, 1978; Ta and Small, 1996; Franke et al, 1994)
- Simplified finite element analyses, usually involving the representation of the foundation system as a plane strain problem (Desai, 1974) or an axi-symmetric problem (Hooper, 1974)
- Three-dimensional finite element analyses (e.g. Zhuang et al, 1991; Lee, 1993; Wang, 1995).

Poulos et al (1997) have compared some of the these methods when applied to a idealized hypothetical problem and to the case of the Westend 1 Building in Frankfurt. The latter case is discussed later in the present paper.

Despite some differences among the various methods, most of those which incorporate nonlinear behaviour were found to give generally similar results, although there were differences among the computed raft bending moments. However, it appeared that, provided the analysis method is soundly based and takes into account the limited load capacity of the piles, similar results may be expected for similar parameter inputs.

Some Characteristics of Piled Raft Behaviour

To illustrate some of the characteristics of piled raft behaviour, the simple hypothetical problem in Figure 3 has been analyzed using the computer program GARP (Poulos, 1994). It involves a rectangular raft foundation 10 m by 6 m in plan, supporting nine column loads, located on a homogeneous clay layer 20 m thick overlying rock. The undrained shear strength of the clay is 50 kPa, while the long-term Young's modulus is 20 MPa and the drained Possion's ratio is 0.3. The foundation design parameters are as follows:

1.	Ultimate bearing capacity below raft	=	300kPa
2.	Ultimate pile shaft friction	=	50 kPa
3.	Ultimate pile base resistance	=	450 kPa.

Attention is focussed on the foundation behaviour under vertical loading only, and the vertical column loadings are as shown in Figure 3. The total vertical load is 7200 kN, thus giving an overall factor of safety against bearing capacity of the raft alone of 2.5.

GARP analyses have been carried out to examine the effects both of raft thickness and of adding various numbers of piles to the raft on the following aspects of foundation behaviour:
1.   Maximum settlement
2.   Maximum differential settlement
3.   Maximum bending moments in the raft.

In particular, attention is concentrated on a comparison of alternative foundations which provide the same maximum allowable differential settlement between the central and corner columns, assumed here to be 5 mm.

For convenience only two types of piles have been considered:
1.   0.5 m diameter and 5 m long
2.   0.5m diameter and 10 m long.

The computed ultimate axial compressive capacity of these two pile types is 480 kN and 873 kN respectively.

In the GARP analyses, use has been made of the symmetry of the raft and the loadings, and the quarter-raft has been divided into 273 nodes and elements.

Figure 4 shows the computed overall factor of safety for five different foundation configurations:
1.   Raft only
2.   The raft with 3 piles 5 m long, the piles being located below the three central

heavily-loaded columns denoted as 'A' in Figure 3
3.   The raft with 3 piles 10 m lcng, one below each column 'A' in Figure 3
4.   The raft with 9 piles 5 m long, one pile below each column
5.   The raft with 9 piles 10 m long, one below each column.
These foundations will be referred to as Configurations 1 to 5 respectively.

Of course, the overall factor of safety increases with increasing number and/or length of piles, from 2.5 for the raft only (Configuration 1) to about 3.5 for Configuration 5. Normal design practice would be to adopt a value somewhere within this range.

Figure 5 summarises the computed maximum differential settlement (between the centre and corner column loads) as a function of raft thickness, for the five foundation configurations. As would be expected, the maximum differential settlement decreases as raft thickness increases. However, it does not necessarily decrease as the number of piles increases; indeed, the least differential settlement occurs for Configuration 3 (raft plus 3 10m long piles). In this case, the piles are strategically located below the largest loads, and serve to reduce the settlements locally while having relatively little effect in suppressing the settlement of the more lightly loaded columns. In contrast, the maximum differential settlement of the foundations with 9 piles tends to be exacerbated because of the piles below the lightly loaded columns further restraining the already small settlements at those locations. Randolph (1994) emphasises the value of careful placement of pile locations in order to obtain the most effective reduction of differential settlements within a piled raft foundation.

If it is assumed that the primary design criterion for this foundation is to limit the maximum differential settlement to 5 mm, then Figure 6 shows the raft thickness required to satisfy this criterion. This figure clearly shows that foundation configuration 3 (raft with three 10 m piles) requires the smallest thickness. Also shown in Figure 6 is the total volume of concrete for each of the foundations giving a 10 mm maximum differential settlement. Again, it is clear that Configuration 3 is by far the most economical design of those considered.

Figure 7 shows the computed performance of the five foundation Configurations for a 0.5 m thick raft. The following points are noted:
1.   Configurations 3 and 5 (each with the longer piles) give the smallest maximum settlements, but the use of nine 10 m piles gives only a modest additional reduction in settlement compared to the case of the three 10 m piles.
2.   Configuration 3, as discussed previously, gives a much smaller differential settlement than the other configurations.
3.   The largest moments occur for the raft without piles. The addition of piles generally reduces the maximum moments, but the most dramatic reduction is obtained for Configuration 3, the raft with three 10 m piles.

Burland's simplified approach has also been used for the analysis of the simplified problem in Figure 3. Configuration 5 has not been considered since the ultimate capacity of the 10 m long piles exceeds the loads on the outer columns, and hence Burland's approach is not directly applicable. Some of the key results are given in Table 1 together with the values obtained from the GARP analysis. For all cases, the raft thickness has been taken as 0.5 m. The following observations are made:

1.   Burland's approach appears to give smaller maximum settlements than GARP
2.   Burland's approach gives maximum moments and differential settlements which are similar to those computed by the GARP analysis.

The smaller settlements obtained from Burland's approach may be a consequence of the soil-pile-raft interactions which are taken into account only approximately in Burland's approach by allowing for only 90% of the pile capacity to be used as a 'negative load'. However, Burland's approach seems to provide a useful and simple means of estimating the bending moments for structural design of the raft.

Table 1.   Comparison Between Results from GARP and from Burland's Approach (0.5 m thick raft)

Configuration	Max Settlement mm		Max. Diffl. Settlement mm		Max. Moment M_{xx} MNm/m	
	GARP	BURLAND	GARP	BURLAND	GARP	BURLAND
2 (Raft + 3*5m piles)	32	27	4.5	3.8	0.553	0.513
3 (Raft + 3*10m piles)	27.3	22	1.5	2.1	0.298	0.336
4 (Raft + 9*10m piles)	32	19	7.3	6.4	0.692	0.644

Geotechnical Parameter Assessment

The design of a piled raft foundation requires an assessment of a number of geotechnical and performance parameters, including:

•   raft bearing capacity
•   pile capacity
•   soil modulus for raft stiffness assessment
•   soil modulus for pile stiffness.

While there are a number of laboratory and in-situ procedures available for the assessment of these parameters, it is common for at least initial assessments to be based on the results of simple in-situ tests such as the Standard Penetration Test (SPT) and the Static Cone Penetration Test (CPT). Typical of the correlations are the following which the author has employed are those based on the work of Decourt (1989, 1995) using the SPT:

Raft ultimate bearing capacity:	p_{ur}	$=$	$K_1.N_r$ kPa (3)
Pile ultimate shaft resistance:	f_s	$=$	$a.[2.8 N_s + 10]$ kPa (4)
Pile ultimate base resistance:	f_b	$=$	$K_2.N_b$ kPa (5)
Soil Young's modulus below raft:	E_{sr}	$=$	2N MPa (6)
Young's modulus along and below pile:	E_s	$=$	3N MPa (7)

where N_r =   average SPT (N_{60}) value within depth of one-half of the raft width

N_s =   SPT value along pile shaft

N_b =   average SPT value close to pile tip

K_1, K_2 =   factors shown in Table 1

a   =   1 for displacement piles in all soils and non-displacement piles in clays,

and   a   =   0.5 - 06 for non displacement piles in granular soils.

Table 2: Correlation factors K_1 and K_2

Soil Type	K_1 (Raft)	K_2 Displacement Piles	K_2 Non-Displacement Piles
Sand	90	325	165
Sandy silt	80	205	115
Clayey silt	80	165	100
Clay	65	100	80

An Application to the Westend Tower, Frankfurt

The Westend 1 tower is a 51 storey, 208 m high building in Frankfurt, Germany, and has been described by Franke et al. (1994), and Franke (1991). A cross-section and foundation plan of the building is shown in Figure 7. The foundation for the tower consists of a piled raft with 40 piles, each about 30 m long and 1.3 m in diameter. The central part of the raft is 4.5 m thick, decreasing to 3 m at the edges. While full details of the geotechnical profile are not available in the published literature, it appears that the building is located on a thick deposit of relatively stiff Frankfurt clay. On the basis of pressuremeter tests, an average reloading soil modulus of 62.4 MPa has been reported by Frank et al. (1994).

Calculations have been reported by Poulos et al. (1997) to predict the behaviour of the building, using a number of different analysis methods:

- a finite element analysis (Ta and Small, 1996)
- the GARP analysis described earlier in this paper
- a piled strip analysis (Poulos, 1991)
- the simple hand calculation method described by Poulos and Davis (1980)
- the approximate linear method developed by Randolph (1983, 1994)
- the combined finite element and boundary element method developed by Sinha (1997)
- the combined finite element and boundary element method described by Franke et al (1994).

In addition Burland's approach has been used.

Table 2 compares the predictions of performance for the above methods, together with the measured values. The calculations have been carried out for a total load of 968 MN, which is equivalent to an average applied pressure of 323 kPa.

The following points are noted:

- the measured maximum settlement is about 105 mm, and most methods (other than Burland's) tend to over-predict this settlement. However, most of the methods provide an acceptable design prediction
- the piles carry about 50% of the total load. Most methods tended to over-predict this proportion, but from a design viewpoint, most methods give acceptable estimates
- all methods capable of predicting the individual pile loads suggest that the load capacity of the most heavily loaded piles is almost fully utilized; this is in agreement with the measurements
- there is considerable variability in the predictions of minimum pile loads. Some of the methods predicted larger minimum pile loads than were actually measured.

This case history clearly demonstrates that the design philosophy of fully utilizing pile capacity can work successfully and produce an economical foundation which performs satisfactorily. The available methods of performance prediction appear to provide a reasonable, if conservative, basis for design in this case.

Table 3: Westend Building - Comparison of Analysis Methods with Measurements

Method	Central Settlement	% Pile Load	Max. MN	Min. Pile Load MN
Poulos & Davis (1990)	131	-	-	-
Randolph (1994)	173	56	-	-
Ta & Small (1996)	117	65	16	16
GARP (Poulos, 1994)	106	61	16	13
GASP (Poulos, 1991)	139	65	16	16
Franke et al (1994)	111	50	15	8
Sinha (1997)	150	55	16	8
Burland (1995)	100	48	-	-
Measured	105	50	15	9

Conclusions

This paper demonstrates that the design of piled raft foundation systems can be carried out as a two-stage process, involving a preliminary design phase to obtain an approximate assessment of the required number of piles, and a detailed design phase to refine piling requirements and locations and provide information for the structural design of the foundation.

Alternative design strategies for the design of the piles have been discussed, and it has been demonstrated that effective and efficient foundations can be designed by utilizing a significant part (if not all) of the available capacity of the piles. This philosophy of designing piles as settlement reducers can lead to foundations with fewer piles than in a conventional design, but which can still satisfy the specified design criteria with respect to ultimate load capacity and settlement. The conventional approach of assuming that all the load should be carried by the piles can often lead to an over-conservative and uneconomical design.

Substantial economy can also be achieved by careful location of the foundation piles so that they are concentrated below heavily loaded areas and not below lightly loaded areas.

The piled raft foundation solution is not suitable for every circumstance. It is unlikely to be appropriate if soft clays or loose sands exist near the surface, and it is generally not a suitable option if ground movements are likely to occur below the raft. However, in cases where the soil conditions allow the raft to develop adequate capacity and stiffness, the piled raft solution may be very suitable.

The main obstacles to increased use of this type of foundation appear to be two-fold:
- a inherent conservatism in foundation design by some foundation engineers
- restrictions imposed by some building codes on minimum factors of safety that may be employed in pile design.

On the positive side, there are an increasing number of examples of successful use of piled rafts. Also, there is a rapidly increasing understanding of the mechanics of behaviour of piled rafts, and a number of design methods and tools now exist to facilitate analysis and calculations for piled rafts. It is to be hoped that these positive aspects will assist foundation design engineers to overcome the obstacles and that piled rafts will become a more commonly employed foundation type.

Acknowledgements

The author is grateful to Dr John Small of the University of Sydney, for many fruitful discussions on methods of analysis, Professor W. Van Impe for his leadership in focussing the efforts of Technical Committee TC18 of the International Society of Soil Mechanics and Foundation Engineering in the period 1994-7 towards piled rafts and professor Michael O'Neill for his contributions to the collection of case histories.

References

Burland, J.B. (1995). Piles as Settlement Reducers. Keynote Address, 18ᵗʰ It. CSM, Pavia

Burland, J.B., Broms, B.B. and de Mello, V.F.B. (1977). 'Behaviour of Foundations and Structures'. Proc. 9 ICSMFE, Tokyo, 2, 495-546.

Butterfield, R. and Banerjee, P.K. (1971). 'The Elastic Analysis of Compressible Piles and Pile Groups'. Geotechnique, 21 (1): 43-60.

Clancy, P. and Randolph, M.F. (1993). 'Analysis and Design of Piled Raft Foundations'. Int. J. NAM Geomechs.

Davis, E.H. and Poulos, H.G. (1972). 'The Analysis of Piled Raft Systems'. Aust. Geomechs. J., G2: 21-27.

Decourt, L. (1989).' SPT - State of the Art Report'. Proc. 12 ICSMFE, Rio de Janeiro, 4.

Decourt, L. (1995). 'Predictions of Load-Settlement Relationships for Foundations on the Basis of SPT-T'. Ciclo de Conf. Int. "Leonardo Zeevaert", UNAM, Mexico, 85-104.

Desai, C.S. (1974). 'Numerical Design analysis for Piles in Sands'. J. Geot. Eng. Div., ASCE, 100 (GT6): 613-635.

El-Mossallamy and Franke (1997). 'Piled Rafts-Numerical Modelling to Simulate the Behaviour of Piled Rat Foundations'. The Authors, Darmstadt, Germany.

Franke, E. (1991). 'Measurements Beneath Piled Rafts'. Keynote Lecture, ENPC Conf., Paris, 1-28.

Franke, E., Lutz, B. and El-Mossallamy, Y. (1994). 'Measurements and Numerical Modelling of High-Rise Building Foundations on Frankfurt Clay'. Geot. Spec. Pub. 40, ASCE, 2: 1325-1336.

Hain, S.J. and Lee, I.K. (1978). 'The Analysis of Flexible Raft-Pile Systems'. Geotechnique, 28(1): 65-83.

Hansbo, S. (1993). 'Interaction Problems Related to the Installation of Pile Groups'. Sem. on Deep Founds. on Bored and Auger Piles, BAP2, Ghent, 59-66.

Hooper, J.A. (1973). 'Observations on the Behaviour of a Piled-Raft Foundation on London Clay'. Proc. Inst. Civ. Engrs., 55 (2): 855-877.

Hooper, J.A. (1974). 'Review of Behaviour of Piled Raft Foundation'. Rep. No. 83, CIRIA, London.

Kuwabara, F. (1989). 'An Elastic Analysis for Piled Raft Foundations in a Homogeneous Soil'. Soils and Foundations, 28 (1): 82-92.

Lee, I.K. (1993). 'Analysis and Performance of Raft and Raft-Pile Systems'. Keynote Lect., 3 Int. Conf. Case Hist. In Geot. Eng., St. Louis (also Res. Rep. R133, ADFA, Univ. NSW, Australia).

O'Neill, M.W., Caputo, V., De Cock, F., Hartikainen, J. and Mets, M. (1996). 'Case Histories of Pile-Supported Rafts'. Rep. For ISSMFE Tech. Comm. TC18, Univ. of Houston, Texas.

Poulos, H.G. (1990). 'DEFPIG Users Manual'. Centre for Geot. Res., Univ. of Sydney, Australia.

Poulos, H.G. (1991). 'Analysis of Piled Strip Foundations'. Comp. Methods & Advances in Geomechs., ed. Beer et al., Balkema, Rotterdam, 1: 193-191.

Poulos, H.G. (1994a). 'An Approximate Numerical Analysis of Pile-Raft Interaction'. Int. J. NAM Geomechs., 18: 73-92.

Poulos, H.G. (1994b). 'Alternative Design Strategies for Piled Raft Foundations'. 3rd Int. Conf. Deep Founds., Singapore, 239-244.

Poulos, H.G. (1998). 'A Simplified Preliminary Design Procedure for Piled Raft Foundations'. Submitted for publication.

Poulos, H.G. and Davis, E.H. (1980). 'Pile Foundation Analysis and Design'. Wiley, New York.

Poulos, H.G., Small, J.C., Ta, L.D., Sinha, J. and Chen, L. (1997).' Comparison of Some Methods for Analysis of Piled Rafts'. Proc. 14 ICSMFE, Hamburg, 2: 1119-1124.

Price, G. And Wardle, I.F. (1986). 'Queen Elizabeth II Conference Centre: Monitoring of Load Sharing Between Piles and Raft'. Proc. Inst. Civ. Engrs., 80 (1): 1505-1518.

Randolph, M.F. (1983). 'Design of Piled Foundations'. Cambridge Univ. Eng. Dept., Res. Rep. Soils TR143.

Randolph, M.F. (1994). 'Design Methods for Pile Groups and Piled Rafts'. S.O.A. Report, 13 ICSMFE, New Delhi, 5: 61-82.

Sinha, J. (1997). 'Piled Raft Foundations Subjected to Swelling and Shrinking Soils'. PhD Thesis, Univ. of Sydney, Australia.

Sommer, H., Wittmann, P. and Ripper, P. (1985). 'Piled Raft Foundation of a Tall Building in Frankfurt Clay'. Proc. 11 ICSMFE, San Francisco, 4: 2253-2257.

Sommer, H., Tamaro, G. and DeBenedittis, C. (1991). 'Messe Turm, Foundations for the Tallest Building in Europe'. Proc. 4 DFI Conf., Stresa, Italy, 139-145.

Ta, L.D. and Small, J.C. (1996). 'Analysis of Piled Raft Systems in Layered Soils'. Int. J. NAM Geomechs., 2: 57-72.

Tamaro, G.J. (1996). 'Foundation Engineers: Why do we need them?' 1996 Martin S. Kapp Lecture, ASCE, New York.

Van Impe, W.F. and Lungu, I. (1996). 'Technical Report on Settlement Prediction Methods for Piled Raft Foundations'. Ghent Univ., Belgium.

Wang, A. (1995). 'Private Communication'. From PhD thesis, Univ. of Manchester, U.K.

Zeevaert, L. (1957). 'Compensated Friction-Pile Foundation to Reduce the Settlement of Buildings on Highly Compressible Volcanic Clay of Mexico City'. Proc. 4 ICSMFE, London, V.2.

Zhuang, G.M., Lee, I.K. and Zhao, X.H. (1991). 'Interactive Analysis of Behaviour of Raft-Pile Foundations'. Proc. Geo-Coast '91, Yokohama, 2: 759-764.

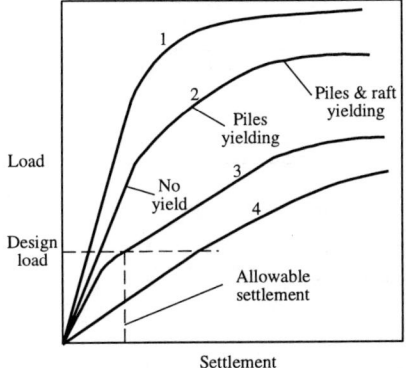

Curve 4:
raft only (settlement excessive)

Curve 1:
raft with pile designed for
conventional safety factor

Curve 2:
raft with piles designed for
lower safety factor

Curve 3:
raft with piles designed for
full utilization of capacity

Fig. 1 Alternative design strategies for piled rafts

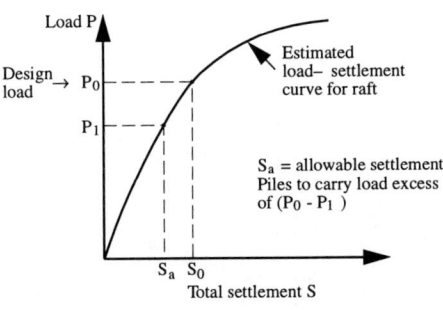

(a) Load settlement curve for raft

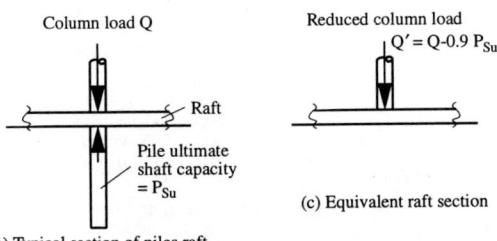

(b) Typical section of piles raft

(c) Equivalent raft section

Fig. 2 Burland's simplified design concept

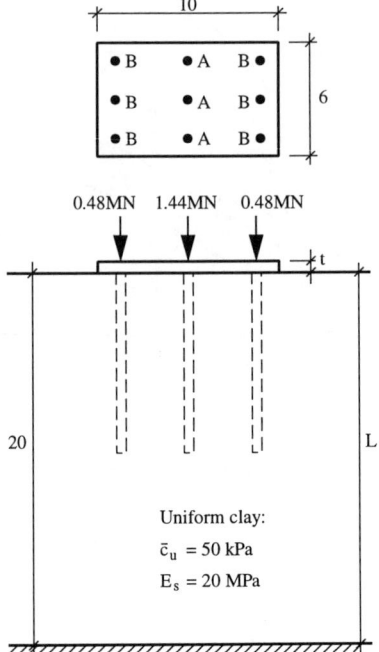

Note: When 3 piles are used, they are located below columns A
When 9 piles are used, they are located below columns A & B

Fig. 3 Problem analysed

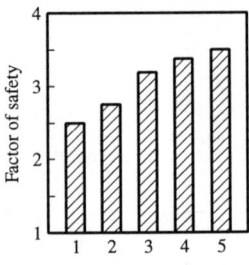

Foundation configuration

Configuration pile details
1 Raft only
2 Raft + 3x5 m piles
3 Raft + 3x10 m piles
4 Raft + 9x5 m piles
5 Raft + 9x10 m piles

Fig. 4 Overall factor of safety for various
foundation configurations

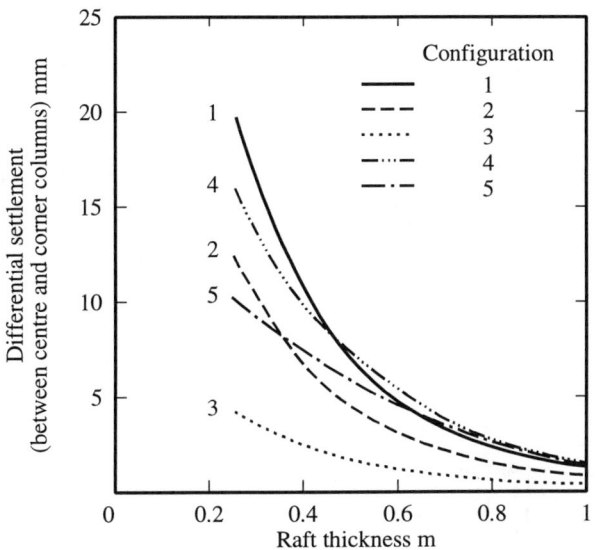

Fig. 5 Maximum differential settlement vs raft thickness

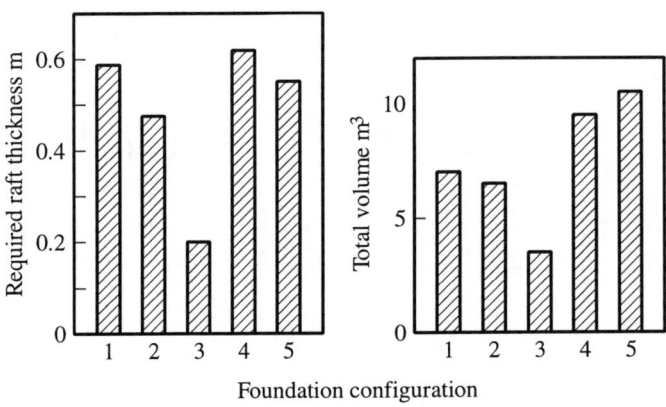

Configuration	Pile details
1	Raft only
2	Raft + 3x5 m piles
3	Raft + 3x10 m piles
4	Raft + 9x5 m piles
5	Raft + 9x10 m piles

Fig. 6 Foundation requirements for 5 mm differential settlement

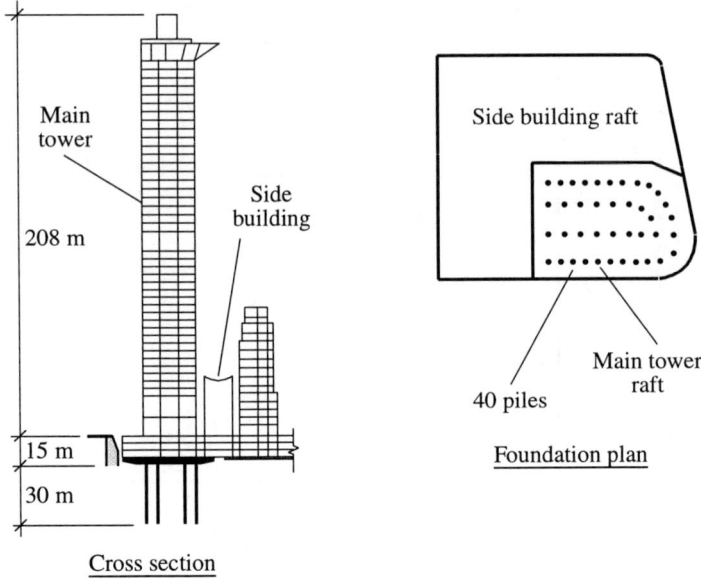

Fig. 7 Westend 1 Building (Franke et al. 1994)

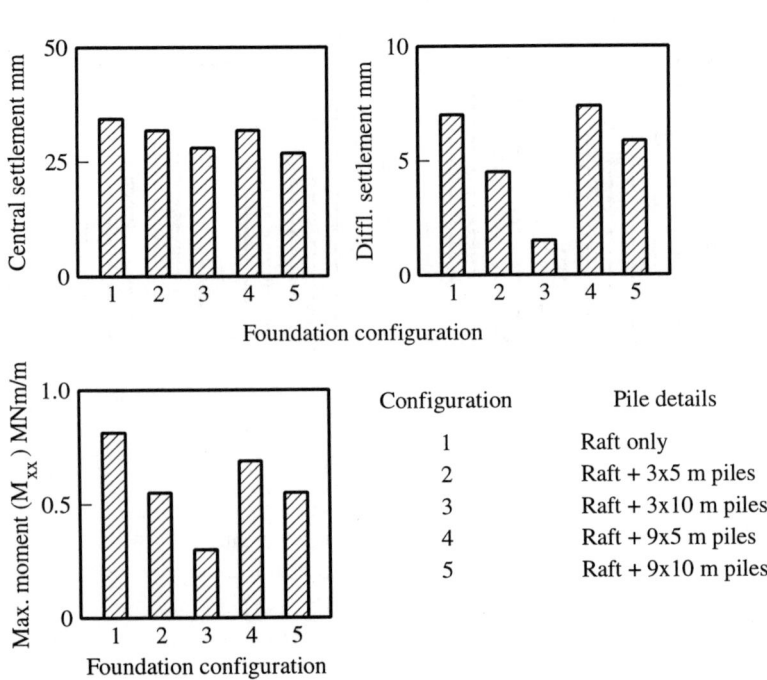

Fig. 8 Performation of various foundation configurations

Three Dimensional Analysis of Deep Foundations

Ian M. Smith[1]

Abstract

Deep foundations are of such geometrical complexity that "analysis" has often meant a formulaic approach, or gross geometrical simplification. As computational power and accessibility have increased, it has been possible to analyse more and more situations as they actually are. The present paper assesses how far this process has advanced, in particular taking into account computations carried out in parallel.

Introduction

Figure 1 shows a computer model (finite element) of the foundation of a milling machine used for the manufacture of the wings of a large modern

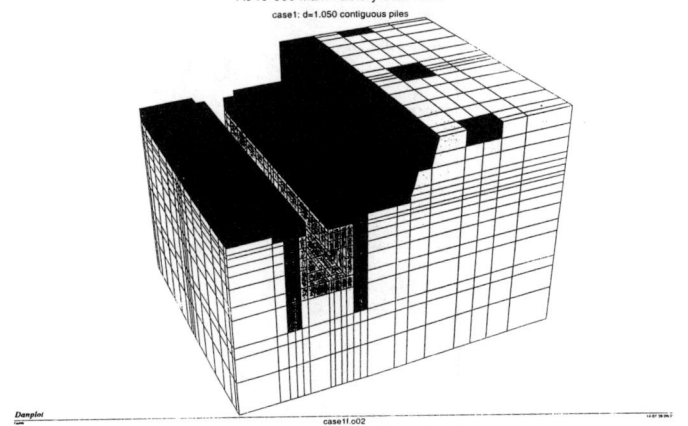

Figure 1 Model of Milling Machine Foundation

[1]Professor of Geotechnical Engineering, University of Manchester, Manchester M13 9PL, UK.

commercial aircraft. A downward-projecting part of the machine passes along the slot in the ground and various options (contiguous or discrete piles for example) can be used to support the base slab. Shown towards the upper right hand corner of the Figure are three foundation pads supporting a travelling crane. The designer needs to know what movements will occur as the various excavations and concrete placements occur during construction and what will be the movements of the machine under working conditions.

It is hard to see how answers to these questions could be given without recourse to computer modelling. Later in this paper it will be shown that analyses using the model shown in Figure 1 can be carried out on a present day "PC" computer but before this some historical perspective will be given to show how analyses of deep foundations have progressed to their current state.

First, a brief history of the process of analysis of single piles, including drivability, will be given. Attention will then be devoted to a specific case history, reported about 10 years ago, to see what changes in analytical procedure might be deployed were the problem to arise today. Finally, strategies for analysing problems such as that described by Figure 1, within the constraints of Civil Engineering budgets, will be given.

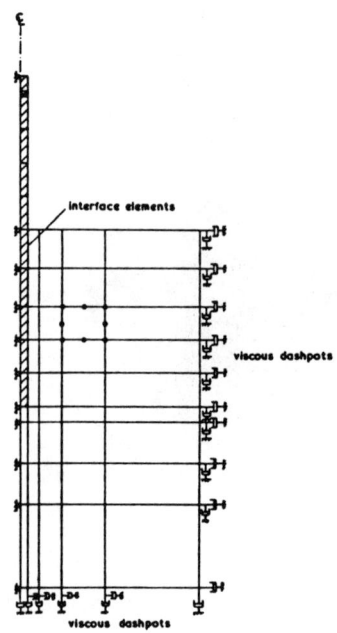

Figure 2 Model of Pipe Pile

Analysis of Single Piles

Figure 2 shows a cross-section of a (finite element) model of a single pipe pile, embedded in the soil. The elements are really three-dimensional, although axisymmetric, and there are about 50 of them.

The situations which can be analysed include drivability and subsequent static load-displacement behaviour using the same model for both situations and the conventional physical properties of stiffness moduli for pile and soil and shear strength parameters (c and φ for example) for the soil and soil/pile interface.

When these results were first reported, at a Conference at Austin, Texas (Smith and Chow, 1982) they had been obtained on a "mainframe" computer, and it could have been argued that such analysis capabilities were not readily available to practising engineers. It is a measure of progress that any two-dimensional analysis of this type, using meshes consisting of thousands of elements, could be completed today in a matter of a few minutes on a PC available on the desk of any practising engineer.

Figure 3 shows the tip displacement versus time computed by the three-dimensional finite element analysis and by the classical one-dimensional approach (Smith, 1960). The three-dimensional analysis results are much more realistic physically, showing an unambiguous residual plastic displacement ("set") of the pile tip.

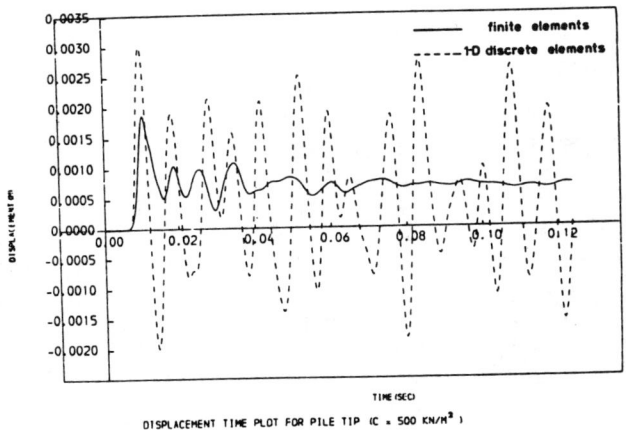

Figure 3 Tip Displacement versus Time for Pile Driving

Following the publication of the above results modifications to the classical one-dimensional drivability analyses were made, for example by Randolph and Simons (1986) and Lee et al (1988) to take account of the radiation damping automatically present in the three-dimensional finite-element models.

It should also be reported that the three-dimensional model, without alteration, was used to study plugging of pipe piles (Smith et al, 1986) and vibratory driving (Smith and To, 1988).

It can be concluded that analysing piles as they really are constitutes a major advance on the "Engineering News Formulae" approach of the pre-computer age.

A Deep Foundation in Calcareous Soil

The writer first encountered Professor Lymon Reese's work while a graduate student at Berkeley in the early 1960s. Professor Reese's doctoral studies had involved measurement of porewater pressures around driven piles in soft clays (Seed and Reese, 1957) and when the writer was taking large diameter piston samples of San Francisco Bay mud at Hamilton Field, Professor Reese's instrumented piles were still in evidence. Indirectly, these experiences carried over to the deep piled foundations for the North Rankin 'A' gas platform, offshore North West Australia. This case history has been extensively reported (Jewell and Khorshid, 1988, 2 volumes) and essentially involved the driving of piles into a very deep deposit of calcareous soil and soft rock.

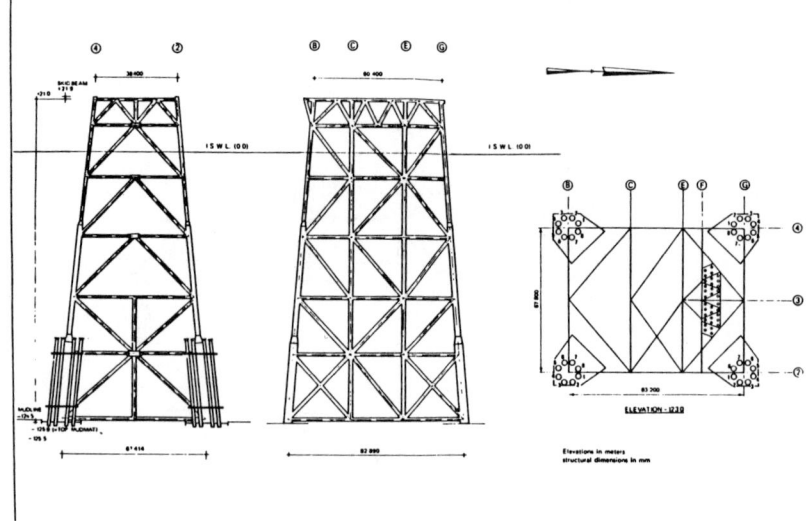

Figure 4 Piled Offshore Structure

Figure 4 shows the structure and Figure 5 that the foundation materials had

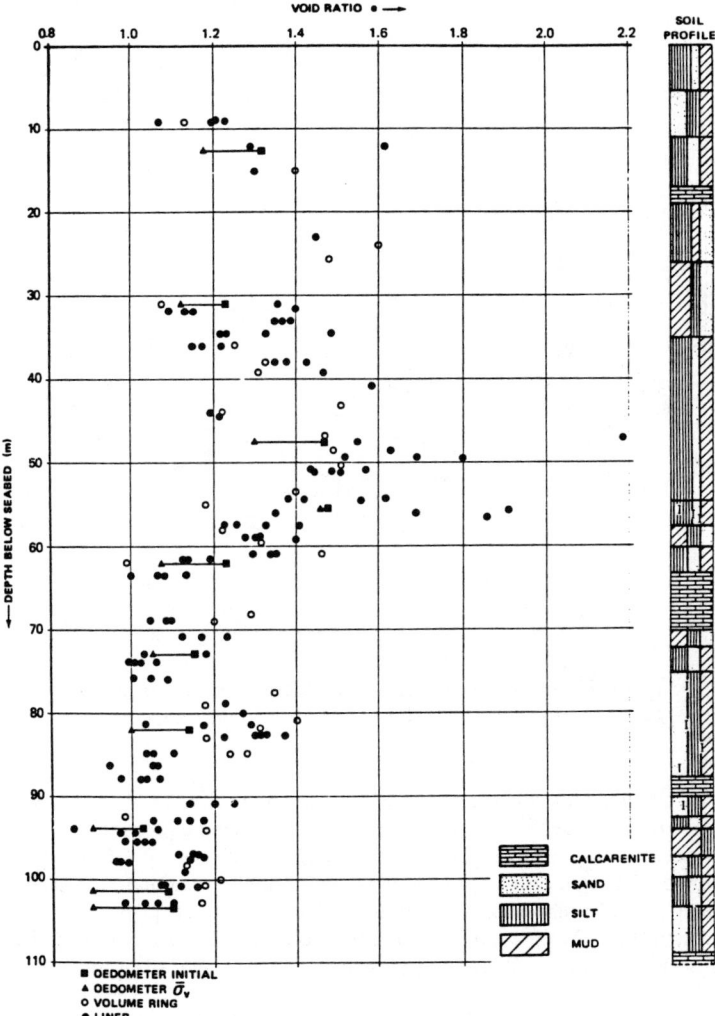

Figure 5 Void Ratios with Depth

high void ratios. However, Figure 6 shows that they also had high friction angles, 40° or more. This Figure also shows "super-critical" generation of porewater pressures during shearing, the stress paths curving downwards from a peak towards the origin indicating a tendency towards "liquefaction".

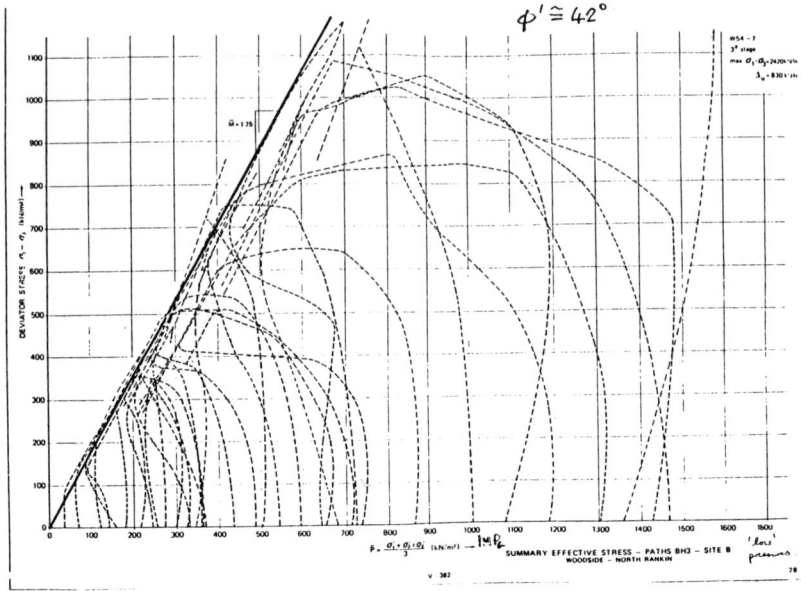

Figure 6   Stress Paths for Foundation Soils

Figure 7 shows the results of oedometer tests, emphasising that these materials were very stiff on unloading, scarcely demonstrating any elastic rebound at all.

One may speculate that the pile design was based on the observation that the materials to a depth of about 100m below seabed were predominantly soft, that high positive excess porewater pressures would be generated on driving but that these would dissipate with time allowing the frictional strength of the material to be at least partially mobilised (as Professor Reese's tests on piles in soft clay had demonstrated). At approximately 120m below seabed there was a hard layer of "calcarenite" which would provide additional end-bearing resistance.

Figure 8 shows the driving record for a typical pile. It penetrated to 70m below seabed under its own weight and exhibited only minor resistance to penetration through harder layers at 70 and 85m before reaching the design depth of about 120m. Pullout tests up to a year after driving indicated very little "set up" due to dissipation of excess porewater pressures.

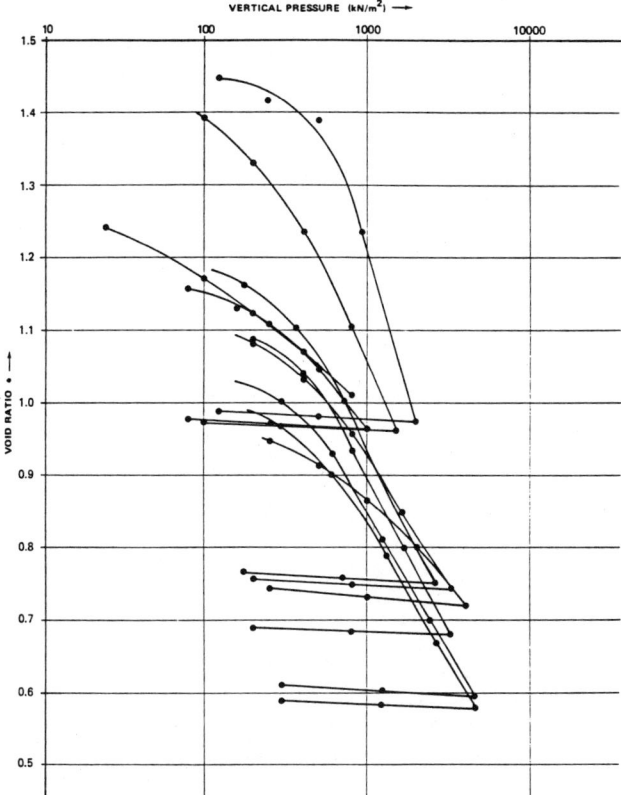

Figure 7 Oedometer Tests on Foundation Soils

With hindsight one may speculate that the high void ratio materials liquefied as the pile passed through them, and that the absence of elastic rebound exacerbated the lack of development of skin friction.

At any rate, the analysis of this deep foundation for vertical loading became concentrated on end-bearing resistance and on ground improvement (Smith et al, 1988; Cuckson et al, 1988). Professor Reese contributed to the analysis of the foundation under lateral loading (Reese et al, 1988; Reese, 1988). The present purpose is to review briefly the vertical loading analyses that were done, and to question what might be done differently were the problem to arise today.

Analyses for Drained, Undrained and Partially Drained Conditions

In a formulaic approach to end-bearing resistance there are thought to be bearing capacity factors based in the "short term" on undrained strength and in the

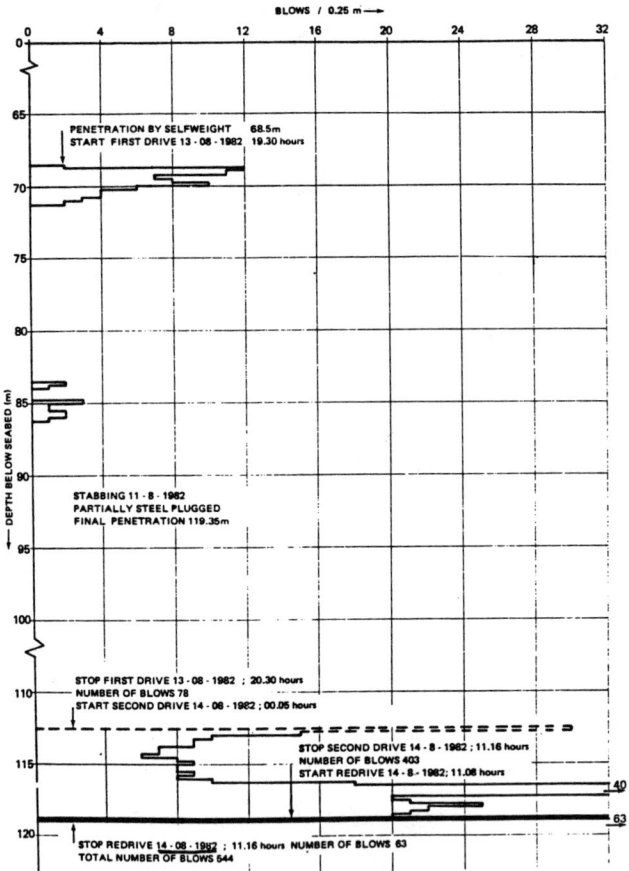

Figure 8 Driving Record for Typical Pile

"long term" on effective strength parameters such as ϕ'. Figures 9 and 10 illustrate the difficulties in applying such concepts in this case. Figure 9 shows that the design (cyclonic) storm load builds up over about 9 to 12 hours, while superimposed on that is a higher frequency wave load which builds up to its peak value in a few (3 to 6) seconds.

Axisymmetric finite element analyses, using a sophisticated constitutive model for the five layer system shown in Figure 10, led to the results shown in Figure 11. Different drained load-displacement responses for different foundation diameters and different depths of founding were computed, while undrained collapse (curve B) could occur at a quarter or less of drained capacity. It therefore became imperative to link the rate of load application to the permeabilities of the underlying soils in order to calculate a likely response. Figure 12 shows a result typical of many foundation analyses carried out since, namely that two orders of magnitude in ground permeability (in this case from 10^{-4} to 10^{-6} m/s) separate essentially drained from essentially undrained behaviour.

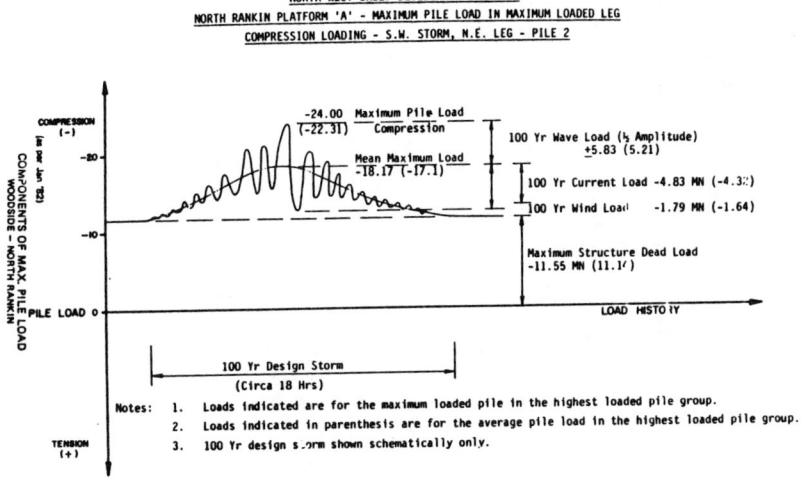

Figure 9 Wave Loading

Geometrical Simplifications

By present day standards, Figure 10 is an extremely crude mesh and indeed the adoption of axisymmetry was a considerable approximation to the true field geometry. Figure 13(a) shows the surveyed locations of tips of the 8-pile groups under two of the legs of the structure and the situation after "bells" had been under-reamed beneath four of the pile tips. Figure 13(b) shows annular idealisations of the treated foundation. Figure 14 shows various "equivalent pier" (i.e. solid axisymmetric) idealisations that were made at the time.

For the remainder of this paper, attention will be devoted to analysing foundation problems, such as those shown in Figures 13 and 14, in their full geometric complexity.

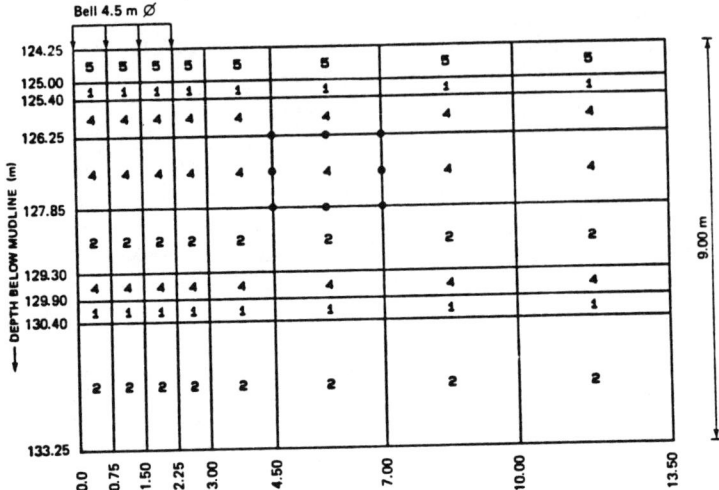

Figure 10 Model of Pile End Bearing

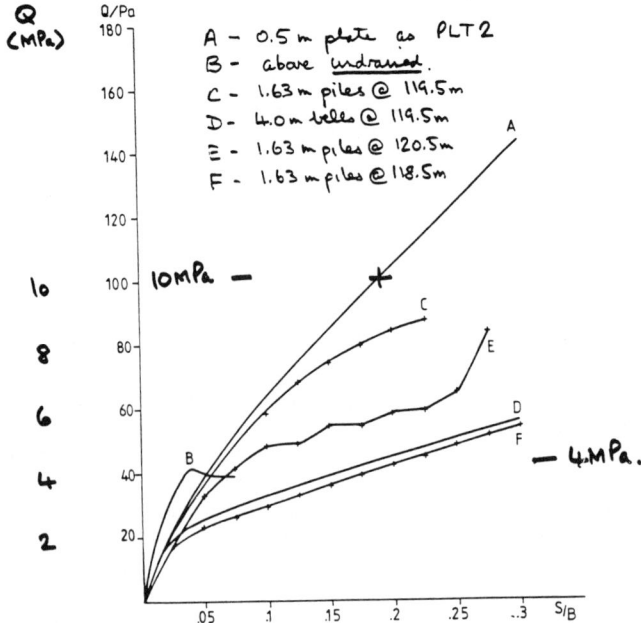

Figure 11 Load-Displacement Behaviour of Different Foundations

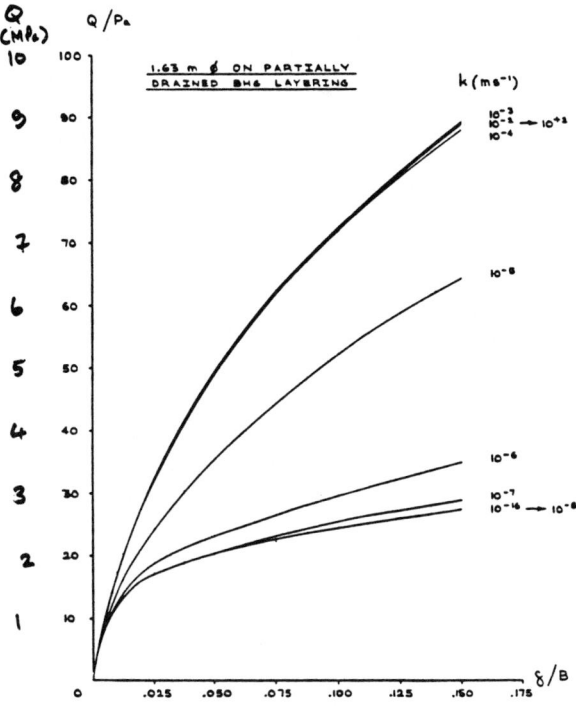

Figure 12 Effect of Ground Permeability on Response

Vector/Parallel, Iterative 3-d Finite Element Analysis

Nearly 25 years ago Ottaviani (1975) published results of three-dimensional finite element analyses of pile groups with pile caps as illustrated in Figure 15. The analyses were carried out using a maximum of 3300 nodes and 2700 eight node brick elements. The (linear elastic) analyses took about 4 hours to complete on what was then a 'mainframe' computer, making substantial use of auxiliary storage. How would such a computation be made today?

First of all, eight-node brick elements would not be used. Smith and Kidger (1992a) showed that as Poisson's ratio increases and the material nears incompressibility, such elements "lock", leading to unreliable results. Instead 14-node (Smith and Kidger, 1992b) or 20-node (Smith and Kidger, 1991) brick elements should be used. In the latter case, "reduced" integration should not be employed, as spurious "zero energy" modes can interfere with the results.

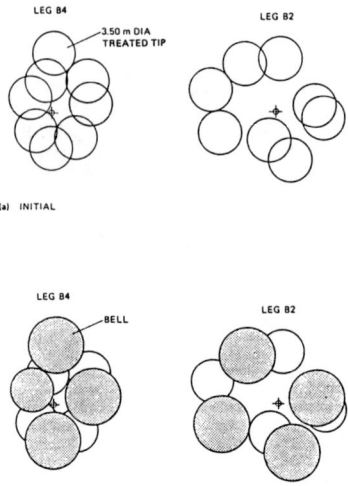

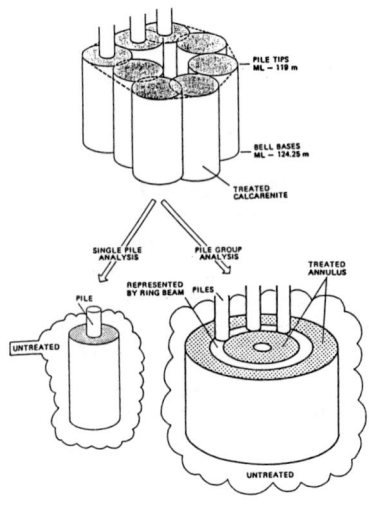

Figure 13 Post-Installation Location of Pile Tips
Annular Idealisation of Treated Tips

LAYOUT / LEG	ACTUAL BELLS		EQUIVALENT PIER*		ESCRIBED FIGURE		COMPACT FIGURE	
	Area A_a (m²)	Perimeter P_a (m)	$\frac{A}{A_a}$	$\frac{P}{P_a}$	$\frac{A}{A_a}$	$\frac{P}{P_a}$	$\frac{A}{A_a}$	$\frac{P}{P_a}$
B2	63.62	56.55	1.75	0.66	1.62	0.66	1.27	0.84
B4	57.33	53.40	1.43	0.60	1.27	0.60	1.08	0.78
G2	63.62	56.55	1.78	0.67	1.68	0.67	1.31	0.87
G4	57.61	53.72	1.34	0.58	1.22	0.58	1.06	0.76

Note :
1. The equivalent pier was chosen to have the same perimeter as the escribed figure for the same bell layout
2. Layout as of May 1986

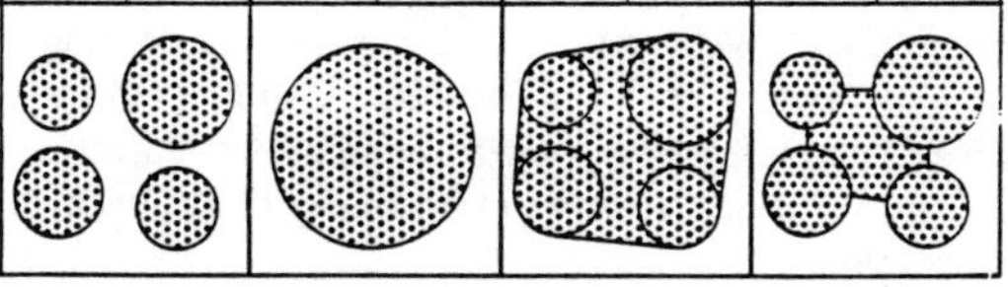

Figure 14 Equivalent Pier Assumptions

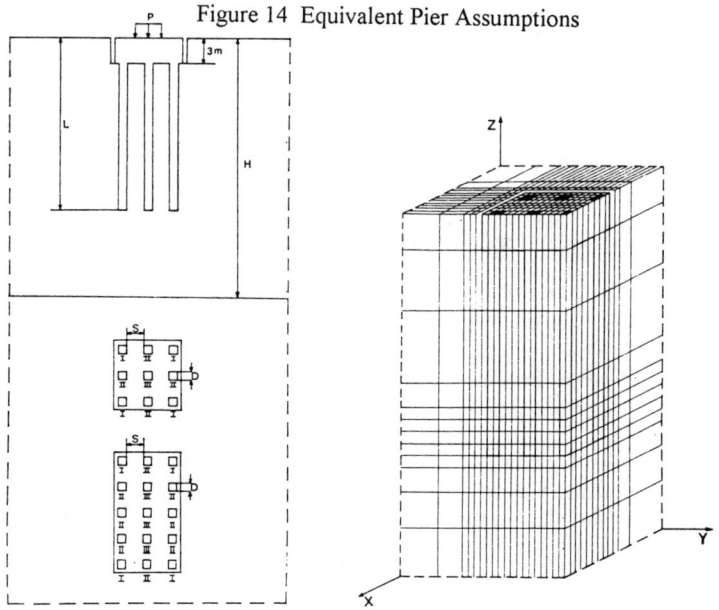

Figure 15 3-D Finite Element Model (after Ottaviani, 1975)

Secondly, elasto-plastic material behaviour would be substituted for the linear elasticity assumed by Ottaviani, thus smoothing out the very high elastic stress concentrations calculated by him. This elastoplasticity might be "non-associated" leading to unsymmetrical element stiffness matrices. Coupling between the solid phase and the porewater phase might be necessary depending on the loading rates relative to the soil permeability (see above). All of these changes lead to a considerable increase in computational complexity, and realistic geometric modelling might involve tens or hundreds of thousands of elements compared with Ottaviani's 2700.

Although computers nowadays have much bigger central storages than in 1975 it will still be found that the large 3-d meshes described above will lead to sets of equation coefficients which cannot be accommodated "in core". In 1975, the universally adopted method for solving the (linearised) equilibrium equations involving these sets of coefficients was "direct", involving a variant of Gaussian elimination. Today, this would still involve the use of auxiliary storage as reported by Ottaviani. But in addition, the penalty to be paid for using a modern computer in this way has increased. Modern computer architectures involve either vector or parallel hardware, or both, and interrupting the flow of such a computation is very time consuming.

For these reasons, present day computation of, for example, the solution of large sets of simultaneous equations is done iteratively rather than directly.

Iterative Solution Methods for Linear Simultaneous Equations

This is an area of intense research activity and although recent texts (Kelley, 1995; Greenbaum, 1997) summarise progress, substantial developments are still occurring, eg. Dongarra et al (1998).

In the present context, at least three distinct types of equation systems have to be considered:

1. Symmetric, positive definite (associated plasticity or where a symmetric calculation matrix can be used).
2. Symmetric, non-positive definite (for coupled problems)
3. Unsymmetric (non-associated plasticity)

Type 1 embraces the standard problems of solid mechanics, Laplacian flow and so on. Type 2 arises in some "structural" applications but also in systems such as the "Stokes" or "Biot" equations. The latter may be written (Smith and Griffiths, 1998) for a timestep in the course of an implicit solution in time:

$$
\begin{bmatrix} \theta \underline{KM} & \theta \underline{C} \\ \theta \underline{C}^T & -\theta^2 \Delta t \, \underline{KP} \end{bmatrix} \begin{Bmatrix} \underline{r}_1 \\ \underline{u}_{w_1} \end{Bmatrix} = \begin{bmatrix} (\theta-1)\underline{KM} & (\theta-1)\underline{C} \\ \theta \underline{C}^T & -\theta(\theta-1)\Delta t \, \underline{KP} \end{bmatrix} \begin{Bmatrix} \underline{r}_0 \\ \underline{u}_{w_0} \end{Bmatrix} \tag{1}
$$

where $\underline{KM}$ is a (symmetric, positive definite) solid "stiffness matrix", $\underline{KP}$ a (symmetric, positive definite) Laplacian fluid matrix, $\underline{C}$ a coupling matrix, Δt the timestep and θ a positive parameter. The coupled variables are displacement, $\underline{r}_1$ and $\underline{r}_0$ and water pressure $\underline{u}_{w1}$ and $\underline{u}_{w0}$ at successive times. The coefficient matrix on the left hand side of eq (1) is clearly symmetric but non-positive definite.

Type 3 may occur in large displacement analyses, non-associated solid plasticity and in many other finite element applications. Quite a severe test of an equation solution algorithm arises from finite element discretisation of the Navier-Stokes equations.

For the steady state condition, Smith and Griffiths (1998) derive the system

$$
\begin{bmatrix}
\underline{C}_{11} & \underline{C}_{12} & \underline{0}_{13} & \underline{0}_{14} \\
\underline{C}_{21} & \underline{0}_{22} & \underline{C}_{23} & \underline{C}_{24} \\
\underline{0}_{31} & \underline{C}_{32} & \underline{C}_{33} & \underline{0}_{34} \\
\underline{0}_{41} & \underline{C}_{42} & \underline{0}_{43} & \underline{C}_{44}
\end{bmatrix}
\begin{Bmatrix}
\underline{u} \\ \underline{p} \\ \underline{v} \\ \underline{w}
\end{Bmatrix}
=
\begin{Bmatrix}
\underline{0} \\ \underline{0} \\ \underline{0} \\ \underline{0}
\end{Bmatrix}
\tag{2}
$$

in which $\underline{u}$, $\underline{v}$, $\underline{w}$ are fluid velocities in the Cartesian directions and $\underline{p}$ is the fluid pressure. Submatrices $\underline{C}_{ij}$ are unsymmetrical.

Most of the methods described in literature, for example by Kelley (1995) or Greenbaum (1997) have been programmed by the author and applied to large finite element systems. Mnemonics are used to distinguish the methods as follows:

Mnemonic	Description
CG	Conjugate Gradient
MINRES	Minimum Residual
GMRES	Generalised Minimum Residual
CGSTAB	Stabilised Conjugate Gradient
BiCGSTAB	Stabilised bi-Conjugate Gradient
BiCGstab(l)	Stabilised hybrid bi-Conjugate Gradient

The prefix "P" means "preconditioned" so that PMINRES refers to a preconditioned minimum residual method.

The methods are supposed to be applied to the three distinct types of equation systems as follows:

System	Method
Symmetric, positive definite	CG, PCG
Symmetric, non-positive definite	PCG?, MINRES, PMINRES
Unsymmetric	GMRES, CGSTAB, BiCGSTAB, BiCGstab(l) and their preconditioned versions.

Vectorisation and Parallelisation

Figure 16 shows an algorithm in Fortran 90 for carrying out the BiCGstab(l) type of iterative solution to a set of linear simultaneous equations $\underline{A} \underline{x} = \underline{b}$. Matrix $\underline{A}$ should be unsymmetrical and (l) represents the (small) number of GMRES-type iterations embedded within the BiCGstab method (Dongarra et al, 1998). The detail is not important here but it should be observed that the (unsymmetrical) equation coefficient matrix $\underline{A}$ occurs only twice in the main iteration loop, on both occasions in the form of a matrix by vector multiplication $\underline{A}*\underline{y}$. In a finite element context $\underline{A}$ has traditionally been assembled from all the property matrices of all the finite elements in an analysis (tens or hundreds of thousands in our case).

```
program bicgstab_l
! Computes solution of Ax = b by the BiCGSTAB(ell) scheme
! No preconditioning at present
implicit none
integer::neq,iters,its,j,k,ell
real:: beta , tol , rho , kappa , gama , omega , norm_r , r0_norm , error ,   &
       cosine , NGamma0 , NGamma1 , alpha
logical::converged
real,allocatable::x0(:),x(:),b(:),A(:,:),r(:,:), rt(:), u(:,:),              &
            y(:),s(:),G(:,:),H(:,:),p(:),q(:),Gamma(:),                       &
            Gamma0(:),Gamma1(:)
open(10,file='bcgstab_l.dat')   ;  open(11,file='bcgstab_l.res')
read(10,*)neq , ell , its , tol , kappa
allocate(x0(neq),x(neq),b(neq),A(neq,neq), r(neq,ell+1), u(neq,ell+1),        &
         y(neq),s(ell+1),G(ell+1,ell+1),H(ell-1,ell-1),p(ell-1),q(ell-1),     &
         Gamma0(ell+1),Gamma1(ell+1),Gamma(ell+1),rt(neq))
read(10,*) A;   read(10,*) b;   read(10,*) x0
!     initialisation phase
     x = x0          ;          y = x    ; y= matmul( A , y)   ; rt = b - y
     r(:,1) = rt     ;          u = .0 ; gama = 1.0  ; omega  = 1.0 ; k = 0
     norm_r = norm(rt)    ;    r0_norm = norm_r  ; error = 1.0  ; iters = 0
!     bicgstab(ell) iterations
        iterations : do
            iters = iters + 1 ; converged = error < tol
            if(iters==its.or. converged) exit
            gama = - omega*gama   ;   y = r(:,1)
            do j = 1 , ell
                rho = dot_product(rt,y)  ;  beta = rho/gama
                u(:,1:j) = r(:,1:j) - beta * u(:,1:j)   ;      y = u(:,j)
                y = matmul(A,y)   ; u(:,j+1) = y ;   gama = dot_product(rt,y)
                alpha = rho/gama            ;    x = x + alpha * u(:,1)
                r(:,1:j) = r(:,1:j) - alpha * u(:,2:j+1)
                y = r(:,j)  ;   y = matmul(A,y) ; r(:,j+1) = y
            end do
            G = matmul(transpose(r),r)   ;  H = -G(2:ell,2:ell)
            call invert(H)
            p = matmul(H,G(2:ell,1))      ;   q = matmul(H,G(2:ell,ell+1))
            Gamma0(1) = 1.0; Gamma0(ell+1) = 0.0; Gamma0(2:ell) = p
            Gamma1(1) = 0.0; Gamma1(ell+1) = 1.0; Gamma1(2:ell) = q
            NGamma0  = dot_product(Gamma0,matmul(G,Gamma0))
            NGamma1  = dot_product(Gamma1,matmul(G,Gamma1))
            omega    = dot_product(Gamma0,matmul(G,Gamma1))
            cosine = abs(omega)/sqrt(abs(NGamma0*NGamma1)); omega=omega/NGamma1
            if(cosine<kappa) omega = (kappa/cosine) * omega
            Gamma = Gamma0 - omega * Gamma1
            s(1:ell) = Gamma(2:ell+1)          ; s(ell+1) = .0
            x = x - matmul(r,s);r(:,1)=matmul(r,Gamma);u(:,1)=matmul(u,Gamma)
            norm_r = norm(r(:,1))  ;  error = norm_r/r0_norm      ; k = k + 1
        end do iterations
print*,"The solution vector is", x
print*,"It took  bicgstab_l ",iters,"iterations to converge"
end program bicgstab_l
```

Figure 16

Figure 17 shows a trivial 2-element example in which it is demonstrated that the product $\underline{A}*\underline{y}$ can be carried out by "element-by-element" summation meaning that $\underline{A}$ need never be assembled at all (Smith and Griffiths, 1998). This has major advantages in terms of storage (only the distinct stiffness matrices need to be stored), vectorisability (the operation matmul $(\underline{A},\underline{y})$ is highly vectorisable) and parallelisability (the main overhead consists in "gathering" the appropriate parts of $\underline{y}$ (denoted by $\underline{y}^1$ and $\underline{y}^2$) in Figure 17 and "scattering" the result into $\underline{A}*\underline{y}$).

The numbers of iterations involved in typical deep foundation and other calculations (Smith, 1999) are shown in Table 1.

Analysis Type	Method	Number of unknowns	Iterations per load step
Elastoplastic (symmetric Calculation matrix)	PCG	787,280	246 (first) then 122
Elastoplastic (Biot-symmetric Calculation matrix)	PMINRES PCG	14,330	Average 332 Average 83
Elastoplastic (unsymmetric Calculation matrix)	BiCGstab(l)	98,000	4 x 531

Table 1 Iteration counts for different analyses types

It can be concluded that:

1. In iterative situations there is a very powerful incentive to use symmetric calculation matrices.
2. The PCG method works better than PMINRES for Biot-type consolidation analyses (Smith, 1998).
3. Iteration counts of several thousand are likely for large problems with unsymmetric calculation matrices (no preconditioning was used in order to gain the greatest advantages of parallelisability).

Element stiffness matrix $\underline{KM} = \begin{bmatrix} \dfrac{EA}{L} & -\dfrac{EA}{L} \\ -\dfrac{EA}{L} & \dfrac{EA}{L} \end{bmatrix}$

Assembled stiffness matrix $\underline{A} = \begin{bmatrix} \dfrac{EA}{L} & -\dfrac{EA}{L} & 0 \\ -\dfrac{EA}{L} & \dfrac{2EA}{L} & -\dfrac{EA}{L} \\ 0 & -\dfrac{EA}{L} & \dfrac{EA}{L} \end{bmatrix}$

$$\underline{A}*\underline{y} = \left\{\begin{array}{l} \dfrac{EA}{L}y_1 \; - \; \dfrac{EA}{L}y_2 \\[2ex] -\dfrac{EA}{L}y_1 \; + \; \dfrac{2EA}{L}y_2 \; - \; \dfrac{EA}{L}y_3 \\[2ex] -\dfrac{EA}{L}y_2 \; + \; \dfrac{EA}{L}y_3 \end{array}\right\}$$

$$\underline{KM_1}*\underline{y}^1 = \left\{\begin{array}{l} \dfrac{EA}{L}y_1 \; - \; \dfrac{EA}{L}y_2 \\[2ex] -\dfrac{EA}{L}y_1 \; + \; \dfrac{EA}{L}y_2 \\[2ex] 0 \end{array}\right\}$$

$$\underline{KM_2}*\underline{y}^2 = \left\{\begin{array}{l} 0 \\[2ex] \dfrac{EA}{L}y_2 \; - \; \dfrac{EA}{L}y_3 \\[2ex] -\dfrac{EA}{L}y_2 \; + \; \dfrac{EA}{L}y_3 \end{array}\right\}$$

$$\underline{KM_1} * \underline{y}^1 \; + \; \underline{KM_2}\,\underline{y}^2 \; = \; \underline{A} * \underline{y}.$$

Figure 17　Assembly versus Element-by-element

The LEGO approach for analysis of deep foundations

Figure 18 shows a differentially-loaded deep-piled raft foundation. Smith and Wang (1998) have described how a three dimensional analysis leads to results of practical importance. For those wedded to a formulaic approach, Poulos and Davis (1980), p 258, say that "no parametric studies of the effects of raft flexibility or pile compressibility have yet been made".

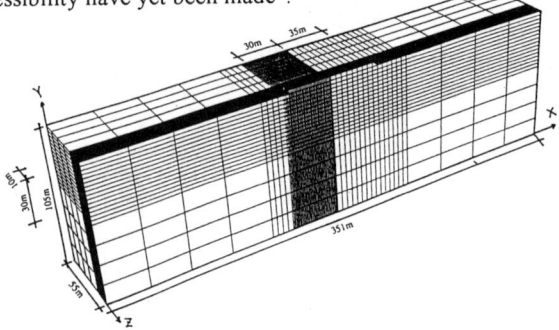

Figure 18　3-D Finite Element Model of Differentially Loaded Raft

Analyses for linear elastic response of the models shown in Figures 1 and 18 can be carried out in reasonable time today on a personal computer (Kidger, 1998). This is because the meshes are quite highly "structured". For example, the model in Figure 18 contains only 26 "types" of element, that is, elements which are geometrically similar but differ only in elastic properties. These 'LEGO' bricks are the only ones whose stiffness matrices need be stored. The linear elastic problem has a symmetric, positive definite $\underline{A}$ matrix (Figure 16) and that the y^I vectors (Figure 17) can be "stacked" up for each element type. This leads to a matrix-matrix rather than matrix-vector multiplication which on some computers is very highly vectorisable.

Analyses for elastoplastic response (especially for the non-associated case – unsymmetrical $\underline{A}$ in Figure 16) and for coupled consolidation response are probably beyond the capacity of personal computers at present. However Smith (1999) presents results obtained on a cluster of PCs running the Linux operating system, and coupled via an ethernet, which show a 5 times speedup on 6 computers. There is every hope therefore of parallelisable methods in the future using hardware within the budgets of Civil Engineering companies.

Conclusion

There is merit in analysing deep foundation problems as they are, without gross oversimplicication.

References

Cuckson, J., van Seters, A.J., Hicks, M.A. and Kenter, C.J. (1988) "Evaluation of response of chemically stabilised calcareous material" in Engineering for Calcareous Sediments, eds Jewell and Khorshid, v2, pp 689-700, Balkema Publishers.
Dongarra, J.J., Duff, I.S., Sorensen, D.C. and van der Vorst, H.A. (1998) "Numerical Linear Algebra for High-Performance Computers", SIAM Publishers.
Greenbaum, A. (1997) "Iterative Methods for Solving Linear Systems", SIAM Publishers.
Jewell, R.J. and Khorshid, M.A. (eds) (1988) "Engineering for Calcarcous Sediments", 2 vols, Balkema Publishers.
Kelley, C.T. (1995) "Iterative Methods for Linear and Nonlinear Equations", SIAM Publishers.
Kidger, D.J. (1998) "A 'LEGO' Based PCG Solver for Finite Elements" Proc. Conf. Parallel Computing, Madrid, ed. Topping.
Lee, S.L., Chow, Y.K., Karunaratne, G.P. and Wong, K.Y. (1988) "Rational Wave-equation model for pile driving analysis" J. Geot.Eng. Proc ASCE, 114, pp 306-325.
Ottaviani, M. (1975) "Three-dimensional finite element analyses of vertically loaded pile groups" Geotechnique, 25, 2, pp 159-174.

Poulos, H.G. and Davis, E.H. (1980) "Pile foundation analyses and design" John Wiley, 397 pp.

Randolph, M.F. and Simons, H.A. (1986) "An improved soil model for one-dimensional pile driving analysis" Proc 3rd Int. Conf. Numerical Methods in Offshore Piling, Nantes, pp 3-17.

Reese, L.C. (1988) "Response of piles in calcareous soils to lateral loading" in Engineering for Calcareous Sediments, eds Jewell and Khorshid, v2, pp 859-868, Balkema Publishers.

Reese, L.C., Wright, S.G., Roesset, J.M., Hayes, L.H., Dobry, R. and Vallabhan, C.V.G. (1988) "Analysis of piles subjected to lateral loading by storm-generated waves" in Engineering for Calcareous Sediments, eds Jewell and Khorshid, v2, pp 647-654, Balkema Publishers.

Seed, H.B. and Reese, L.C. (1957) "The Action of Soft Clay along Friction Piles", Transactions, ASCE, Paper 2882, v22, pp 731-754.

Smith, E.A.L. (1960) "Pile driving analysis by the wave equation", J. Soil Mechanics and Foundations Division, ASCE, 86, SM4.

Smith, I.M. and Chow, Y.K. (1982) "Three-dimensional analysis of pile drivability" Proc 2nd Int. Conf. Numerical Methods in Offshore Piling, Austin, Texas, pp 1-19.

Smith, I.M., To, W.P. and Willson, S.M. (1986) "Plugging of Pipe Piles", Proc 2nd Int. Conf. Numerical Methods in Offshore Piling, Nantes, pp 53-73.

Smith, I.M. and To, W.P. (1988) "Numerical Studies of Vibratory Pile Driving", Int. J. Num. Analytical Meth. Geomechanics, 12, 5, pp 513-531.

Smith, I.M., Hicks, M.A., Kay, S. and Cuckson, J. (1988) "Undrained and Partially Drained Behaviour of End Bearing Piles and Bells founded in Untreated Calcarenite", in Engineering for Calcareous Sediments, eds Jewell and Khorshid, v2, pp 663-680, Balkema Publishers.

Smith, I.M. and Kidger, D.J. (1991) "Properties of the 20-node Brick" Int. J. Num. Analytical Meth. Geomech, 15, 12, pp 871-891.

Smith, I.M. and Kidger, D.J. (1992a) "Eigenvalues and Eigenmodes of 8-node Brick Elements" Comm. App. Num. Meth., 8, pp 193-205.

Smith, I.M. and Kidger, D.J. (1992b) "Elastoplastic Analysis Uisng the 14-node Brick Element Family" Int. J. Num. Meth. Eng., 35, 6, pp 1263-1275.

Smith, I.M. and Griffiths, D.V. (1998) "Programming the Finite Element Method" John Wiley, pp 534.

Smith, I.M. (1998) "Parallel Coupled Analyses in Geotechnical Engineering" Proc 4th European Conf. on Numerical Methods in Geomechanics, pp 25-34, Springer Publishers.

Smith, I.M. and Wang, A. (1998) "Analysis of Piled Rafts" Int. J. Num. Analytical Meth. Geomechanics, 22, 10, pp 777-790.

Smith, I.M. (1999) "A General Purpose System for Finite Element Analyses in Parallel"" submitted to Engineering Computations.

Modeling Spatial Variability in Pile Capacity
for Reliability-Based Design

Samuel J. Gambino[1], Associate Member, and Robert B. Gilbert[2], Member, ASCE

Abstract

The American Petroleum Institute's (API) Recommended Practice for designing pile foundations for offshore structures entails using site-specific data for soil properties. The 1993 version of the code establishes load and resistance factors in order to achieve a desired level of reliability (API 1993). This paper will introduce the development of a site-specific model for predicting axial side capacity, which is the first step in a methodology for achieving the same level of reliability in design at a site whether or not a soil boring has been drilled at that site by using the properties of the offshore field as a whole. This approach is applicable to the re-qualification of older platforms with poor quality or no boring data, the design of new platforms for which the platform location and the boring location do not coincide, and the design of new platforms without drilling a site-specific boring. The model was developed by compiling and analyzing geological and geotechnical data from more than fifty sites from an offshore field. The model provides site-specific profiles of axial side capacity within the field and quantifies the site-specific uncertainty in pile capacity arising from spatial variability in soil properties.

Introduction

The American Petroleum Institute (API) recommends using site-specific soil properties when developing pile capacity profiles for offshore structures (API 1993). However, there are many situations where this site-specific information is not available or would be costly and time consuming to obtain. Examples include re-qualifying older platforms that have poor quality or no boring data, designing new platforms for which the platform location and the boring location do not coincide, and

---

[1] URS Greiner Woodward Clyde, 500 12th Street, Suite 200, Oakland, CA 94607
[2] The University of Texas at Austin, ECJ 9.227, Austin, TX 78705

135

designing new platforms without drilling a site-specific boring. It is therefore advantageous to be able to predict the expected pile capacity at a site that does not have a site-specific soil boring by using the properties of the offshore field as a whole.

This paper describes a process for developing a model to predict site-specific pile capacity profiles with depth and to quantify the uncertainty associated with these predictions. The process of model development consists of the following steps: (1) establish a conceptual geologic model; (2) compile geotechnical data to relate the geologic model to pile capacity; (3) develop a quantitative model describing spatial trends and variability in pile capacity; and (4) calibrate the quantitative model with geotechnical data.

The process of model development will be illustrated using axial side capacity for one particular offshore field, herein referred to as the case-study field. The focus on axial side capacity is considered appropriate because approximately 90 percent of the total axial capacity for a typical pile in this field is derived from side friction. Although not presented in this paper, the process of model development has also been applied to end bearing, total axial capacity, and foundation response under axial and lateral loading conditions. The proprietary information used in this case study included reports describing the geology of the field and a database of consulting reports containing geotechnical and geophysical data for more than 50 sites. The application to this particular offshore field is provided only for illustrative purposes; the process of model development is general and could be applied to any offshore field.

Step 1: Establish the Conceptual Geologic Model

The conceptual geologic model is used to identify the factors that will affect axial pile capacity in the field. For the case-study offshore field, the typical soil profile in the upper 85 meters (the length of most piles) is normally to slightly over-consolidated, marine clay with lithified soil strata and subsurface channels.

This conceptual geologic model is based on the assumption that the mean sea level was relatively constant (at an elevation similar to today's mean sea level) up until the mid-Pleistocene era (750,000 years ago). Since that time, there have been at least six major periods in which the mean sea level has fallen on the order of 120 meters before returning to its current elevation. This drop in sea level has happened roughly every 100,000 years, the duration of each low stand being approximately 15,000 years. During each low stand, the exposed surface was believed to have undergone desiccation, thereby forming an over-consolidated crust. The most recent period of desiccation ended roughly 20,000 years ago, around the beginning of the Holocene era. Since the mid-Pleistocene era, sedimentation rates have remained fairly constant at approximately 0.3 mm per year, and a typical 85-meter deep section is likely to consist of soils deposited during the past 350,000 years. Two to three such desiccated crusts are therefore likely to be encountered in a typical 85-meter deep section. The first crust will be overlain by relatively soft Holocene clay on the order of several meters thick; subsequent crusts will occur within the Pleistocene material at spacings on the order of 30 meters.

Within the lithified soil strata, this conceptual geologic model includes the presence of valley/channel systems that developed during periods of low sea-level stand. Thick channels near the surface have the potential to be filled with weak Holocene-age clay, although this weak surficial clay layer is generally less than 6 meters thick. Deep buried channels are predominantly filled with clay (the result of marine sedimentation). However, the bottom portion of a given channel may have a greater silt content due to alluvial deposition. A key assumption in this conceptual model is that overburden pressure will generally cause deep clayey fill material to have similar strength properties as the soil surrounding the channel.

Step 2: Compile Geotechnical Data

The purpose of compiling geotechnical data is to relate the geologic model to pile capacity. For the case-study field, geotechnical data from more than 50 borings were compiled and analyzed. Example data from one boring are shown on Figure 1. To provide points of reference, the solid line on the unconsolidated-undrained (UU) strength versus depth plot in this figure corresponds to a typical strength profile for a normally consolidated clay. The typical normally consolidated strength profile increases linearly at a rate of 1.9 kPa per meter of depth (or a c/p value of approximately 0.24). The soil profile in Figure 1 is consistent with the conceptual geologic model: a slightly overconsolidated clay with desiccated crusts at an approximate spacing of 30 meters. Each crust is characterized by an increase in UU shear strength, an increase in unit weight, and a decrease in water content (Figure 1).

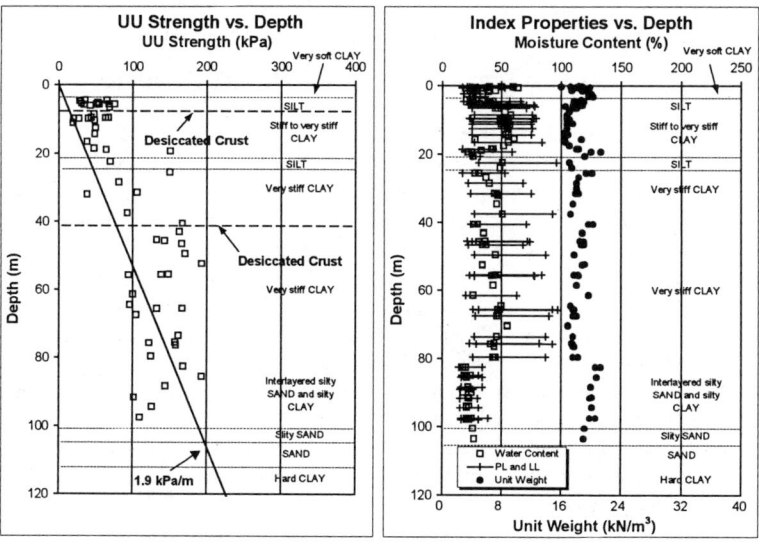

Figure 1. Geotechnical Data for a Typical Site

The boring data shown in Figure 2 provide an example of a site with a deep buried channel. Although the overlying material and channel fill material differ with respect to plasticity, moisture content and unit weight, they exhibit similar strength properties (following the 1.9 kPa/m line) in accordance with the conceptual geologic model.

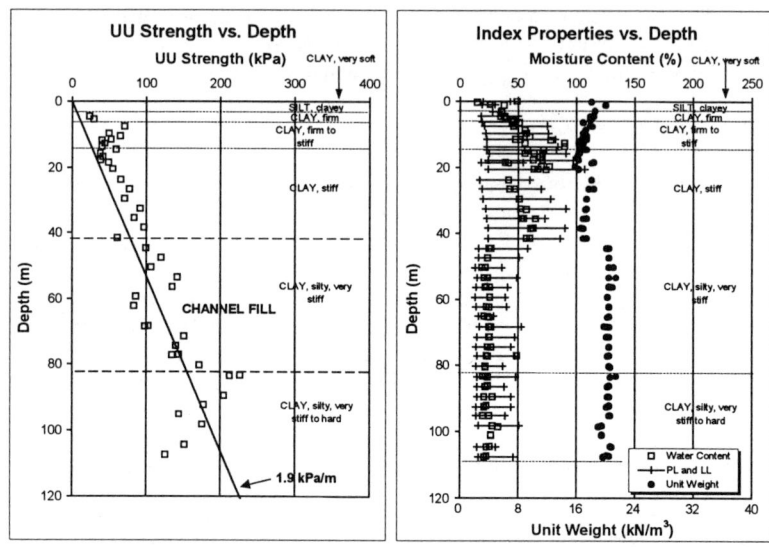

Figure 2. Geotechnical Data for a Site with Channel Fill

The design unit side friction, $f_S(z)$, is plotted as a function depth in Figure 3 for all of the borings in this field that were drilled with pushed sampling methods. Each data point in Figure 3 represents the design unit side friction at a given depth (shown in 3-meter intervals) for one particular site. As expected, based on the conceptual geologic model of a slightly overconsolidated marine clay, the unit side friction tends to increase nearly linearly with depth. It should be noted that the conspicuously vertical sets of data (evident at depths greater than 45 meters) represent the handful of sites with cohesionless soils; limiting side friction values are placed on cohesionless soils as per the API code. A subset of borings that were drilled with driven sampling methods is not included on Figure 3 due to the potential effects of sample disturbance, as discussed below.

While the conventional practice today is to use pushed sampling methods to obtain samples for laboratory strength testing, approximately 25 percent of the borings in the case-study field are older and were drilled using percussion (or driven) sampling methods. Driven samples were retrieved with 75-mm diameter by 600-mm long Shelby tubes driven with wireline hammers, in contrast to pushed samples, which were retrieved with 76-mm diameter by 1-m long Shelby tubes pushed into the ground by hydraulically activated systems for controlled penetration. It is of interest to relate the information from the driven samples to that from the pushed samples because the driven samples represent a significant portion of the total data set and because platforms designed with driven samples may require re-qualification.

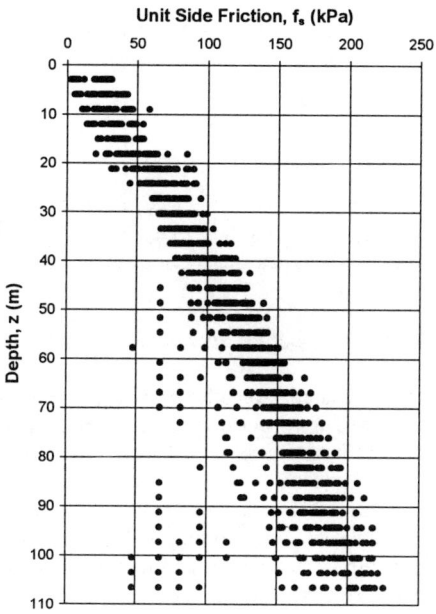

Figure 3. Unit Side Friction versus Depth

The boring data indicate that the measured strengths for the driven samples are, on average, lower than those for the pushed samples, presumably due to sample disturbance (e.g., Quiros et al. 1983). The effect of sample disturbance can be observed by separating the geotechnical data into two categories (pushed versus driven samples), discretizing the data into 15-meter depth intervals, and then comparing the average soil properties for both categories within each depth interval. Table 1 summarizes these results. The relative differences in Table 1 were calculated by subtracting the pushed average from the driven average and then dividing by the pushed average. The average measured strengths for the driven samples tend to be on the order of 20 to 30 percent lower than those from the pushed samples. The difference in the average design unit side friction values is smaller, although the design strength profiles from the older borings are apparently still biased low.

Table 1 Effect of Sample Disturbance for Older Borings

Depth Interval (m)	Avg. Measured UU Strength			Avg. Design Unit Side Friction		
	Driven Samples (kPa)	Pushed Samples (kPa)	Relative Difference (%)	Driven Samples (kPa)	Pushed Samples (kPa)	Relative Difference (%)
0 – 15	58	66	-12	26	25	4
15 – 30	69	95	-27	56	64	-13
30 – 45	115	118	-3	85	96	-11
45 – 60	109	133	-18	109	122	-11
60 – 75	107	148	-28	128	145	-12
75 – 90	120	167	-28	151	166	-9

Step 3: Develop a Quantitative Model for Spatial Trends and Variability

As discussed above, design strength profiles are conventionally developed from unconsolidated-undrained tests performed on specimens obtained via pushed sampling. These strength profiles are used to calculate the axial side friction for piles of specified length as per API (1993). The model for axial side friction described herein attempts to recreate the design profile at a site where unconsolidated-undrained test data are not available from borings with pushed samples.

For various pile lengths, the average side friction over the length of the pile was evaluated. The average side friction over the length of the pile, $F_S(L)$, is given by the following equation:

$$F_s(L) = \frac{1}{L}\int_0^L f_s(z)dz \tag{1}$$

where $f_S(z)$ is unit side friction as a function of depth and L is a specified pile length. The average side friction, when multiplied by the total length and circumference of the pile, can be used to calculate total side friction. This average side friction is plotted as a function of pile length in Figure 4. There is significantly less scatter in $F_S(L)$ (Figure 4) versus $f_S(z)$ (Figure 3) due to the effect of averaging; extremely large and small values of $f_S(z)$ tend to cancel out when they are averaged over depth. The distribution of $F_S(L)$ for a typical 85-meter long pile is also illustrated by the frequency distribution shown in Figure 5.

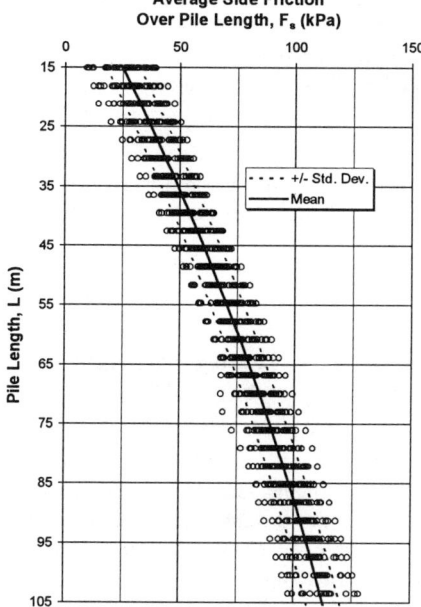

Figure 4. Average Side Friction Over Length of Pile versus Length

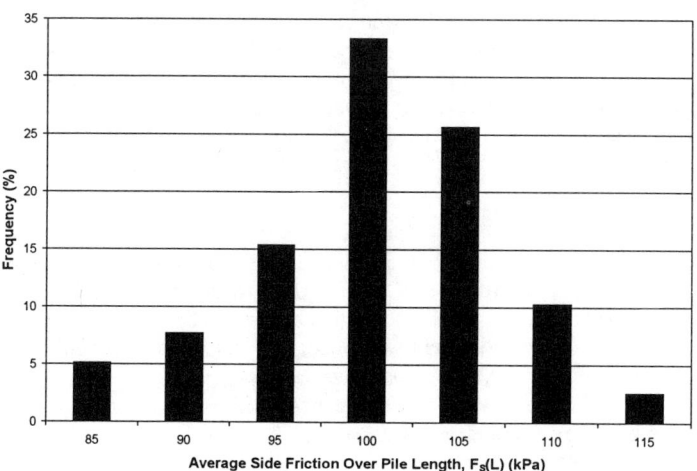

Figure 5. Frequency Distribution of $F_S(L)$ for Piles 85 m in Length

The quantitative model to predict pile capacity consists of three components: a component describing spatial trends, a component describing spatial variability, and a component describing spatial correlations between pile locations. These components are described in the following subsections.

Trend in Average Side Friction

The trend in $F_S(L)$ versus pile length is represented by the mean value, $\mu_{Fs(L)}$. This mean has been modeled as a second order polynomial with respect to pile length via the following equation:

$$\mu_{Fs(L)} = \phi_0 + \phi_1 L + \phi_2 L^2 \tag{2}$$

where L is the pile length and the ϕ_i variables are parameters that describe the model. The model for $\mu_{Fs(L)}$ is indicated in Figure 4 by a solid line.

Variability in Average Side Friction

The scatter in $F_S(L)$ due to spatial variability in soil properties is represented by the standard deviation, $\sigma_{Fs(L)}$. Since the amount of scatter about the mean remains relatively constant as pile length increases, the standard deviation has been modeled as a constant value via the following equation:

$$\sigma_{Fs(L)} = e^{\phi_4} \tag{3}$$

The standard deviation is expressed as an exponential function of the model parameter ϕ_4 in order to prevent the possibility of a negative standard deviation. The model for $\sigma_{Fs(L)}$ is indicated in Figure 4 by the dashed lines shown on both sides of the mean. Notice that while the magnitude of the mean for $F_S(L)$ increases with pile length, the magnitude of the standard deviation for $F_S(L)$ stays the same. Therefore, the ratio of the standard deviation versus the mean (and thus the amount of relative spatial variability) decreases with increasing pile length. Another way of saying this is that the coefficient of variation for $F_S(L)$ decreases with increasing pile length.

Spatial Correlation in Average Side Friction

The expected axial side friction at a new site is affected by the site's proximity to existing borings. The term "generic" will be used to describe the side friction profile at a location relatively far from any existing borings. Thus, the generic profile reflects the properties of the entire field as a whole; it is not spatially conditioned on local borings. For a new site, "A", conditioned on an existing boring, "B", the conditional forms of the equations for the mean at A given B, $\mu_{A|B}$, and the standard deviation at A given B, $\sigma_{A|B}$, are modeled as follows:

$$\mu_{A|B} = \mu_{Fs(L)} + \rho(\sigma_{Fs(L)}/\sigma_{Fs(L)})(F_{SB} - \mu_{Fs(L)}) \tag{4}$$

$$\sigma_{A|B} = [\sigma_{Fs(L)}^2(1 - \rho^2)]^{1/2} \tag{5}$$

where $\mu_{Fs(L)}$ is the generic value for the mean side friction, $\sigma_{Fs(L)}$ is the generic value for the standard deviation in side friction, F_{SB} is the known value of $F_s(L)$ at B, and ρ is the correlation coefficient between the side friction at A and that at B. The correlation coefficient, ρ, indicates the strength of the relationship in side capacity between any two locations: it has a maximum magnitude of 1.0 (indicating a perfect, one to one relationship in capacity between two locations) and a minimum magnitude of 0.0 (indicating no relationship in capacity between two locations). Note that in Eqs. (4) and (5), as ρ approaches 1.0, the mean side friction at site A approaches the known value at site B and the standard deviation in side friction at A approaches zero. These equations can be extended to include multiple borings in the vicinity of the site.

The correlation coefficient tends to decrease with distance between any two locations. The correlation coefficient in side friction between two sites resulting from the physical distance between those two sites, ρ_{Dist}, is modeled as follows:

$$\rho_{Dist} = (e^{-|\tau|/\theta}) \tag{6}$$

where τ is the horizontal distance separating A and B, and θ is a parameter called the correlation distance. The correlation distance, θ, is the spatial distance (in units of length) over which data tend to resemble one another. The correlation distance can be interpreted in the following way: if a site is horizontally within θ of another site, and the first site has a higher than average side capacity profile, then the second site will also be likely to have a higher than average side capacity profile. Mathematically, the correlation distance is the distance at which the correlation coefficient between two locations is reduced to approximately $1/3$.

For the case-study field, the correlation distance tends to increase with increasing pile length. This behavior could be due to greater continuity of strata at depth or it could be an artifact of sample disturbance in deep samples, which tends to reduce variability. In order to capture this behavior, the correlation distance is modeled as follows

$$\theta = e^{\phi_6} + Le^{\phi_7} \tag{7}$$

where, as in Eqs. (2) and (3), ϕ_6 and ϕ_7 are parameters that describe the model.

Inclusion of Driven Samples in Estimating Average Side Friction

Data from borings with driven samples have been incorporated into the model via the following relationships:

$$\mu_{Fs(L)Driv} = e^{\phi_3}\mu_{Fs(L)} \tag{8}$$

$$\sigma_{Fs(L)Driv} = e^{\phi_5}\sigma_{Fs(L)} \tag{9}$$

$$\rho_{Push/Driv} = \exp(-1/e^{\phi_9}) \tag{10}$$

where $\mu_{Fs(L)Driv}$ is the mean of the average side friction at a site where driven sampling methods were used to develop the design strength profile, $\sigma_{Fs(L)Driv}$ is the standard deviation in average side friction at a site with driven samples, and $\rho_{Push/Driv}$ is the theoretical correlation between average side friction if both driven and pushed sampling methods were used at the same location. As before, the ϕ_i variables in Eqs. (8) through (10) are parameters that describe the model. In particular, e^{ϕ_3} is a reduction factor to account for the disturbance in the driven samples. Similarly, e^{ϕ_5} is a reduction factor to account for the notion that a portion of the variability within the driven subset is masked by sample disturbance. The correlation coefficient $\rho_{Push/Driv}$ accounts for the fact that even if a driven sample and a pushed sample were to be taken at the exact same location, the resulting strength data [adjusted for disturbance using Eqs. (8) and (9)] would not be perfectly correlated. In other words, $\rho_{Push/Driv}$ is less than unity.

Step 4: Calibrate the Model

The final step in the process of model development is to calibrate the quantitative model developed in the previous step. Calibration means estimating values for the model parameters, the ϕ_i variables in Eqs. (2), (3), and (7) through (10), based on the database of geotechnical data for this field (e.g., Figure 4). In order to estimate the model parameters, the average side friction values across the field were assumed to follow a multivariate normal distribution. The justification for this assumption is that the effect of averaging will tend to result in a normal distribution based on the Central Limit Theorem (e.g., Ang and Tang 1975). This assumption is consistent with the shape of the frequency distribution shown on Figure 5 for this data set. The set of parameter values that maximized the likelihood of this multivariate normal distribution for the data set was then obtained through optimization. In addition, the magnitude of uncertainty in these parameter estimates was quantified. This uncertainty for this case study was found to be relatively insignificant due to the large size of the data set.

Example Applications

Examples of how this model can be applied as a design tool are presented in the following sections. The design capacity will be calculated for an 85-meter long, 1-meter diameter pile at different locations in the case study field. For illustrative purposes, the design capacity will be defined as the 5-percentile value of capacity, i.e., the value of capacity for which there is a 95-percent probability that the actual capacity (the value given by the API code if a boring were actually drilled at the site) will be greater than the design capacity. The design capacity for a given pile length is calculated as follows:

$$Q_{side}* = \mu_{Qside} - 1.65\sigma_{Qside} \tag{11}$$

where $Q_{side}*$ is the design capacity and μ_{Qside} and σ_{Qside} are respectively the mean and standard deviation in the total side capacity for that pile length. At a site with a soil boring (using pushed sampling methods), μ_{Qside} is obtained from the design unit side friction profile and σ_{Qside} is equal to zero; hence, $Q_{side}*$ is simply the design value used in conventional practice. Note that the conventional factors of safety or resistance factors would then be applied to this value in a standard design. As the platform location moves away from the boring location, σ_{Qside} will increase and a more conservative design will be required to account for uncertainty in the side capacity.

The definition of design capacity in Eq. (11) is used in this paper for illustrative purposes only. The model developed herein for side friction is intended to be combined with one for end bearing as well as with models for uncertainty in the design method and uncertainty in the applied load in order to develop a reliability-based design capacity. In addition, models have been developed to incorporate information from geophysical surveys into this design methodology. However, these extensions of the methodology are beyond the scope of this paper.

Example 1: Calculating the Expected Axial Side Capacity at a Generic Site

Suppose that a new platform is to be installed relatively far from any existing borings. The value of average side friction is calculated as follows from Eq. (2):

$$\mu_{Fs(L)} = 9.0 + 1.25(275) - 0.0024(85)^2 = 98 \text{ kPa}$$

where the estimated model parameters have replaced the variables in Eq. (2) (for example, when the model was calibrated with actual data, the parameter ϕ_0 was estimated to have a value of 9.0 kPa). Similarly, the standard deviation in $F_S(L)$ is obtained from Eq. (3) as follows:

$$\sigma_{Fs(L)} = e^{1.97} = 7.2 \text{ kPa}$$

The average axial side capacity, μ_{Qside}, is found by multiplying $\mu_{Fs(L)}$ by the length and circumference of the pile:

$$\mu_{Qside} = (98 \text{ kPa})(\pi)(1 \text{ m})(85 \text{ m}) = 26,200 \text{ kN}$$

Since the standard deviation in $F_S(L)$ is 7.2 kPa, the resulting uncertainty in axial side capacity expressed as a standard deviation, σ_{Qside}, is:

$$\sigma_{Qside} = (\pi)(1 \text{ m})(85 \text{ m})(7.2 \text{ kPa}) = 1,920 \text{ kN}$$

Finally, the design capacity for axial side capacity is calculated as follows:

$$Q_{side}* = 26,200 - (1.65)(1,920) = 23,000 \text{ kN}$$

Example 2: Conditioning the Generic Profile on an Adjacent Boring

The following example illustrates how to horizontally condition the model on a single adjacent boring. Suppose that the new platform discussed in Example 1 is to be located 350 meters from an existing boring. It is known from the design strength profile that $F_s(350)$ is 95 kPa for 85-meter long, 1-meter diameter piles at the site of the existing boring, and the samples from the existing boring were obtained using pushed sampling techniques.

Using the spatial conditioning procedure described above, the expected profile can be tailored such that it reflects the influence of the existing boring. From Example 1, the generic (i.e., not conditioned on any nearby borings) mean and standard deviation in $F_S(L)$ for a 85-meter pile are 98 kPa and 7.2 kPa, respectively. If ϕ_6 and ϕ_7 are replaced by calibrated values in Eq. (7), the correlation distance for 85-meter long piles is

$$\theta = e^{6.64} + 85e^{2.66} = 1980 \text{ m}$$

From Eq. (6), the correlation coefficient between the two sites is then

$$\rho_{Dist} = (e^{-|1200|/6446}) = 0.83$$

Applying Eqs. (4) and (5) gives the following conditioned mean and standard deviation

$$\mu_{A|B} = 98 \text{ kPa} + (0.83)(7.2 \text{ kPa}/7.2 \text{ kPa})(95 \text{ kPa} - 98 \text{ kPa}) = 95.5 \text{ kPa}$$

$$\sigma_{A|B} = [(7.2 \text{ kPa})^2(1 - 0.83^2)]^{1/2} = 4.0 \text{ kPa}$$

Note that the standard deviation of 4.0 kPa reflects a decrease of 44% from the generic value of 7.2 kPa. Uncertainty reduction in this example, expressed as the ratio of $\sigma_{A|B}$ to $\sigma_{Generic}$, is plotted versus dimensionless horizontal distance in Figure 6. When the new site is at the location of the existing boring (i.e., where τ/θ is zero) there is zero uncertainty in F_S. As the new site is moved farther and farther from the existing boring, the uncertainty asymptotically increases back to the generic value. Consequently, the reduction ratio goes to unity.

The remaining calculations are the same as in Example 1. The average axial side capacity, μ_{Qside}, is 25,500 kN. The standard deviation in axial side capacity, σ_{Qside}, is 1,070 kN. The design value for axial side capacity, $Q_{side}*$, is therefore 23,700 kN. Note that this simple approach can be extended to include both vertical and horizontal correlation relationships as well as multiple adjacent borings.

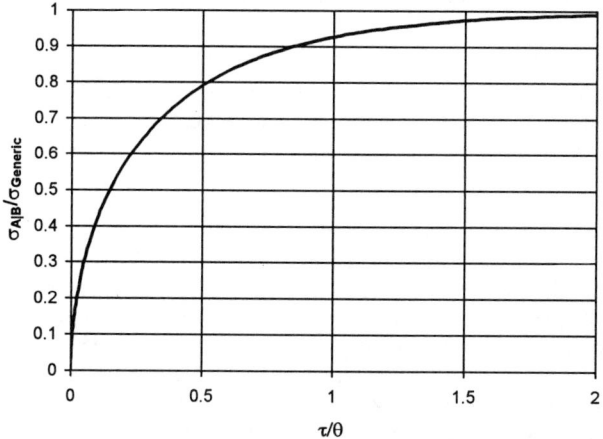

Figure 6. Uncertainty Reduction Due to Spatial Conditioning

Example 3: Conditioning the Generic Profile on an Adjacent Driven Boring

The following example illustrates how to horizontally condition the model on a single adjacent boring characterized by driven sampling techniques. Suppose that the samples from the existing boring in Example 2 were obtained using driven sampling techniques. For comparison, assume that the new platform is still 350 meters from the existing boring and $F_s(85)$ is still 95 kPa for 85-meter long, 1-meter diameter piles at the existing site. The values for $\mu_{Fs(L)}$, $\sigma_{Fs(L)}$, θ, and ρ_{Dist} are therefore the same as in Example 2.

If ϕ_3 is replaced by its calibrated value in Eq. (8), the generic mean $F_s(L)$ for driven sampling is:

$$\mu_{Fs(L)Driv} = e^{-0.10}(98) = 89 \text{ kPa}$$

If ϕ_5 is replaced by its calibrated value in Eq. (9), the standard deviation in the generic $F_s(L)$ is:

$$\sigma_{Fs(L)Driv} = e^{-0.15}(7.2) = 6.2 \text{ kPa}$$

If ϕ_9 is replaced by its calibrated value in Eq. (10), the correlation coefficient between pushed and driven sampling techniques is:

$$\rho_{Push/Driv} = \exp(-1/e^{0.60}) = 0.58$$

Since the correlation coefficient between the two sites based solely on separation distance, ρ_{Dist}, is 0.83, the composite correlation coefficient is $(0.83)(0.58) = 0.48$.

Applying Eqs. (4) and (5) then gives the following conditioned mean and standard deviation

$$\mu_{A|B} = 98 \text{ kPa} + (0.48)(7.2 \text{ kPa}/6.2 \text{ kPa})(95 \text{ kPa} - 89 \text{ kPa}) = 101 \text{ kPa}$$

$$\sigma_{A|B} = [(7.2 \text{ psf})^2(1 - 0.48^2)]^{1/2} = 6.3 \text{ kPa}$$

The average axial side capacity, μ_{Qside}, is 27,000 kN and the standard deviation, σ_{Qside}, is 1,682 kN. Finally, the design value for axial side capacity, Q_{side}^*, is 24,200 kN.

Summary

A process was presented to develop a model of spatial variability in pile capacity for offshore fields. The model provides a methodology to predict site-specific pile capacity profiles with depth and to quantify the uncertainty associated with these predictions. The process of model development consists of the following steps: (1) establish a conceptual geologic model; (2) compile geotechnical data to relate the geologic model to pile capacity; (3) develop a quantitative model describing spatial trends and variability in pile capacity; and (4) calibrate the quantitative model with geotechnical data. This process was demonstrated in this paper using axial side capacity for one particular offshore field.

The research presented here is part of a larger study in which field-specific models are also developed for predicting end bearing, total axial capacity, and foundation response under axial and lateral loading conditions. The models are part of a reliability-based methodology for designing offshore pile foundations without site-specific geotechnical data. The essence of the methodology is to quantify the additional level of conservatism required in design as a function of the amount of site-specific information. The methodology can be applied to: the re-qualification of existing platforms that either have no geotechnical data from a site-specific boring or poor quality data from an older boring; the design of new platforms using site-specific geophysical information and regional geotechnical data from soil borings in the site vicinity; and the design of new platforms using only regional geotechnical data from soil borings in the site vicinity. The methodology can also be used to quantify the expected value of additional information at a site before obtaining that information; this expected value can then be input into a cost-benefit analysis to decide whether or not the information should be obtained and to design optimal site investigation programs.

Acknowledgments

The authors wish to acknowledge Rick Dupin and Jared Black with the Unocal Corporation for their support of this research.

References

API (1993), "Recommended Practice for Planning, Designing, and Constructing
    Fixed Offshore Platforms – Load and Resistance Factor Design," RP 2A-LRFD,
    American Petroleum Institute, Washington D. C.
Ang, A. H-S. and Tang, W. H. (1975), *Probability Concepts in Engineering
    Planning and Design: Vol. I – Basic Principles*, John Wiley & Sons, New York
    City.
Quiros, G. W., Young, A. G., Pelletier, J. H. and Chan, J. H-C. (1983), "Shear
    Strength Interpretation for Gulf of Mexico Clays," ASCE Conf. on Geotechnical
    Practice in Offshore Engineering, Austin, 144-165.

by William M. Isenhower,[1] Member, ASCE

ABSTRACT: Pile groups may be loaded by soil displacements generated by slope instability, seepage forces, nearby pile driving, and other construction activities. Methods used to analyze pile groups subjected to deep-seated soil displacements are presented. An example analysis is presented to demonstrate the proposed methods.

INTRODUCTION

The *p-y* method for analyzing individual piles subjected to lateral loading was originally conceived by McClelland and Focht (1958) and subsequently developed by Reese, Matlock, and others in later years. Later developments include the development of *p-y* models for most commonly encountered soils. Additional developments have included the application of the *p-y* method for the analysis of pile groups and the inclusion of effects of nonlinear bending stiffness for piles and drilled shafts.

The topic of this paper is an extension of the *p-y* method for the analysis of pile groups in which the piles are embedded in soils that are displaced after the piles are installed. Soil displacements can be due to a number of causes. A few of the more common causes of soil movements are:

- shear deformations due to slope instability and/or seepage forces,
- lateral spread of soils due to liquefaction during earthquakes, and
- volume change of soil due to consolidation or swelling,
- soil displacements resulting from adjacent construction, such as pile driving or excavation.

---

[1] Project Manager, Ensoft, Inc./Lymon C. Reese & Associates, P. O. Box 180348, Austin, Texas 78718, ensoft@ensoftinc.com.

The numerical techniques presented herein are suitable for pile groups with both vertical and battered piles. Methods are presented for both axial and lateral load-transfer analysis that are capable of handling soil displacements in both vertical or horizontal directions without difficulty. These methods also allow unlimited rigid body motion resulting from soil displacements. An example analysis reported by Isenhower (1996) of a large, pile-supported dam monolith subjected to deep seated soil displacements resulting from seepage forces is presented as an application of the proposed procedure.

ANALYTICAL PROCEDURES

The analytical solution presented in this paper was originally developed for making first-order, second-moment reliability analyses of pile groups subjected to seepage force displacements (Isenhower and Mosher, 1996). Thus, it was necessary to develop a solution technique that was convergent for a wide variation of input parameters. The variations of input data allowed by the solution algorithms include the following:

- The foundations could be composed of any combination of battered and plumb piles.
- The direction of soil displacements could be in any direction.
- Soil displacements causing rigid body motion in any direction were allowable.
- No zone of soil could be restricted in its magnitude for direction of soil displacements.

Computation of Soil Displacements

When a pile is embedded in a displaced soil profile, the soil displacement usually acts at an inclined angle with a pile. Consequently, it is necessary to compute the components of the soil displacement vector that act in the axial and transverse directions of the pile. An illustration of the soil displacement vectors acting on battered piles is shown in Figure 1.

The best method for computing soil displacements is to use a finite element analysis that includes the effects of nonlinear stress-strain properties and interface elements. Experience has found that when developing the finite element mesh for an analysis, the mesh should be sized so that it extends to three times the maximum depth of the piles to avoid mesh boundary effects.

The transformation of the soil displacement vector to the soil displacements along the axial and transverse directions of a pile are computed using the following procedure. The transformation of the soil displacement vector from the x-y coordinate system to the u-y' coordinate system of an individual pile shown in Figure 1 is based on a transformation matrix multiplication. The transformation matrix is made up of the direction cosines of the angles between the x-axis (positive to the right) and y-axis (positive upwards) and the u-axis (axial direction, positive downwards) and the y'-axis (transverse, positive to the right). Defining the pile batter angle as α, the transformation equation is

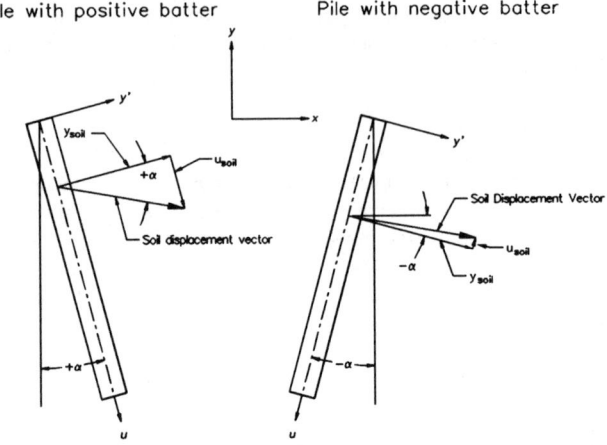

Figure 1. Soil Displacement Vectors Acting on Battered Piles.

$$
\begin{bmatrix} \cos(\alpha - \pi/2) & \cos(\pi + \alpha) \\ \cos(\alpha) & \cos(\alpha - \pi/2) \end{bmatrix} \begin{Bmatrix} dx \\ dy \end{Bmatrix} = \begin{Bmatrix} u_{soil} \\ y_{soil} \end{Bmatrix}
\tag{1}
$$

The values of dx and dy are input at every point computed in the soil displacement analysis. This is usually a finite element analysis, but could be data derived from slope inclinometer measurements or some other technique. Values of soil displacement at pile station locations between the input values are found by linear interpolation. Values of u_{soil} and y_{soil} are computed at all pile stations for all piles. Values of u_{soil} are used in computing of axial load-settlement curves for each pile and values of y_{soil} are used in computing the lateral loading response of each pile.

Lateral Loading

As part of the solution, the differential equation governing the behavior of a laterally loaded pile is solved. The conventional form of this differential equation is

$$
EI\, y'''' + P_x\, y'' + E_s\, y - W = 0
\tag{2}
$$

This differential equation is an expression of the static equilibrium of the pile in the lateral direction. The quantity $EI\, y''''$ is the lateral load intensity due to elastic bending in the pile. The quantity $P_x\,y''$ is the lateral load due to the moment magnifier effect for bending. The quantity $E_s\,y$ is the product of soil modulus, defined using a p-y curve, and pile deflection y. The last term, W, is the distributed lateral load intensity due to an external distributed loading along some length of the pile.

In the case where the soil is moving relative to the pile, the soil reaction is determined by the relative movement of soil and pile. In this case, the term in Equation 2 for soil response may be rewritten as

$$E_s\,(y - y_{soil}) \tag{3}$$

The value of soil modulus is determined at a relative pile movement of y-y_{soil} as shown in Figure 2.

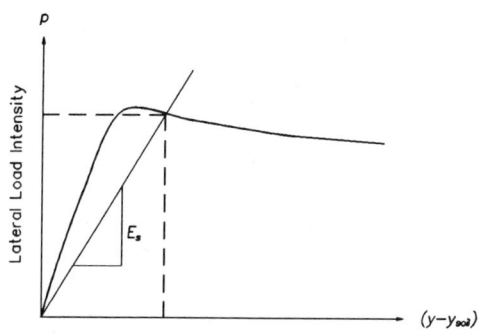

Figure 2. Soil Modulus as Function of Relative Pile Movement

Earlier studies by Reese and Wang (1991) used a method in which the value of W was varied to account for the effect of moving soil. Experience found that the former solution algorithm was unstable and was not unconditionally convergent because the form of the differential equation could behave either as "soft" or "stiff" depending on the magnitudes of y and y_{soil}. The differential equation will be soft whenever the value of E_s decreases with increasing values of $y - y_{soil}$ and is "stiff" whenever the opposite occurs. This was the cause of the numerical instability because the finite difference solution algorithm will only work for "soft" equations. Thus, it was necessary to write the differential equation in a form that is always "soft." The following method was developed to overcome this problem.

The modified formulation of the beam-column equation is now

$$EIy'''' + P_x y'' + E_s(y - y_{soil}) - W = 0 \tag{4}$$

where the primes denote differentiation versus depth and:

EI =   the bending stiffness of the pile,
y =   lateral pile displacement,
P =   the axial thrust force acting on the pile,
E_s =   the "soil modulus" computed at displacement y on the p-y curve,
y_{soil} =   soil displacement transverse to the axis of the pile computed from
           the soil displacement values, and
W =   external distributed loading on the pile.

In the case where the cause of soil movement is lateral spread due to liquefaction of strata of sand, it will be necessary to reformulate the p-y curves for the liquefied soils. One approach that has been recommended by Wang and Reese (1998) is to evaluate the residual strength of the liquefied soil and to use this strength to compute a p-y curves using Matlock's (1970) criteria for soft clay. However, this approach will not include the inertia effects of soil flowing around the foundation pile. Instead, a better approach is to determine the density and viscosity of the liquefied soil and to use Morrison's equation (Morrison et al., 1950) to compute the ultimate value of lateral load transfer to include effects due to inertia of the flowing soil.

Solution Algorithm for Lateral Loading

The methods used to solve the above differential equation use a finite difference solution algorithm and p-y curves. In the finite difference solution algorithm, the pile is divided into n discrete segments of length, h. Stations along the length of the pile are numbered from 0 at the pile head to n at the pile tip. Any pile stations outside of this range are imaginary and are used to obtain solutions only. It is possible to express the differentials in Equation 4 in finite difference form, which yields a set of equations that can be assembled into a banded system of simultaneous equations. These equations are expressed as:

$$(EIy'''')_i = [y_{i-2} R_{i-1} + y_{i-1}(-2R_i - 2R_{i-1}) + y_i(4R_i + R_{i-1} + R_{i+1})$$
$$+ y_{i+1}(-2R_i - 2R_{i+1}) + y_{i+1}R_{i+1}]\frac{1}{h^4} \tag{5}$$

$$(P_x y'')_i = \frac{P_x(y_{i-1} - 2y_i + y_{i+1})}{h^2} \tag{6}$$

where R is the bending stiffness of pile (EI_i) at pile station i and the subscripts indicate the pile station number relative to pile station i. Note that Equations 5 and 6 contain terms related to the bending of the pile only and do not contain any terms related to the soil response.

Equations 3, 5, and 6 can be substituted into Equation 2 and rearranged to the form

$$A_i \, y_{i+2} + B_i \, y_{i+1} + C_i \, y_i + D_i \, y_{i-1} + E_i \, y_{i-2} = F_i \tag{7}$$

where:

$$A_i = R_{i-1} \tag{8}$$

$$B_i = -2\,R_{i-1} - 2\,R_i + P\,h^2 \tag{9}$$

$$C_i = R_{i-1} + 4\,R_i + R_{i+1} - 2\,P_x\,h^2 + E_s\,h^4 \tag{10}$$

$$D_i = -2\,R_i - 2\,R_{i+1} + P_x\,h^2 \tag{11}$$

$$E_i = R_{i+1} \tag{12}$$

$$F_i = -W_i\,h^2 + E_s\,y_{soil}\,h^2 \tag{13}$$

It is necessary to make two adjustments when solving the above system of equations. First, the initial values of pile displacement along the pile are set proportional to the soil displacement values, y_{soil}, obtained from a finite element analysis of soil displacements. Secondly, it is necessary to compute the soil modulus along the p-y curve at point i, using an abscissa value of $y - y_{soil}$ as shown in Figure 2.

After the system of equations is solved, bending moment is computed using

$$M_i = R_i \left(\frac{y_{i+1} - 2y_i + y_{i-1}}{h^2} \right) \tag{14}$$

Axial Loading

The axial capacity of a pile in compression is computed using the following formulae:

$$Q_{ult} = Q_s + Q_b \tag{15}$$

$$Q_s = f_s\,A_s \tag{16}$$

$$Q_b = q_b\,A_b \tag{17}$$

where

Q_{ult} = ultimate pile capacity
Q_s = shaft resistance of pile due to skin friction
Q_b = base resistance of the pile due to end bearing

f_s = average unit skin friction
A_s = surface area of pile in contact with the soil
q_b = unit end bearing capacity
A_b = effective (gross) area of the tip of the pile in contact with the soil

The axial capacity of a pile in tension is computed using

$$Q_{ult} = Q_{st} \tag{18}$$

where Q_{st} is the axial capacity in tension due to skin friction only.

It is necessary to describe the development of both axial load-transfer and end bearing as they vary with axial pile displacements. The curve of axial load-transfer versus displacement is called a $T\text{-}z$ curve and the curve of end bearing versus tip displacement is called a $Q\text{-}z$ curve. Procedures have been developed for computing these curves for different types of piles in sands and clays. These curves can also be measured experimentally in an axial pile load test using a instrumented pile.

Solution Algorithm for Axial Loading

The following numerical algorithm used to compute axial pile settlement is based on an algorithm that was originally developed by Seed and Reese (1957). The algorithm described herein has been modified to account for any soil displacements acting in the axial direction. The principal advantage of using this algorithm is that any nonlinear load-transfer relations for skin friction or end bearing can be used. The difference between the following algorithm and the original Seed and Reese algorithm is that load-transfer values are based on the relative pile displacements, not the absolute pile displacements.

In this algorithm, the pile is divided into elements as shown in Figure 3. Skin friction is modeled using the $T\text{-}z$ curves computed for each pile element and end bearing is modeled using the $Q\text{-}z$ curve computed for the tip of the pile.

The solution starts by first computing the load-transfer in skin friction acting along the length of pile using the methods discussed above. Next, a value for the pile-tip displacement is assumed and the pile tip movement relative to the soil displacement is computed. This relative tip movement is

$$u_{tip,r} = u_{tip} - u_{soil} \tag{19}$$

The force acting at the tip is obtained from the $Q\text{-}z$ curve at the relative pile tip movement. The force acting on the bottom of the lower most element is now known for the assumed pile-tip displacement.

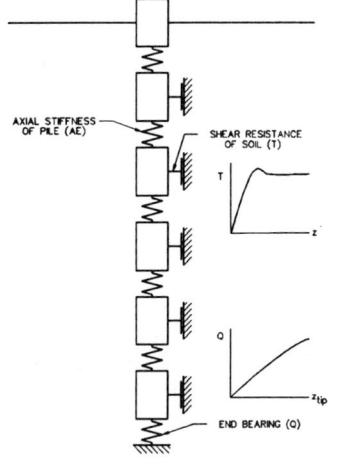

The force distribution along the length of the shaft is now solved incrementally, working from the pile tip up to the ground surface. The solution algorithm is as follows: (1) First, set the force on the bottom of an element, F_{bot}, equal to the force on the top of the element below and set the pile displacement, u_{bot}, equal to the displacement of the element below . For the lower most element, this means that

$$F_{bot} = F_{tip} \qquad (20)$$

and

$$u_{bot} = u_{tip} \qquad (21)$$

Figure 3. Model for Axial Load-Transfer and Pile Settlement Analysis

(2) Next, assume that the pile displacement at the middle of the element, u_{mid}, equals the pile displacement at the bottom of the element minus the average soil displacement in the axial direction for the pile increment.

$$u_{mid} = u_{bot} - u_{soil,avg} \qquad (22)$$

where $u_{soil,avg}$ is the average soil displacement acting over the pile increment.

(3) The force due to skin friction is obtained from the T-z curve computed for the pile element in question for a pile movement equal to u_{mid}.

$$F_{side} = T(u_{mid}) \qquad (23)$$

(4) The force acting on the top of the element is found from equilibrium of forces.

$$F_{top} = F_{bot} + F_{side} \qquad (24)$$

(5) Once the force acting on the top of the element is known, the pile displacement at the middle of the element can be computed using

$$u_{mid} = u_{bot} + \left(\frac{3}{8} \cdot F_{bot} + \frac{1}{8} \cdot F_{top} \right) \cdot \frac{h}{AE} \qquad (25)$$

where

h = length of the pile increment
A = area of the piles structural cross-section

E = modulus of elasticity of the pile

Steps 3, 4, and 5 are repeated until the computed value of u_{mid} converges. Usually, three to five iterations is sufficient for convergence.

(6) The pile displacement at the top of the element is computed using the average axial force in the element.

$$u_{top} = u_{bot} + \frac{(F_{bot} + F_{top})}{2} \cdot \frac{h}{AE} \tag{26}$$

The above algorithm is repeated in turn for each element on the pile from the pile tip to the upper most element of the pile. The pile-head load-settlement curve is generated by assuming a series of values of pile tip movements and computing the resulting pile-head forces and displacements.

Group Analysis

Several algorithms are available for the solution of the response of piles in a group. The choice of algorithm depends on the requirements for the solution. Two basic approaches are available. These choices are algorithms based on the stiffness method and on the flexibility method. The effectiveness and efficiency of the solution algorithm will vary depending on the geometric configuration of the pile group and whether hysteretic load-transfer curves are used for cases involving cyclic loading.

In addition to the numerical method used to solve for group action, analyses may consider either two-dimensional or three-dimensional group action. For many problems, two-dimensional analysis is an appropriate choice. It is usually simpler to characterize the foundation loading in two-dimensions. When three-dimensional analyses are made, characterization of foundation loadings may be quite time consuming.

The solution algorithm used in the example application presented later in this paper uses the flexibility method. The procedure is the following:

1. Specify an initial displacement for the pile cap.
2. Compute the corresponding pile-head displacements.
3. Compute the pile reactions for the given pile-head displacements.
4. Sum the pile reactions to obtain the group reaction for the given displacement.
5. Compute the difference between the applied loads and the pile reactions to obtain a force correction vector.
6. Give a virtual displacement to obtain the stiffness matrix.
7. Invert the stiffness matrix to obtain the flexibility matrix.

8. Multiply the flexibility matrix by the force correction vector to get the displacement correction vector.
9. Correct the pile-cap displacement by adding the displacement correction vector to the pile displacement vector.
10. Repeat the above steps until the displacement-correction vector becomes acceptably small.

Sequence of Analyses for Cases Involving Seepage Forces

The general procedure for problems involving seepage forces is illustrated in Figure 4. A brief discussion of the analysis is as follows:

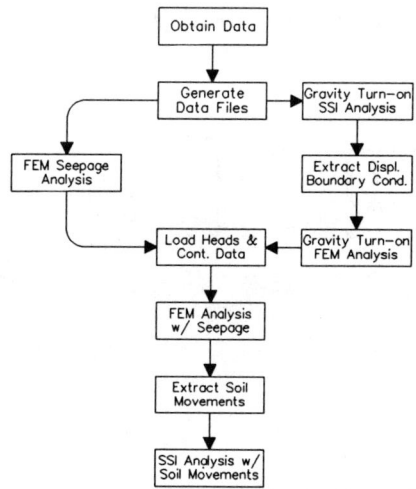

Figure 4. Flow Chart for Analyses of Involving Seepage Forces

1. If the piles are widely spaced then no pile-soil-pile interaction will take place and the interaction factors may be omitted from the input files. If pile spacing is close enough for interaction factors to be computed then compute interaction factors and check the magnitude of lateral group displacement in the analysis. Studies by Brown et al. (1988) have found if the pile group movements are less than 1.0 inch (25 mm) then no interaction takes place. Omit the interaction factors from the input file if the computed lateral group displacement is less than 1.0 inch.
2. Compute structural loadings for gravity turn-on analyses, FEM gravity turn-on analysis, FEM seepage force analysis, and SSI analysis including seepage forces.
3. Determine nodal point numbers in finite element mesh for support displacement boundary conditions and create data for later use.
4. Perform seepage analysis for all random variable cases related to soil properties used in seepage analyses.

5. Perform analyses as illustrated in Figure 4.

EXAMPLE ANALYSES

The structure chosen for the example analyses is the Lindy C. Boggs Lock and Dam on the Red River Waterway in Louisiana, shown in Figure 5. This structure is a navigation control structure with dam composed of 11 monoliths supporting one tainter gate each, a single 800 ft by 105 ft (244m by 32 m) lock, and the adjacent flood walls. Each dam monolith is founded on 188 H-piles arranged in 12 rows of alternating batter direction across the monolith and 15 or 16 per row along the flowline. Piles battered downstream are battered 2.5 vertical to 1 horizontal. Piles battered upstream are battered 4 vertical to 1 horizontal. Typical spacing between rows is 5 feet (1.52 m) center-to-center. The length of piles varies between 88 to 110 feet (26.8 to 33.5 m).

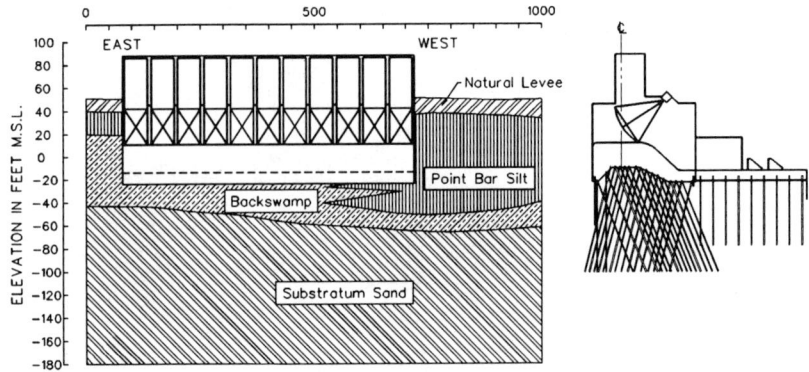

Figure 5. Profile and Cross-section of Lindy C. Boggs Lock & Dam

This study considered two soil profiles. These soil profiles are shown in Figure 5. These soil profiles are those found under the east and west abutments of Lindy C. Boggs Lock and Dam. Four soil strata are found at this location. From bottom to top these strata are the substratum sand, backswamp soil, point bar silts, and compacted backfill placed around the structure during construction. The soil profiles of the east and west abutments differ in that the point bar silts are absent under the east abutment.

The finite element mesh developed for Lindy C. Boggs Lock and Dam is shown in Figure 6. This mesh has 518 elements and 546 nodal points. Interface slip elements are provided between the soil and the structure.

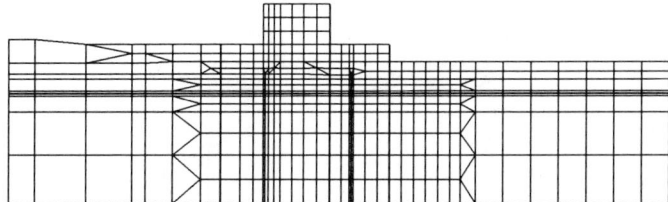

Figure 6. Finite Element Mesh for Lindy C. Boggs Lock and Dam

The same basic finite element mesh was used for both the east and west abutments. The two cases differed only by the inclusion of the point bar silts (replacing the backswamp deposit) for the west abutment.

The influence of soil displacement on pile deflection and bending moment is illustrated in Figures 7 and 8 for two piles located near the centerline of the dam. One pile is battered upstream and the other is battered downstream. The pile battered upstream develops larger lateral deflections and bending moments than the pile battered downstream which supports the dam primary via axial loading. Please note that the lateral pile deflections plotted in the figures are perpendicular to the axis of pile and are not horizontal deflections because the piles are battered. Also shown in the figures is the horizontal component of soil displacement along the centerline of the dam. Not shown in the figures is the vertical component of soil displacement, which is acting in a downward direction. This downward soil displacement adds an additional loading on the piles due to downdrag.

The backswamp deposit was not present at the east abutment. Thus, the magnitude of bending moment developed in the piles is smaller, as shown in Figure 9. This is due to the smaller soil displacement developed by the seepage forces at this location.

In general, the largest values of bending moment are located near the interfaces between soil layers where variations in curvature of the soil displacement versus depth profile are largest. At this site, the backswamp deposit is both softer and lower in permeability that the point bar silt above. Thus, the largest shear strains due to seepage forces occur in the backswamp deposit, resulting the development of the largest bending moments in that zone. It is interesting to note for this problem that bending moment resulting from soil displacements is larger than that resulting from structural loading alone.

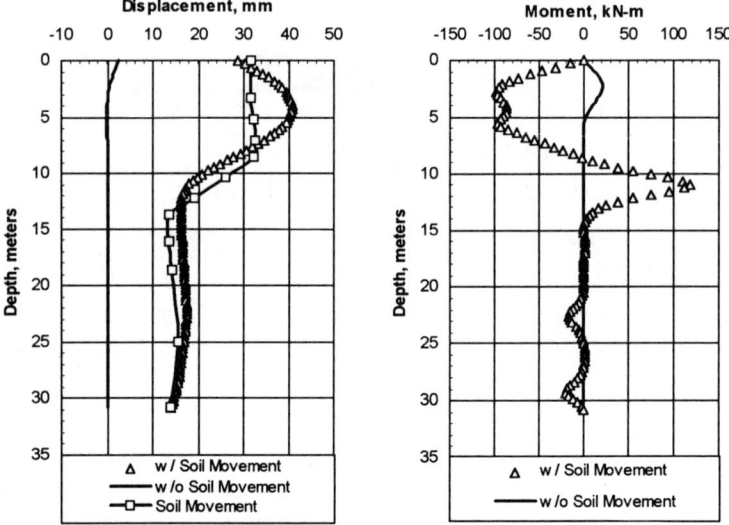

**Figure 7. Computed Results for West Abutment
(Pile 11, Upstream Batter, Near Centerline of Dam)**

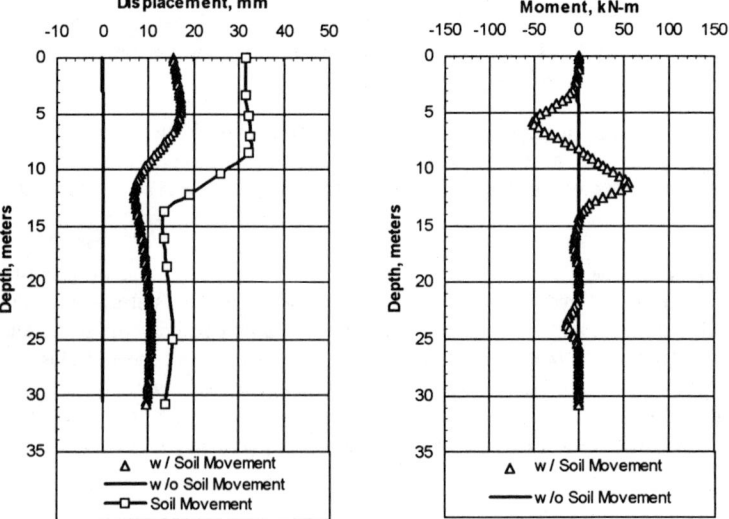

**Figure 8. Computed Results for West Abutment
(Pile 12, Downstream Batter, Near Centerline of Dam)**

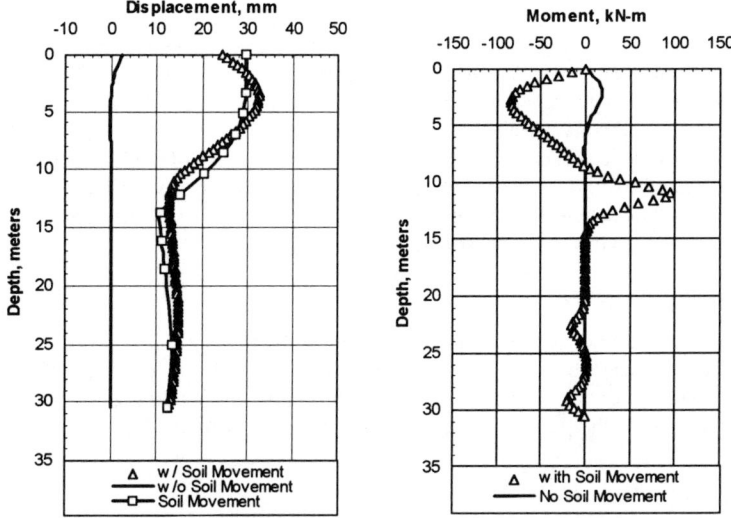

**Figure 9. Computed Results for East Abutment
(Pile 11, Upstream Batter, Near Centerline of Dam)**

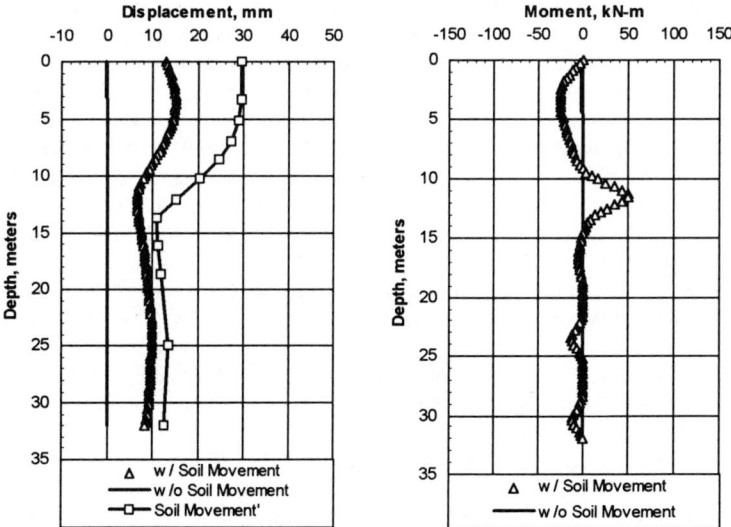

**Figure 10. Computed Results for East Abutment
(Pile 12, Downstream Batter, Near Centerline of Dam)**

CONCLUSIONS

Methods for analyzing the effects of soil displacements on the response of pile groups have been presented. The techniques presented can deal with the effects of soil displacements acting in any direction on lateral and axial load-transfer response of pile groups. The methods permit rigid body motion in both horizontal and vertical directions. An example was presented of the methods for the case of a pile-supported structure subjected to soil displacements resulting from seepage forces. The analysis found that for this example the bending moments resulting from soil displacements were larger than those from external structural loading alone.

ACKNOWLEDGMENTS

The research described in this paper was supported by the US Army Engineer Waterways Experiment Station under the auspices of the US Army Research Office Scientific Services Program administered by Battelle. The views of the author does not purport to reflect the position of the Department of the Army or Department of Defense.

REFERENCES

Brown, D. A., and Reese, L. C. (1988). "Behavior of a Large-Scale Pile Group Subjected to Cyclic Lateral Loading," Miscellaneous Paper GL-88-2, US Army Waterways Experiment Station, Vicksburg, MS.

Isenhower, W. M., "Reliability Analysis of a Pile-Support Navigation Structure," Report to US Army Waterways Experiment Station, Contract No. DAAL03-91-C-0034, TCN 96-076, Scientific Services Program, August, 1996, 106 pp.

Isenhower, W. M., and Mosher, R. L., "Evaluation of Reliability of Pile-Supported Structures," *Proceedings,* Uncertainty in the Geologic Environment: From Theory to Practice, ASCE, Geotechnical Special Publication No. 58, 1996, pp. 666-684.

Matlock, H. "Correlations for Design of Laterally Loaded Piles in Soft Clay," *Proceedings*, Offshore Technology Conference, 1970 Vol. 1, pp. 577-594.

McClelland, B., and Focht, J. A., Jr., "Soil Modulus for Laterally Loaded Piles," *Transactions*, ASCE, Vol. 123, 1958, pp. 1049-1086.

Morrison, J. R., O'Brien, M. P., Johnson, J. W., and Schaff, S. A., "The Force Exerted by Surface Waves on Piles," *Transactions of AIME*, Vol., 189, 1950, pp. 149-154.

Reese, L. C., and Wang, S-T., "Study of the Stability of Pile Foundations for Lock Structures, Phase Three Report," Report to US Army Waterways Experiment Station, Contract DACW39-88-C-0034.

Seed, H. B., and Reese, L. C., "The Action of Soft Clay Along Friction Piles," *Transactions*, ASCE, 1957, Vol. 122, pp. 731-754.

Wang, S-T., and Reese, L. C., "Design of Pile Foundation in Liquefied Soil," *Proceedings*, Geotechnical Earthquake Engineering and Soil Dynamics III, ASCE, Geotechnical Special Publication No. 75, 1998, Vol. 2, pp. 1331-1343.

Lateral Resistance of Piles in Liquefying Sand

By Daniel W. Wilson[1], Ross W. Boulanger[1], and Bruce L. Kutter[1]
Members, ASCE

Abstract

This paper provides an overview of a study on the dynamic response of pile foundations in liquefying sand during seismic shaking. The study included a series of dynamic centrifuge tests of pile supported structures in liquefying sand and the back-calculation of time histories of p-y behavior. The experimentally determined dynamic p-y curves provide insight into the mechanisms of soil-pile interaction in liquefying sand, showing characteristics that are consistent with the undrained behavior of liquefying sand. The study also included pseudo-static analyses that account for the effects of liquefaction using an apparent p-y scaling factor.

Introduction

The potential significance of liquefaction-related damage to piles has been clearly demonstrated during past earthquakes, from the 1964 Alaskan earthquake (e.g., Youd and Bartlett 1989) to the 1995 Kobe earthquake (e.g., Tokimatsu and Asaka 1998). Observations during past earthquakes have shown that modern pile foundations in firm soils generally perform well, while the performance of piles in liquefied ground can range from excellent to poor (e.g., structural damage or excessive deformations).

Analysis and design procedures for piles in liquefying ground have generally had large uncertainties due to both a lack of physical data against which they can be evaluated and a lack of understanding of the mechanisms involved in soil-pile-structure interaction in liquefying soils. The documentation of case histories from Kobe have been invaluable in providing field data against which engineering procedures can be evaluated (e.g., JGS 1996, 1998). In addition, recent physical

---

[1] Post-Graduate Research Engineer, Associate Professor, and Professor, respectively, Department of Civil and Environmental Engineering, University of California at Davis, One Shields Avenue, Davis, CA 95616

modeling studies (e.g., Tokida et al. 1992, Liu and Dobry 1995, Abdoun et al. 1997, Horikoshi et al. 1998, Wilson et al. 1998) also have provided insight into the mechanisms of soil-pile-structure interaction in liquefying soils.

In the study presented herein, the lateral resistance of liquefying sand was examined through: (1) a series of dynamic centrifuge tests of pile-supported structures in saturated sands; (2) the back-calculation of p-y time histories from a single-pile-supported structure; and (3) the evaluation of pseudo-static analysis methods that account for the effects of liquefaction using an apparent p-y scaling factor. This paper provides an overview of the above research study and summarizes some of the conclusions, while detailed descriptions of these analyses and more discussion on the conclusions can be found in Wilson (1998) and Wilson et al. (1999).

Experimental Setup

Tests were performed using the servo-hydraulic shaking table on the 9 m radius centrifuge at UC Davis (Kutter et al. 1994). Models were tested in a Flexible Shear Beam (FSB) container at a centrifugal acceleration of 30 g. All results presented herein are in prototype units. See Kutter (1992) for a discussion of the applicable scaling laws.

A typical partial model layout is shown in Figure 1. Two models (containers) are examined herein. The soil profile consisted of two horizontal layers of saturated fine, uniformly graded Nevada sand ($C_u = 1.5$, $D_{50} = 0.15$ mm). The lower layer was at a relative density (D_r) of about 80% in both models. The upper layer was pluviated to $D_r \approx 35\%$ in one model (Csp2), and to $D_r \approx 55\%$ in another (Csp3). Structural models included single-pile- and pile-group-supported structures, but this paper will focus on results obtained with one heavily instrumented single-pile-supported structure. The single pile approximated a steel pipe pile with a diameter (D) of 0.67 m and wall thickness of 19 mm supporting a superstructure load of 480 kN at 3.8 m above grade. The piles were installed at 1 g (prior to spinning the centrifuge), and remained elastic during all earthquake events.

Each model configuration was shaken with up to 15 simulated earthquake events. Each event was a scaled version of a strong motion accelerogram from Port Island in the Kobe Earthquake or UC Santa Cruz in the Loma Prieta Earthquake, with some slight

▶ acceleration   ▪ pore pressure

▭- displacement   ‥ strain gauge

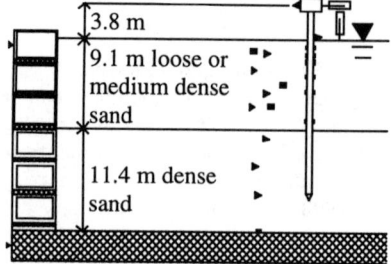

Figure 1. Schematic of layout and
instrumentation

modifications to their frequency content. Detailed documentation of the tests, including all recorded time histories, are available in Wilson et al. (1997a-b). An evaluation of the modeling techniques and limitations are in Wilson et al. (1997c).

Model Behavior

Typical recordings within the soil profile and on the pile-supported structure are presented in Figure 2 for Csp2 event H and Figure 3 for Csp3 event J, both scalar multiples of the Kobe motion with significant excess pore pressure generation. Phase transformation behavior (i.e., transition from contractant to dilatant behavior) in the upper soil layer is evident in the soil profile time histories. Here sharp decreases in pore pressure coincide with sharp acceleration peaks. These pore pressure decreases and acceleration peaks were more pronounced in Csp3 than in Csp2, which is consistent with the medium dense sand of Csp3 having a lower cyclic mobility (i.e., experiencing phase transformation at lower strain levels) than the loose sand of Csp2. Note, however, that the loose sand in one large shaking event (Kobe motion, peak base acceleration of 0.62 g) did dilate long enough during the strong early pulses of motion to transmit relatively large accelerations through the ground to the superstructure (Wilson 1998).

For a given earthquake (i.e. Kobe or Santa Cruz motion), superstructure accelerations for the single pile supported structure in the medium dense sand (Csp3) were typically greater than or about equal to those in the loose sand (Csp2). This is illustrated in Figure 4, where peak superstructure accelerations are plotted versus peak base accelerations. At small levels of shaking ($a_{max,base} < 0.05$ g), the structural models in Csp2 and Csp3 responded similarly to a given earthquake motion because excess pore pressures were small. At intermediate levels of shaking ($a_{max,base} \approx 0.1$ g), excess pore pressures were high in Csp2 but were still low in Csp3. And at higher levels of shaking ($a_{max,base} > 0.2$ g), high excess pore pressures were generated in the upper sand layer in both containers. As excess pore pressures increased, the loose sand in Csp2 "softened" more than the medium dense sand in Csp3, which affected the ground motions and caused a greater lengthening of the "fundamental" period of the structure. It should be noted that the term fundamental period may not be strictly appropriate for a nonlinear system, but is commonly used to indicate the "apparent" fundamental period for a given level of loading. The generally lower peak superstructure accelerations in Csp2 are consistent with the effects liquefaction had on the ground motions and structural period in these tests. Note, however, that the peak superstructure accelerations for the 0.6 g Kobe events in both Csp2 and Csp3 were approximately the same (≈ 1.5 g), due to dilation in the loose sand (and in the medium dense sand) during the strong early pulses of motion as discussed above.

Also clear in Figure 4 is that the Kobe base motion excited a much larger superstructure response than did the Santa Cruz base motion. This was due to the higher energy content of the Kobe motion near the 0.9 to 1.5 second range of fundamental periods of the structure for these shaking events (Wilson 1998).

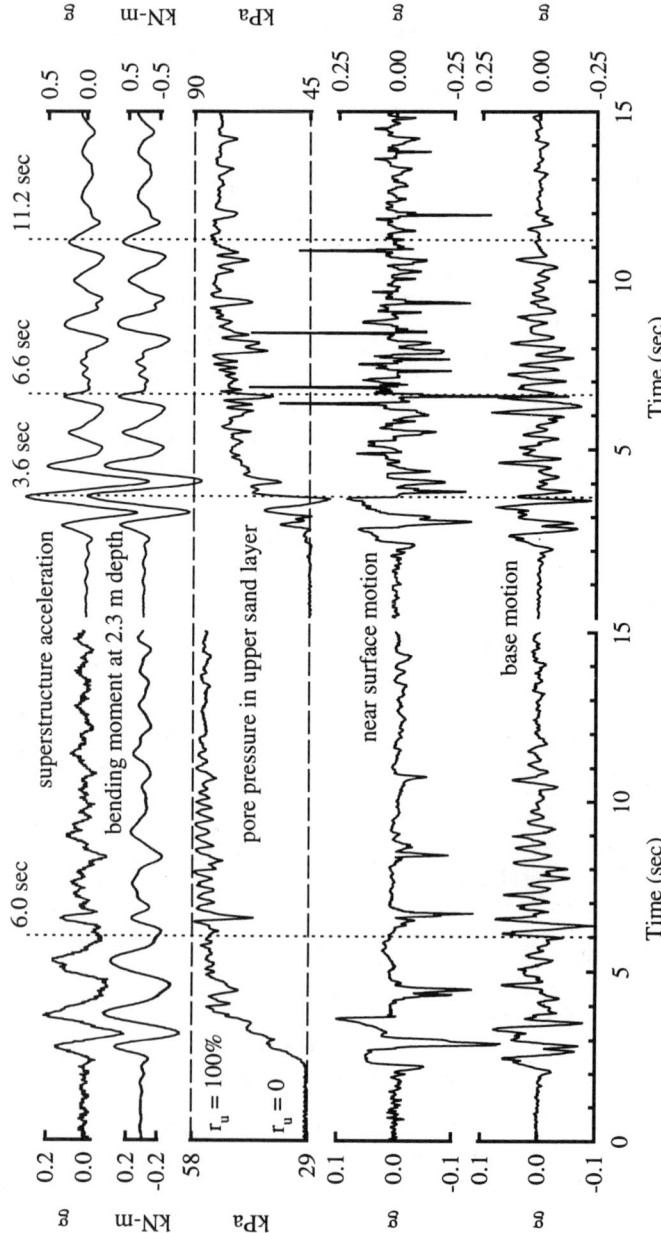

Figure 2. Time histories in Csp2 event H: upper sand layer at
$D_r \approx 35\%$, Kobe motion with $a_{max,base} \approx 0.1$ g

Figure 3. Time histories in Csp3 event J: upper sand layer at
$D_r \approx 55\%$, Kobe motion with $a_{max,base} \approx 0.22$ g

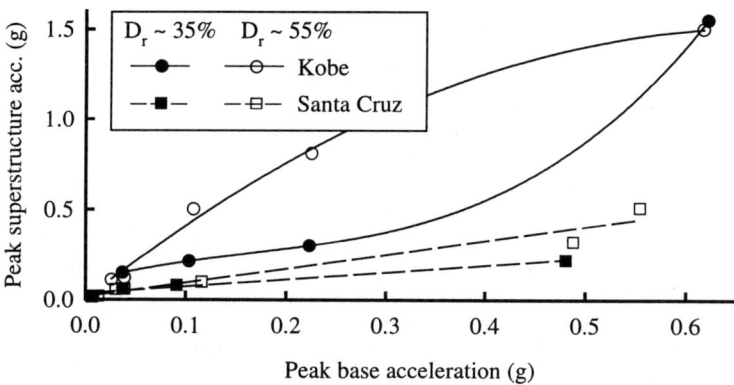

Figure 4. Peak superstructure versus peak input acceleration for the single-pile supported structure

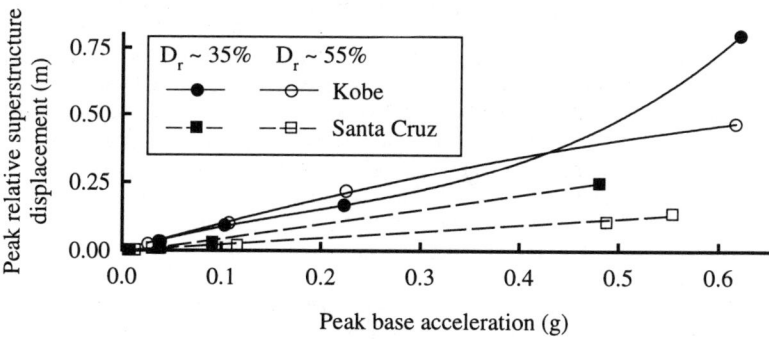

Figure 5. Peak superstructure displacement versus peak input acceleration for the single-pile supported structure

Analyses of the bending moment and superstructure acceleration time histories showed that the peak bending moments could be strongly correlated to the inertial loads from the superstructure for these model configurations, while kinematic loads from the soil profile had significant effects on bending moments deeper in the soil profile (i.e., below the depth of peak bending moment for these models). Thus the same general trends in peak superstructure acceleration were observed in peak bending moments.

Peak superstructure displacements, however, did not follow the same trends as the peak superstructure accelerations. Peak superstructure displacements relative to the pile tip for the single-pile-supported structure are plotted versus peak base acceleration in Figure 5. For a given earthquake (i.e., Kobe or Santa Cruz motion), peak superstructure displacements in Csp2 (loose sand) ranged from slightly smaller to substantially larger than those in Csp3 (medium dense sand). The occurrence of greater peak superstructure displacements despite lower peak superstructure accelerations was due to the greater "softening" of the looser sand in Csp2 and the resulting greater lengthening of the structure's "fundamental" period.

Measuring p-y Behavior

Dynamic soil-pile interaction is often represented numerically using a Beam on Nonlinear Winkler Foundation (BNWF) model, where interaction is represented through parallel, independent, nonlinear p-y springs. Much of the early work developing the p-y representation of soil-pile interaction was performed by Professor Reese and his colleagues, for example Reese (1958), Matlock and Reese (1960), and Reese et al. (1974).

In this study, experimentally observed p-y curves during earthquake shaking were back-calculated using the data collected in these tests. A detailed description of the methods used in these calculations is given in Wilson et al. (1999).

The soil profile displacement (y_{soil}) was found by fitting an interpolation function to the displacements measured at each accelerometer in the soil profile vertical array. Pile displacements (y_{pile}) were calculated by double-integrating the recorded bending moment distribution with respect to distance along the pile, subject to global constraints of the measured displacement at the pilehead and the net displacement between the pile and soil being "small" (or zero) at 9 m depth. Only transient displacements are obtained from these calculations and thus the resulting values of $y = y_{pile} - y_{soil}$ represent the transient component of relative displacements.

The distribution of lateral resistance (p) was obtained by double-differentiating the recorded bending moment distribution with respect to depth. A new approximation to the second derivative of bending moment distribution was developed using the concept of weighted residuals (WR), as typically applied in finite element analyses. The WR method of differentiation provides a means of obtaining a derivative from discrete data that has the same smoothness properties as the original interpolation of the data. This makes it especially useful for unevenly spaced data that exhibits noise. The method is described in Wilson et al. (1999).

Resulting p-y Curves

Time histories of p and y at a depth of 2-D are plotted for Csp2 event H, Csp3 event L, and Csp3 event J in Figures 6, 7, and 8, respectively. Select p-y loops from these same events at depths of 2-, 3-, and 4-D are plotted in Figures 9 and 10. Monotonic p-y curves based on American Petroleum Institute (API 1993) recommendations for drained loading are shown on the plots for reference. To develop the API curves, values of $\phi' = 30°$ and $k = 8.1$ MN/m^3 were used for Csp2 ($D_r \approx 35\%$) and $\phi' = 35°$ and $k = 21.7$ MN/m^3 for Csp3 ($D_r \approx 55\%$).

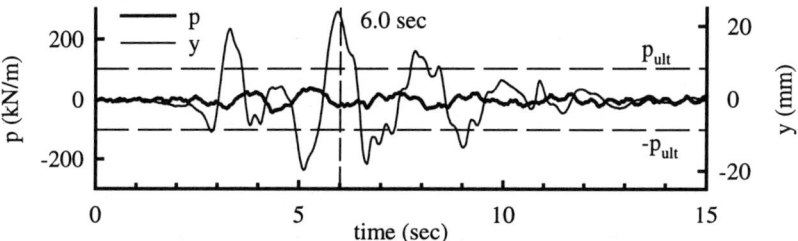

Figure 6. P and y time histories, Csp2 event H ($D_r \approx 35\%$) at depth 3-D.
- liquefaction early in shaking (p_{ult} based on API 1993)

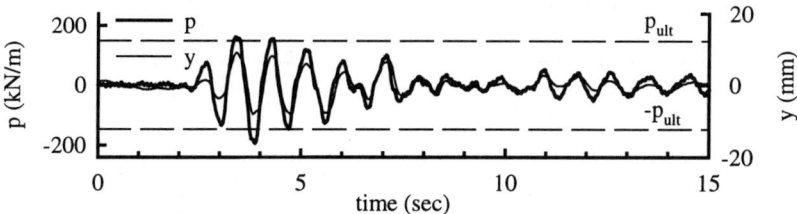

Figure 7. P and y time histories, Csp3 event L ($D_r \approx 55\%$) at depth 3-D.
- low r_u throughout shaking (p_{ult} based on API 1993)

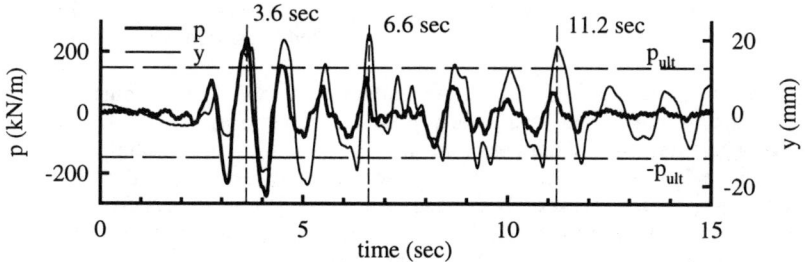

Figure 8. P and y time histories, Csp3 event L ($D_r \approx 55\%$) at depth 3-D.
- r_u increasing during shaking (p_{ult} based on API 1993)

The upper soil layer in Csp2 was loose ($D_r \approx 35\%$) and liquefied early in shaking for most events. In Csp2 event H, liquefaction occurred early in shaking (see Figure 2) and the back-calculated p-y results shown in Figures 6 and 9 indicate there was very little lateral resistance on the pile even under large relative displacements (>25 mm, or $y/D \approx 0.035$, @ $z = 2$ m). This behavior was consistent for all earthquake events which generated high pore pressures in Csp2 except the largest

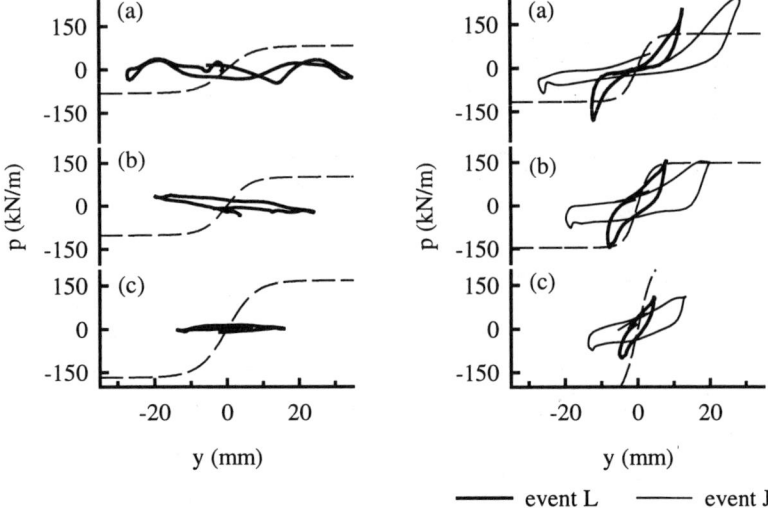

Figure 9. P-y loops for Csp2 event H
from 4.5 to 6.5 seconds ($D_r \approx 35\%$,
liquefaction early in shaking) at
depths (a) 2-D, (b) 3-D, and (c) 4-D
(D = 0.67 m). Dashed lines from
API (1993)

Figure 10. P-y loops for Csp3 events L
and J, both from about 4 to 5
seconds ($D_r \approx 55\%$, low r_u in L,
moderate r_u in J) at depths (a) 2-D,
(b) 3-D, and (c) 4-D (D = 0.67 m).
Dashed lines from API (1993)

Kobe event where the displacements and strains were large enough to cause phase
transformation during the strong early pulses of motion, as discussed earlier. In
events where liquefaction did not occur (i.e., low shaking levels), lateral resistances
were well below the API (drained) p_{ult} values (because inertial load demands were
small) and showed no signs of cyclic degradation or softening in time. That is, p and
y remained relatively proportional throughout shaking, similar in most respects to the
p and y time histories shown in Figure 7 for Csp3 event L.

The upper sand layer in Csp3 was medium dense ($D_r \approx 55\%$) and had a
greater resistance to liquefaction than in Csp2. The observed p-y behavior in these
events degraded with cyclic loading and increasing pore pressures but showed
hardening as the relative displacement approached or exceeded maximum past
values, especially near the surface. The effect of r_u ($= \Delta u/\sigma_{vo}'$) on p-y behavior is
examined by comparing p and y time histories for event L and event J, both scalar
multiples of the Kobe motion. In event L ($a_{max,base} \approx 0.11$ g), there was no significant
excess pore pressure generation and p and y remained relatively proportional
throughout shaking (see Figure 7). In event J ($a_{max,base} \approx 0.22$ g), pore pressures
increased during shaking (see Figure 3) and p values peaked at about 4 seconds, with

only lower p values developing afterward (see Figure 8). In contrast, several peaks of similar magnitude in the y values occur between 4 and 11 seconds.

The softening of the p-y relationship with increased pore pressures is examined more closely in Figure 10 by comparing loading cycles at similar levels of lateral resistance (i.e. similar mobilized strengths) from events L and J. The lateral resistance mobilized from about 4 to 5 seconds was similar for each event. However, the p-y loop from event J is significantly softer than that from event L. The average pore pressure ratio during these time frames was negligible in event L, and was approximately 40% at 4.5 m depth in event J (Figure 3). P-y loops for later times in event J show even greater softening as r_u continued to increase, as evidenced in the p and y time histories in Figure 8.

Pseudo-Static BNWF Analyses - Background

Current design methodologies often involve analyzing the soil-pile-structure system as two uncoupled problems, the superstructure and the foundation, and then finding solutions to each that are compatible with the expected response of both parts. In the first step of the analysis, the foundation for the superstructure may be replaced with a set of springs that represent the effective foundation stiffness. The displacement demand on the superstructure can then be estimated based on either a linear elastic dynamic response analysis or linear elastic response spectra, plus some correction for the effects of inelastic superstructure behavior. The load demand transmitted to the foundation may be limited by inelastic behavior in the superstructure, and hence may be smaller than the elastic force demand (as calculated from linear elastic response analyses).

The displacement capacity of the superstructure and foundation system may then be evaluated using a nonlinear pushover analysis that incorporates nonlinear behavior of both the superstructure and the foundation. The pseudo-static response of the foundation may be modeled using a BNWF method as previously described with the soil parameters appropriately modified to account for the effects of seismic loading. The design load on the foundation is obtained from the pushover analysis with the superstructure statically pushed to the estimated displacement demand. Finally, the foundation loads and displacements from this step must be consistent with the effective foundation stiffness used in the linear elastic response analysis of the superstructure above. Note that this approach does not account for kinematic loading of the piles if the p-y supports are fixed. Kinematic loads can, however, be approximated in the BNWF analysis by imposing free-field soil displacements on the p-y supports.

One approach to account for the effects of liquefaction on the lateral resistance of soil for a pseudo-static analysis would be to apply an apparent scaling factor to static p-y curves, as suggested by JRA (1980), AIJ (1988), and Liu and Dobry (1995). This approach was initially evaluated against the data from this study by Boulanger et al. (1997), and then more extensively by Wilson (1998), as described below.

Pseudo-static p-y analyses of the highly instrumented single pile system in tests Csp2 ($D_r \approx 35\%$) and Csp3 ($D_r \approx 55\%$) were performed using the program PAR (PMB 1988). The program is one of several programs that were used in a study by Wang et al. (1998) to perform dynamic soil-pile-superstructure interaction analyses based on the BNWF model. The pseudo-static p-y analyses were done at snapshots in time, with the time closely corresponding to a peak inertial load (i.e., peak superstructure acceleration). Baseline sets of p-y curves were established for each test using API (1993) recommendations for static loading with the properties listed earlier. The baseline curves provide a fixed reference and are not implied to be necessarily correct for the static loading condition for these tests. In an analysis, the measured inertial loads and free-field displacements were applied to the numerical model, and the calculated bending moment distribution and pile deformed shapes were compared to recorded values. The analysis was repeated with different scalar multiples of the baseline p-y curves (scalar multiplier applied to the p values) in the upper soil layer until a reasonable match between calculated and measured responses was obtained.

Pseudo-Static BNWF Analyses - Typical Results

Typical results are illustrated in Figures 11 and 12. Note that these figures represent only two analyses from the suite of analyses performed by Wilson (1998). The results in Figure 11 are for a time of 6.0 seconds in Csp2 event H ($D_r \approx 35\%$), the same event as represented by the time histories in Figures 2 and 6. At the particular snapshot in time shown in Figure 11, the soil profile and pile are deformed in opposing directions, a condition that accentuates the effects of the p-y resistance. The excess pore pressure ratio averaged about $r_u \approx 85\%$ in the upper sand layer at this time. The comparison between the calculated and recorded deformed pile shape and bending moment distribution is very poor if the baseline p-y curves are used (i.e., APIx1.0), but quite reasonable if the baseline p-y curves are scaled (multiplied) by a factor of 0.1. A slightly better fit is possible with an even smaller scaling factor in this one case, but 0.1 is a reasonable value obtained from several such analyses. Note that a p-multiplier of 1.0 to 2.0 was shown to provide reasonably good agreement for peak loading cycles during events without liquefaction in Csp2.

The results in Figure 12 are for a time of 11.2 seconds in Csp3 event J ($D_r \approx 55\%$), the same event as represented by the time histories in Figures 3 and 8. At this time, r_u averaged about 82% in the upper sand layer. The deformed shape and bending moment distribution were reasonably matched with a scaling factor of 0.2-0.25 applied to the baseline p-y curves (i.e., API/4 to API/5). However, good agreement at 3.6 seconds in this event was obtained with about 2.0 times the baseline p-y curves. The need for stiffer/stronger p-y curves at 3.6 seconds is consistent with the back-calculated p-y curves in Figure 10. Hence the results in Figure 12 indicate that the p-y curves softened during shaking by a scaling factor as small as 0.1 (i.e.,

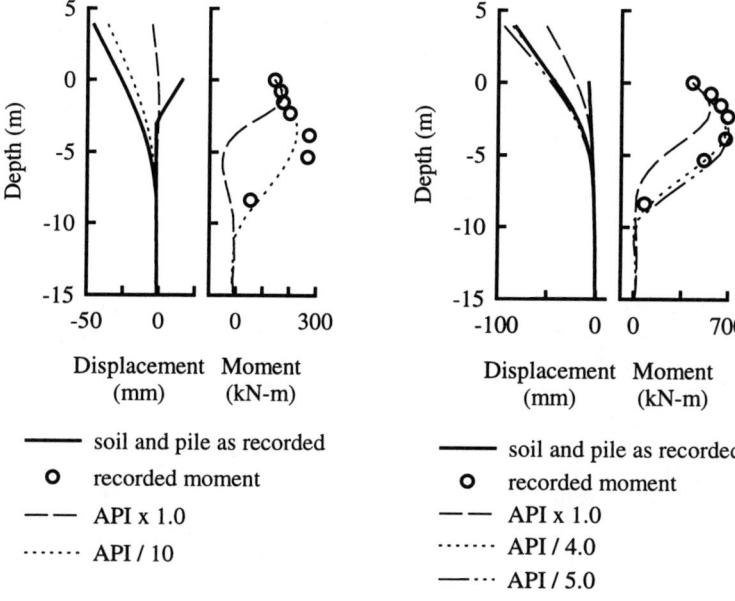

Figure 11. Pseudo-static analysis at 6.0 s in Csp2 event H (Dr ≈ 35%, r_u ≈ 85%)

Figure 12. Pseudo-static analysis at 11.2 s in Csp3 event J (Dr ≈ 55%, r_u ≈ 82%)

2.0xAPI early vs. 0.2xAPI late). Similar analyses were performed for the peak load near 6.6 seconds in this event (r_u ≈ 65%), with the best agreement obtained with 0.5 times the baseline p-y curves (i.e., a scaling factor of about 0.25 relative to early in shaking). These results illustrate the progressive softening of the p-y resistance of the liquefying sand during earthquake shaking due to the combined influences of increasing r_u, prior displacement history, and cyclic degradation.

Other Factors Affecting Lateral Resistance

The actual p-y resistance of liquefying sand in the field is undoubtedly dependent on numerous factors that are not yet understood or are not readily quantifiable. Recent physical modeling studies have illustrated some of these influences. Liu and Dobry (1995) performed centrifuge tests with sand at D_r ≈ 55% and recommended that a scaling factor be applied to static drained p-y curves with the scaling factor decreasing more or less linearly with r_u and reaching a minimum value of about 0.1 when r_u = 100%. Tank experiments by Tokida et al. (1992) showed that the lateral resistance of liquefying sand varied significantly with loading rate (which may be attributed to the effects of partial drainage and/or viscous behavior), and with pile group dimensions and geometry.

The study presented herein also did not address the problem of soil-structure interaction in laterally spreading ground. Large kinematic loads can develop due to lateral spreading of liquefying soils or due to high strain gradients in soft clays, and may be particularly damaging under certain stratigraphies (e.g., strong crusts overlying soft or liquefying soils). For example, Tokimatsu et al. (1998) described cases of pile damage caused by liquefaction-induced lateral spreading at Kobe and the results of analyses that accounted for the kinematic loads imposed by the spreading ground. Abdoun et al. (1997) and Dobry and Abdoun (1998) describe centrifuge test results for piles embedded in various profiles that included layers of loose Nevada sand ($D_r \approx 40\%$) with overlying and/or underlying layers of non-liquefiable cemented sand. These results showed that the kinematic loads imposed on the piles were dependent on the free field soil displacements, the stratigraphy, and the properties of the non-liquefied and liquefied soils. Results were presented for the lateral pressures exerted by the liquefied soil on the pile, and these pressures were found to be important for piles "floating" in liquefied soil. In cases where the pile spanned across a liquefied layer into overlying and underlying non-liquefied soils, the loads from lateral spreading of the overlying layer became dominant and the pressures from the liquefied soil became relatively unimportant. Horikoshi et al. (1998) describe centrifuge test results for piles embedded near a quay wall in a profile having a non-liquefied crust overlying a layer of saturated sand ($D_r \approx 50$-60%), and also showed that kinematic loads from the non-liquefied crust can have a dominating effect.

Summary and Conclusions

The back-calculated p-y curves across many shaking events (as described by Wilson 1998) showed characteristics that are consistent with the stress-strain response of liquefying sand. The p-y resistance of the $D_r \approx 35\%$ upper sand layer in Csp2 was much smaller and softer than for the $D_r \approx 55\%$ upper sand layer in Csp3. This observation is consistent with the effects of D_r on the undrained shear resistance (or cyclic mobility) of saturated sand, as commonly observed for example in undrained cyclic triaxial tests.

The ultimate lateral resistance in loose sand (Csp2, $D_r \approx 35\%$) was generally small when the soil liquefied, even when displacements were fairly large (i.e., y/D > 0.035 @ 3-D depth). Larger resistances were observed, however, in the largest shaking event due to dilation of the soil during the strong early pulses of motion.

In medium dense sand (Csp3, $D_r \approx 55\%$), the p-y behavior progressively softened with time during shaking as pore pressures, strains, and number of load cycles increased. The observed p-y behavior was found to be displacement hardening when relative displacements approached or exceeded past values, especially near the surface. This behavior may be attributed to the nearly undrained loading conditions and the tendency for the soil to dilate under these loading conditions (i.e., large enough strains to move the sand through a phase transformation). The observed behavior was similar in some respects to results reported by Dou and Byrne (1996) and Kagawa et al. (1994).

For the conditions presented herein, the pseudo-static BNWF analyses appeared to provide reasonable estimates of pile deformation and bending moment distribution based on known (or estimated) superstructure inertia loads, known (or estimated) soil displacements, and p-y curves scaled from established static p-y curves. Based upon the analyses by Wilson (1998), of which only two typical examples were presented herein, a reasonable p-multiplier for representative peak loading cycles on a single pile in liquefied sand may be about 0.1-0.2 for $D_r \approx 35\%$ and about 0.25-0.35 for $D_r \approx 55\%$. Caution is warranted in design, however, because the apparent p-y resistance can be expected to progressively soften during shaking as pore pressures increase. It should also be noted that the estimation of inertial and kinematic loads for input to these analyses requires the challenging task of estimating the effects of liquefaction on the dynamic response of the soil profile and soil-pile-structure system.

In a nonlinear push-over (pseudo-static) analysis, it is not necessarily conservative to assume an overly soft p-y resistance for liquefying soils in evaluating foundation performance. Instead, the questions are how the p-y resistance will affect the structure's "fundamental" period and how liquefaction will affect the ground motions. Only then can the effect of liquefaction on superstructure force or displacement demand be rationally evaluated. In general, the moment and displacement demands will depend on a combination of the earthquake motion and structural response characteristics and the effects of liquefaction on site response and soil-structure interaction.

Large kinematic loads can develop due to lateral spreading of liquefying soils or due to high strain gradients in soft clays, and may be particularly damaging under certain stratigraphies (e.g., strong crusts overlying soft or liquefying soils). In situations where the kinematic loading is dominated by strong crusts overlying liquefiable soil, the analyses may be insensitive to the p-y representation of the liquefied soil, provided it has been sufficiently softened relative to the other non-liquefied soils.

Acknowledgments

Funding was provided by CALTRANS (contract number 65V495) under contract manager Abbas Abghari. However, the contents of this paper do not necessarily represent a policy of that agency nor endorsement by the State government. The shaker was designed and constructed with support from the NSF, Obayashi Corporation, and Caltrans. The authors would like to acknowledge the contributions of Dr.'s M.M. Rashid, who suggested the WR derivative, A. Abghari, and I.M. Idriss. The above support and assistance is greatly appreciated.

References

Abdoun, T., Dobry, R., and O'Rourke, T.D. (1997). "Centrifuge and numerical modeling of soil-pile interaction during earthquake induced soil liquefaction and lateral spreading." *Observation and Modeling in Numerical Analysis and Model Tests in Dynamic Soil-Structure Interaction Problems*, Geotechnical Special Publication 64, ASCE, pp. 76-90.

AIJ (1988). *Recommendations for design of building foundations*. Architectural Institute of Japan. (in Japanese).

API (1993). *Recommended Practice for Planning, Designing and Constructing Fixed Offshore Platforms*. API RP 2A-WSD, 20th ed., American Petroleum Institute

Boulanger, R.W., Wilson, D.W., Kutter, B.L., and Abghari, A. (1997). "Soil-pile-superstructure interaction in liquefiable sand." *Transportation Research Record No. 1569*, TRB, NRC, National Academy Press, 55-64.

Dobry, R., and Abdoun, T. (1998). "Post-triggering response of liquefied sand in the free field and near foundations." *Geotechnical Earthquake Engineering and Soil Dynamics III*, Geotechnical Special Pub. No. 75, ASCE, Vol. 1, pp. 270-300.

Dou, H., and Byrne, P.M. (1996) "Dynamic response of single piles and soil-pile interaction." *Canadian Geotechnical Journal*, Vol. 33, pp. 80-96.

Horikoshi, K., Tateishi, A., and Fujiwara, T. (1998). "Centrifuge modeling of a single pile subjected to liquefaction-induced lateral spreading." *Soils and Foundations*, Special Issue No. 2, pp. 193-208.

JGS (1998). *Soils and Foundations* Special Issue on Geotechnical Aspects of the January 17, 1995 Hyogoken-Nambu Earthquake, No. 2, Japanese Geotech. Soc.

JGS (1996). *Soils and Foundations* Special Issue on Geotechnical Aspects of the January 17, 1995 Hyogoken-Nambu Earthquake, Japanese Geotech. Soc.

JRA (1980). *Specifications for highway bridges*. Japan Road Association. (in Japanese).

Kagawa, T., Minowa, C., Mizuno, H., and Abe, A. (1994). "Shaking-table tests on piles in liquefying sand", *Proceedings, Fifth US National Conference on Earthquake Engineering*, Chicago, Illinois, Vol. IV, 107-113.

Kutter, B.L. (1992). "Dynamic centrifuge modeling of geotechnical structures." *Transp. Res. Rec. 1336*, Foundation Engineering: Seismic Design, Drilled Shaft and Other Issues. TRB, NRC, Wash., D.C. 24-30.

Kutter, B.L., Idriss, I.M., Kohnke, T., Lakeland, J., Li, X.S., Sluis, W., Zeng, X., Tauscher, R., Goto, Y., and Kubodera, I. (1994). "Design of a large earthquake simulator at UC Davis." *Centrifuge 94*, Balkema.

Liu, L. and Dobry, R. (1995). "Effect of liquefaction on lateral response of piles by centrifuge model tests." *National Center for Earthquake Engineering Research (NCEER) Bulletin*, Vol. 9, No. 1, January, pp. 7-11.

Matlock, H., and Reese, L.C. (1960). "Generalized solutions for laterally loaded piles." *Journal of the Soil Mechanics and Foundations Division*, ASCE, Vol. 88, No. SM 5, pp. 63-91.

PMB Engineering Inc. (1988). PAR (Pile Analysis Routine) Manual, 500 Sansome Street, San Francisco, CA 94111.

Reese, L.C. (1958). Discussion of "Soil modulus for laterally loaded piles." by B. McClelland and J.A. Focht, Jr., Transactions, ASCE, Vol. 123, pp. 1071-1074.

Reese, L.C., Cox, W.R., and Koop, F.D. (1974). "Analysis of laterally loaded piles in sand." *Proceedings, Sixth Annual Offshore Technology Conference*, Houston, Texas, pp. 473-483.

Tokida, K.-I., Matsumoto, H., and Iwasaki, H. (1992). "Experimental study of drag acting of piles in ground flowing by liquefaction." *Proc. Fourth Japan-U.S. Workshop on Earthquake Resistant Design of Lifeline Facilities and Countermeasures for Soil Liquefaction*, NCEER 92-0019, Vol. 1, SUNY, Buffalo, N.Y.

Tokimatsu, K., and Asaka, A. (1998). "Effects of liquefaction-induced ground displacements on pile performance in the 1995 Hyogoken-Nambu earthquake." *Soils and Foundations*, Special Issue No. 2, pp. 163-178.

Wang, S., Kutter, B.L., Chacko, J., Wilson, D.W., Boulanger, R.W., and Abghari, A. (1998). "Nonlinear seismic soil-pile-structure interaction." *Earthquake Spectra*, Earthquake Engineering Research Institute, Vol. 14, No. 2.

Wilson, D.W., Boulanger, R.W., and Kutter, B.L. (1999). "Seismic lateral resistance of liquefying sand." In preparation for *Journal of Geotechnical Engineering*, ASCE.

Wilson, D.W. (1998). *Soil-pile-superstructure interaction at soft and liquefying soil sites*. Ph.D. dissertation, Dept. Civil and Env. Engrg., UC Davis.

Wilson, D.W., Boulanger, R.W., and Kutter, B.L. (1998). "Signal processing for and analyses of dynamic soil-pile interaction experiments." *Centrifuge 98,* Balkema, pp. 135-142.

Wilson, D.W., Boulanger, R.W., and Kutter, B.L. (1997a). *Soil-pile-superstructure interaction at soft or liquefiable soil sites - centrifuge data report for Csp2*. UCD/CGMDR-97/03, Center Geotech. Modeling, UC Davis.

Wilson, D.W., Boulanger, R.W., and Kutter, B.L. (1997b). *Soil-pile-superstructure interaction at soft or liquefiable soil sites - centrifuge data report for Csp3*. UCD/CGMDR-97/04, Center Geotech. Modeling, UC Davis.

Wilson, D.W., Boulanger, R.W., Kutter, B.L., and Abghari, A. (1997c). "Aspects of dynamic centrifuge testing of soil-pile-superstructure interaction." *Observation and Modeling in Numerical Analysis and Model Tests in Dynamic Soil-Structure Interaction Problems*, Geotechnical Special Publication 64, ASCE, pp. 47-63.

Youd, T.L. and Bartlett, S.F. (1989). "Case histories of lateral spreads from the 1964 Alaskan earthquake," *Proceedings, Third Japan-U.S. Workshop on Earthquake Resistant Design of Lifeline Facilities and Countermeasures for Soil Liquefaction*, Report NCEER-91-0001, National Center for Earthquake Engineering Research, SUNY, Buffalo, N.Y.

ACCURACY OF METHODS FOR PREDICTING AXIAL CAPACITY OF DEEP FOUNDATIONS

James H. Long[1] and Michael H. Wysockey[2]

ABSTRACT

Several design methods are available for determining the axial capacity of drilled shafts and driven piling, yet there is considerably less information on how accurately these methods predict capacity and how to determine the impact accuracy has on cost. A procedure is outlined herein for comparing predictive methods. The accuracy and precision for each design method is quantified and discussed. In addition, the impact of accuracy and precision on foundation cost is presented in terms of a Wasted Capacity Index.

Examples of how well these methods predict capacity are provided for both drilled shafts and driven piling. The methods investigated for drilled shafts include LPC cone, the Meyerhof cone, FHWA, a modified FHWA, Chen and Kulhawy Alpha, and Chen and Kulhawy Beta. The methods investigated for driving piling include the Engineering News formula, the Gates Formula, WEAP, Measured Energy approach, PDA, and CAPWAP.

INTRODUCTION

A predictive method must be shown sufficiently accurate before the method can be considered generally applicable. To obtain comparable results of prediction methods it is necessary to standardize the procedures of statistical evaluation. The procedures outlined here offer a consistent basis for comparison. Methods for predicting axial capacity are evaluated by their ability to predict the behavior measured from results of full-scale load tests. The ratio of predicted to measured capacity (Q_p/Q_m) is used as a measure of a methods ability to predict capacity. The distribution of Q_p/Q_m is log-normal and simple statistical parameters (mean and standard deviation) are used to quantify accuracy and precision for the predictive methods.

When foundation support is provided by drilled shafts or piling, the engineer must select from predictive methods that differ in their ability to predict accurately the axial capacity, but also differ in the cost to use the method. Thus, the engineer must consider the accuracy of the method, as well as the expense required to use the method. For example, load tests provide an excellent means to calibrate a method at a site, but can be costly and time consuming. Estimates of capacity for driven piling can be determined using simple dynamic formulae or by monitoring pile accelerations and strains during pile driving, but each method

----

[1] Associate Professor of Civil Engineering, University of Illinois, Department of Civil Engineering, 205 North Mathews, Urbana, Illinois 61801
[2] Engineer, Thatcher Engineering Corporation, Gary, Indiana 46406

requires a different investment of resources to obtain predictions. Accordingly, the engineer is required to select a technique that provides the greatest reliability for a reasonable cost. Two collections of load test data, one on drilled shafts and one on driven piles, are used to compare predicted versus measured capacity.

PREDICTION OF CAPACITY FOR DRILLED SHAFTS

As a part of a research effort at the University of Illinois, results of axial load tests on drilled shafts were collected with the purpose of assessing predictive methods. Basic information recorded for each load test includes details of the shaft geometry, construction procedures, maximum load, maximum deflection, and general soil description. This study focuses on a filtered database satisfying the following criteria:
- embedded length > 4.5m
- tip diameter > 0.3m
- tests in end bearing must reach a deflection of at least 4% of the tip diameter, or display a "plunging" load-deflection relationship
- tests with side resistance only must reach a deflection of at least 13mm
- soil strength information must be documented and classify as soil
- shafts fail by exceeding soil bearing capacity (not structural capacity)

The first two criteria exclude model tests and shaft behavior dominated by near-surface soils. The second two require a test to reach sufficient displacement to study bearing capacity. The final two criteria are self-explanatory.

The authors have collected and reviewed over 1200 axial load tests and over 200 have met the criteria above. Using different datasets is necessary to illustrate the performance of the several predictive methods investigated in this study. For the study reported in this paper, 6 subset of this database are used. These are as follows:

177: consists of 177 axial load tests on drilled shafts meeting criteria above

122: consists of the 177 database with tests removed in which problems with construction or soil investigation were reported or were apparent

44: only instrumented shafts of the highest quality

Alpha: shafts in clay in which side resistance could be determined and alpha calculated

Beta: shafts in clay in which side resistance could be determined and beta calculated

Cone: axial load tests in which results from static cone penetrometer results were available.

Traditional Methods Investigated

A traditional method refers to the state-of-practice (SOP) and is defined herein by the FHWA88 method (Reese and O'Neill, 1988) described later. Typically, axial capacities are estimated with SOP methods with knowledge of the shaft geometry and soil information including depth to groundwater, standard penetration blow counts, unconfined compressive strength or unconsolidated undrained (UU) triaxial tests, index properties, water contents, and a visual description of the soil.

Axial capacity is calculated as the sum of side resistance and end bearing. The soil conditions are then idealized as either clay or sand. Each component of resistance (e.g., end bearing in sand) was analyzed individually, comparing the FHWA88 method with other existing methods (Wysockey, 1999). Space limitations preclude inclusion of the analysis in full, but as an example, side resistance in clay is described below. The results are summarized in the section entitled modified FHWA88 method.

Side resistance in cohesive soil. Two methods that have been proposed for estimating the side resistance for drilled shafts embedded in clay soils are termed α-methods and

β–methods. The α-method was originally developed for driven piles (Tomlinson, 1957), but has been extended to drilled shafts. The method relates side resistance, f_s, with undrained strength, s_u, using the empirical reduction factor, α. Chen & Kulhawy (1994) offer several correlation equations, which are similar in form to Tomlinson's original curve. The following equation gave the best statistical results based on the methods described later in this paper:

$$\alpha = 0.31 + 0.17(p_a / s_u) \leq 1 \tag{1}$$

where $f_s = \alpha \cdot s_u$, p_a is atmospheric pressure. Chen & Kulhawy (1994) recommend using s_u values from isotropically consolidated, undrained triaxial compression tests. While this may be a worthwhile recommendation, such test results were available for few sites in the database. The important issue is that a consistent and relevant measure of strength be used. The standard undrained strength used herein is the s_u(UU).

The FHWA88 method simply uses a constant α of 0.55, but neglects any contribution from side resistance along the top 1.5 m (5 ft) and for a distance of one diameter up from the tip. The FHWA88 method limits the side resistance to 260 kPa.

The two α–methods exhibit similar scatter in their predictions of capacity for drilled shafts in the database. The similarity is because drilled shafts installed in clays are typically installed in stiff or hard clays where both methods will predict relatively constant α values. Shown in Figure 1a is a plot of predicted capacity versus measured capacity for the Chen and Kulhawy α-method. The solid 45° line represents a line of perfect agreement between predicted capacity (Q_p) and measured capacity (Q_m). The 45° dashed lines plotted above and below represent $Q_p = 2Q_m$ and $Q_p = \frac{1}{2}Q_m$, respectively.

Beta methods estimate the side resistance (f_s) as equal to the product of the vertical effective stress (σ_v') and the beta value:

$$f_s = \sigma_v' \cdot \beta \tag{2}$$

where $\beta = K_s \cdot \tan \phi'$, K_s is the ratio of horizontal to vertical effective stress, and ϕ' is the effective stress friction angle. While the concept of using a β-method for cohesive soil has been around for a long time (Chandler, 1968), the lack of specific usable recommendations for assessing the necessary soil properties has limited its use in standard practice. Chen & Kulhawy (1994) offer easy to follow procedures requiring only the vertical effective stress, water content, and Atterberg limits. This simplified procedure requires some imperfect correlations, which are reflected in the results. Results for predicted versus measured capacity are shown in Figure 1b. The plot shows more scatter than with the alpha method.

FHWA88 method. Specific details for the FHWA88 method are presented in the original reference (Reese and O'Neill, 1988). The method employs an α-method for predicting side resistance of the shaft in clay (described above). End bearing (q_{eb}) in clay is estimated as $9s_u$. End bearing in sand is estimated from results of the standard penetration test:

$$q_{eb}(kPa) = 58N \tag{3}$$

where N is the standard penetration value. N is limited to values less than 75. The side resistance in sand is found from a β-method for which β depends on depth only:

$$\beta = 1.5 - 0.245\sqrt{z} \tag{4}$$

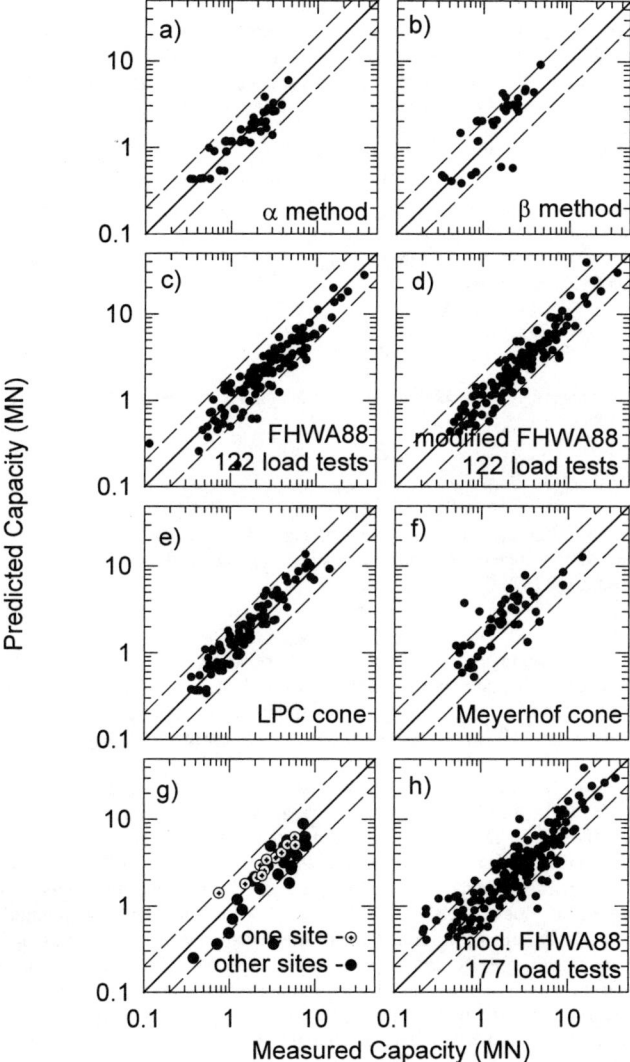

Figure 1. Plots of predicted capacity versus measured capacity

where z is the depth in metres. The value of f_s for depths greater than 26 m is equal to f_s at 26 m. The value of β has upper and lower limits which are $1.2 \geq \beta \geq 0.25$. The method was intended by the original authors to be conservative in sand. Predicted versus measured capacity is shown for the FHWA88 method in Figure 1c using the 122 database. Most of the data fall within the $Q_p = 2$ to $\frac{1}{2} Q_m$ range identified with the dashed lines. The data tend to plot slightly below the $Q_p = Q_m$ line, and therefore, the method underpredicts capacity.

Modified FHWA88. Modifications to the FHWA88 method were developed which result in a decrease to the scatter in Q_p/Q_m (Wysockey, 1999). The modified method predicts side resistance in clay using the entire shaft surface and the variable α of Chen & Kulhawy, 1994. No change is made for end bearing in clay. Side resistance in sand is predicted using standard penetration test results N_{60} (where N_{60} represents an attempt made to correct the delivered energy to 60% theoretical) as:

$$f_s(kPa) = 4N_{60} \tag{5}$$

where $N_{60} \leq 75$. End bearing in sand is calculated in a manner similar (but with a different constant) to FHWA88 as

$$q_{eb}(kPa) = 67N_{60} \tag{6}$$

with a limit of $N_{60} \leq 75$. Predicted versus measured capacities are shown in Fig. 1d. The modified FHWA88 method appears to predict measured capacity well with most of the data plotting between the $Q_p = 2$ to $\frac{1}{2} Q_m$ lines.

Cone Methods

There are many advantages for correlating cone resistance with ultimate capacity of drilled shafts that include: 1) the cone penetrometer is a simple, fast, and economical test, 2) it provides a continuous measurement of in-situ soil resistance with depth (good profiling tool), 3) the cone is particularly valuable in cohesionless soils where retrieving undisturbed samples is difficult.

A cone database (Alsamman, 1995) was used to assess the accuracy of predictive methods that use cone results directly for estimating capacity. Ninety-five full-scale load tests were collected that satisfied the general criteria and reported results of static cone penetration. Results of predicted versus measured capacity are shown for the Meyerhof cone method (Meyerhof, 1976) in Fig. 1f. The data exhibit a fair amount of scatter and there appears to be a tendency for the data to plot slightly above the $Q_p = Q_m$ line. The Meyerhof method was developed only for sands. Results of predicted versus measured capacity for the LPC method (Bustamante and Gianeselli, 1982) are shown in Fig. 1e. The LPC method includes relationships for side resistance and end bearing as a function of cone tip resistance for both sand and clay. Accordingly, more load tests are plotted in the figure than for the Meyerhof method. The LPC method shows excellent agreement between predicted and measured capacity.

Site Specific Variability

Load tests are conducted to improve the economy for a foundation, or to improve estimates of safety for a specific site. While one drilled shaft has its capacity measured, it must be assumed that the other shafts at the site will have approximately the same capacity. The site variability can be assessed in cases with several load tests at a single site.

Less scatter is expected when results of predicted versus measured capacities are plotted for a specific site because many of the parameters that affect capacity, such as soil conditions, construction details, etc. vary over a smaller range than when considering a complete database of load tests. Shown in Fig. 1g are the results of predicted (using the FHWA88 method) versus measured capacity for a database on high-quality load tests (Long and Shimel, 1989). The solid circles represent predicted versus measured capacity for the entire collection of load tests, and the hollow circles represent drilled shaft tests at a single site in London (Whitaker and Cooke, 1966). The decrease in scatter from the general database to the site specific group shows the effect of calibrating a prediction method for a given site with a load test. In this example the site conditions are uniform, but the shafts varied in length and shape (shafts were belled or straight sided).

Construction Effects

The 177 database includes all drilled shafts load tests in which information reported is adequate to allow an estimate of capacity using the modified FHWA88 method. The database contains all load tests meeting this requirement, regardless of the quality of construction or the quality of soil information. Accordingly, the 177 database contains drilled shafts in which poor construction methods were reported (e.g., shafts without cleanout, excavated hole left open for days, etc.). Results of predicted versus measured capacity are shown in Fig. 1h. This figure should be compared with Fig. 1d which uses the 122 database (the 122 database excludes shafts with obvious problems with poor construction or soil information). The scatter in Fig. 1h is greater and the majority of the data indicate overprediction, whereas the 122 database exhibits less scatter, with a slight tendency to for the method to underpredict capacity.

COMPARING PREDICTED AND MEASURED CAPACITIES

Inspection of Figs 1a to 1h can provide the reader with a subjective assessment of the accuracy for the methods presented above. The plots can be used to visually determine trends for the method such as a method's tendency to over- or under-predict capacity, and a method's tendency to exhibit scatter. The ability for a method to predict axial capacity accurately is important, but a degree of uncertainty (scatter) is associated with every method. Accordingly, the accuracy of a method is important to objectively quantify the uncertainty, and to allow comparison of the uncertainty associated with several different methods. In this section, the accuracy of these predictive methods are quantified and compared.

Statistical Interpretation

Bias and precision are two statistical parameters useful for defining a method's ability to predict capacity. Bias is a systematic error between the average ratio of Q_p/Q_m and the ideal ratio of Q_p/Q_m (which is unity). Statistically, the bias can be estimated with a sample mean. Precision is the scatter or "variability of a large group of individual test results obtained under similar conditions" (ASTM, C670-90a). Statistically, precision can be estimated with a sample standard deviation. The two terms, mean and standard deviation are defined in detail below. The distribution of Q_p/Q_m is log-normal (Cornell, 1969). A log-normal distribution means that the values of $\ln(Q_p/Q_m)$ are normally distributed. Accordingly, we can estimate the mean and standard deviation for the $\ln(Q_p/Q_m)$ for the predictive measures as a means to assess the bias and precision.

The mean value (μ_{ln}) of $\ln(Q_p/Q_m)$ is calculated as

$$\mu_{\ln} = \frac{1}{n}\sum_{i=1}^{n}\ln\left(\frac{Q_p}{Q_m}\right) \tag{7}$$

where n is the number of observations. The mean can be converted to an arithmetic equivalent (μ) by the following:

$$\mu = e^{\mu_{\ln}} \tag{8}$$

A mean value equal to zero ($\mu = 1$) represent that, on the average, predicted capacity equals measured capacity. For $\mu < 1$, the method, on the average, underpredicts capacity, and the method, on the average, overpredicts capacity if $\mu > 1$. A measure for scatter exhibited by a predictive method can be quantified with a standard deviation ($\sigma_{\ln}$). The equation for standard deviation for the $\ln(Q_p/Q_m)$ is as follows:

$$\sigma_{\ln}^2 = \frac{1}{n-1}\sum_{i=1}^{n}\left(\ln\left(\frac{Q_p}{Q_m}\right)_i - \mu_{\ln}\right)^2 \tag{9}$$

Values of mean and standard deviation for drilled shafts are given for each of the predictive methods in Table 1. Mean values vary from 0.82 to 1.32 and $\sigma_{\ln}$ values vary from 0.2 to 0.5.

Table 1. Bias and precision for predictive methods

Method	μ	$\sigma_{\ln}$	# of tests	WCI (99%)	Database
alpha	0.98	0.269	43	1.8	Alpha
beta	1.32	0.504	35	3.4	Beta
FHWA88	0.82	0.398	122	2.5	122
mod.FHWA88	0.93	0.331	122	2.1	122
LPC Cone	1.11	0.287	95	1.9	cone – all
Meyerhof cone	1.30	0.485	48	3.2	cone – sand only
FHWA88	0.80	0.460	44	3.0	44
Load test – Whitaker	1.09	0.204	12	1.6	44
mod. FHWA88	0.99	0.495	177	3.3	177

Cumulative Distribution

An additional plot is used to help compare and quantify the ability of a method to predict capacity. The plot is constructed for each predictive method by sorting the Q_p/Q_m data from smallest ratio to largest ratio and numbering each Q_p/Q_m value from $i=1$ to n. A cumulative probability value (CP) for each Q_p/Q_m ratio is determined as

$$CP_i = \frac{i}{n+1} \tag{10}$$

Values of Q_p/Q_m versus cumulative probability are plotted on probability paper as shown in Fig. 2 for the modified FHWA88 method using the 177 database. It can be seen that the data follow approximately a straight line which indicates a log-normal distribution. The plot can also provide a relationship between Q_p/Q_m and shaft reliability. For example, it can be seen that a cumulative probability of 50 percent corresponds to a Q_p/Q_m of 0.95. This means 50 percent of the time, the predictive formula underpredicts capacity by more than a factor of 0.95. In addition, a factor of 3.0 corresponds to a cumulative probability of approximately 97 percent. Thus, if a shaft with a 97 percent reliability is desired, the pile would require a predicted capacity three times greater than needed.

Cumulative distributions are plotted for all the methods presented above (Fig. 3a-h). These plots allow assessment of the distribution of Q_p/Q_m, determination of the value of Q_p/Q_m at 50% probability, and assessment of the method's precision (steeper slopes reflect greater precision). It can be observed that each of these methods results in a different mean (value at CP=50%). Estimated probabilities at Q_p/Q_m=3 vary from 95 percent to 99.99 percent. It can also be seen that values of Q_p/Q_m required for a cumulative probability equal to 99 percent vary significantly for each method range from 1.7 to 4.3. This large range illustrates the importance of the bias and precision of a method in determining the relationship between Q_p/Q_m and cumulative probability.

Observations

Not surprisingly, the steepest slope (least scatter) in Figs. 3a-3h is exhibited by load tests at a single site (Fig. 3g). Since the tests are associated with specific soil conditions, construction conditions, etc, the result is a low degree of scatter, and a high degree of confidence in predicted values. The LPC method exhibits less scatter than the rest of the methods. The good predictions are because the cone is an excellent profiling tool, but also because the soil conditions are limited to the types of soils through which a cone can be penetrated. Very stiff, hard soils that pose difficulties in characterizing strength, are not considered in the cone method. By contrast, the Meyerhof method exhibits a more significant scatter and a significant bias. On the average, the method overpredicts capacity by 30 percent.

The cumulative distribution plot for the alpha method exhibits a much steeper slope than the beta method, meaning the beta method exhibits significantly greater scatter. These results confirm the observation of greater scatter made when comparing Q_p versus Q_m plots in Fig. 1. Comparison of cumulative distribution plots for FHWA88 methods (Fig. 3c) with modified FHWA88 (Fig. 3d) show the modified FHWA88 method to have slightly less scatter.

A comparison of the cumulative distribution plots for the 177 versus 122 database illustrates the value of obtaining

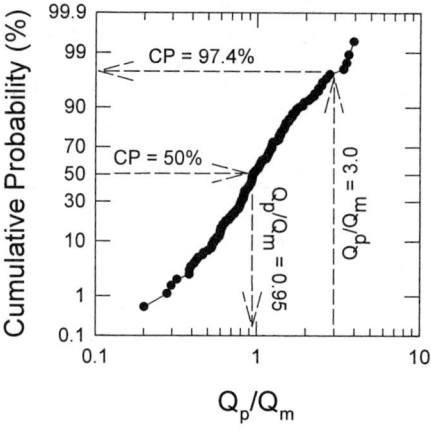

Figure 2. Cumulative probability plot for FHWA88 method with 177 database

good quality soil information and good construction procedures. The 122 database results plot significantly steeper than for the 177 database. For cumulative probabilities less than 50 percent, the results of the 122 database plot below the 177 database. This means predicted capacities using the 177 database were more wasteful than predictions with the 122 database. This difference is due to poor quality soil information where strengths of soil were

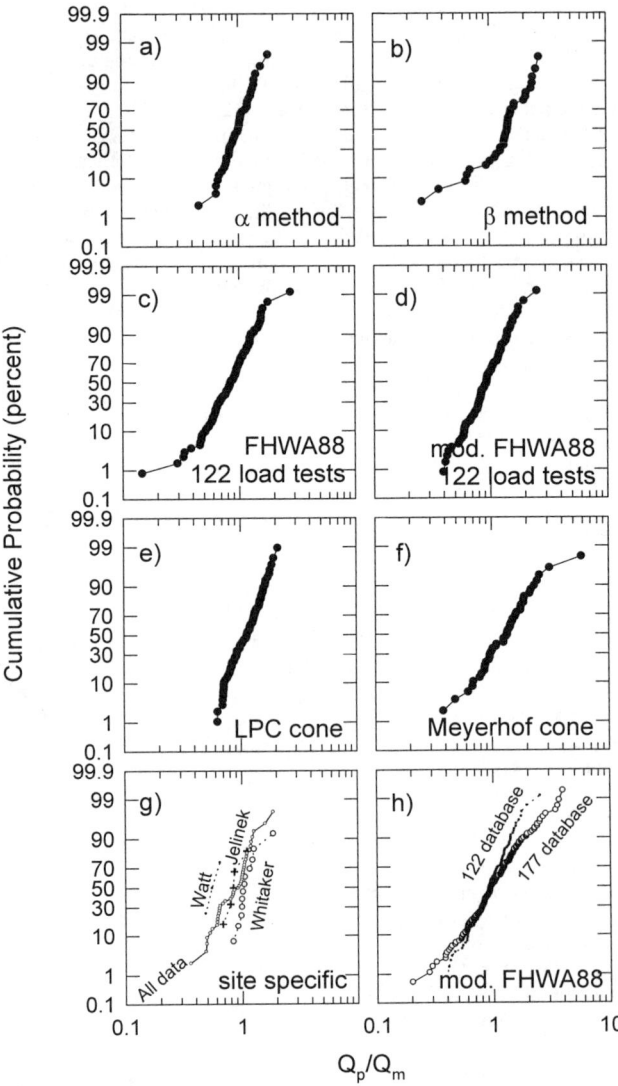

Figure 3. Cumulative distribution plots for Q_p/Q_m

underestimated in the load tests reports, and accordingly, resulted in underpredicting capacity. For cumulative distributions greater than 50 percent, the results from the 177 database plot to the right of the 122 database. The 177 database includes shafts constructed poorly, with a correspondingly reduced capacity, and resulting in the tendency for overestimation of axial capacity.

EFFICIENCY OF PREDICTIVE METHODS

It is important to quantify the cost effectiveness for a specific predictive method so that decisions can be made whether to load test, invest in additional soil tests, or remain with a traditional method. Methods that exhibit less scatter require a smaller factor of safety to achieve a target foundation reliability, and therefore can provide better economy. However, additional costs may be associated with using methods that are more accurate. For example, load tests provide the most reliable method but also require a significant investment and time to conduct the load test. The LPC cone method also provides reliable estimates, but requires results of cone penetration tests. Some methods require similar soil information, but exhibit very different reliability (e.g. LPC and Meyerhof cone methods require the same information, but LPC results in more accurate predictions).

It is impossible to compare the actual cost of these methods in a generally applicable way because the expense of conducting a load test, cone tests, getting more detailed soil information depends on factors that vary with location, availability, and time. Furthermore, the cost for acquiring additional capacity is impractical to estimate in general, since it too is site specific. Additional capacity may be acquired by requiring additional foundations, or the same foundation extended to a greater depth, or a myriad of other solutions available to the engineer. Therefore, the approach taken here is to estimate "wasted capacity" for a method, which is an inefficiency term that will be quantified later. So while the increased cost of requiring more capacity is very site dependent, on the average, wasted capacity can serve as a general indicator of increased cost.

Wasted capacity refers to the extra capacity for which a foundation must be designed to account for uncertainties. The concept of wasted capacity is illustrated with an example for a prediction method with a value of Q_p/Q_m equal to unity ($\mu=1.0$) and standard deviation ($\sigma_{ln}=0.46$). A cumulative distribution curve with these statistical parameters would show $Q_p/Q_m = 2.5$ at a cumulative probability of 97.7 percent. Thus, a pile predicted to carry a load of $2.5Q$ has a 97.7 percent chance of resisting the load Q without failing. However, there is a continuous distribution of Q_p/Q_m. Thus, some piles exhibit a $Q_p/Q_m = 0.2$ while other piles may exhibit a $Q_p/Q_m = 2.0$. Piles plotting on the distribution curve with $Q_p/Q_m = 2$ are overdesigned by a factor of 1.25 (=2.5/2). Piles with $Q_p/Q_m = 0.75$ are overdesigned by a factor of 3.3 (=2.5/0.75). Thus, overdesign, or Wasted Capacity for a particular Q_p/Q_m can be calculated as the ratio of $(Q_p/Q_m)_{required}$ divided by the value of Q_p/Q_m.

The overall measure of wasted capacity must consider both the probability associated with Q_p/Q_m and the wasted capacity associated with each value of Q_p/Q_m. A Wasted Capacity Index (WCI) is determined by summing the product of probability ($p(x)$) and wasted capacity from zero to Q_p/Q_m required for a given foundation reliability. Mathematically, WCI is determined as:

$$WCI = \int_0^{(Q_p/Q_m)_{reqd}} p(x) \frac{(Q_p/Q_m)_{reqd}}{x} dx \qquad (11)$$

The distribution of probability for Q_p/Q_m ($p(x)$) is controlled by μ_{ln} and σ_{ln}. The plot of Wasted Capacity is determined from $(Q_p/Q_m)_{reqd}$, which is governed by μ_{ln}, σ_{ln}, and the selected reliability. However, the Wasted Capacity Index is a function only of σ_{ln} and reliability. The relationship of WCI versus standard deviation (for a predictive method) is shown in Fig. 4 for pile reliabilities of 50 percent, 99 percent and 99.9 percent.

The WCI can be used to quantify wasted capacity associated with a method, and therefore allows comparisons of different methods. For example, assume a foundation requires a reliability equal to 99 percent and we want to compare the difference in wasted capacity between a traditional method and a cone-based (LPC) method. The standard deviation for the

the cone-based method is approximately 0.29 (Table 1). Starting at a standard deviation of 0.29 (Fig. 4) and moving vertically until intersecting the 99 percent reliability curve, we determine the WCI to be 1.9. For comparison with the FHWA88 method, the standard deviation is approximately 0.4, which corresponds to a WCI equal to 2.5. Thus, on the average, the FHWA88 method wastes about 1.3 times the capacity (=2.5/1.9) than the LPC cone-based method.

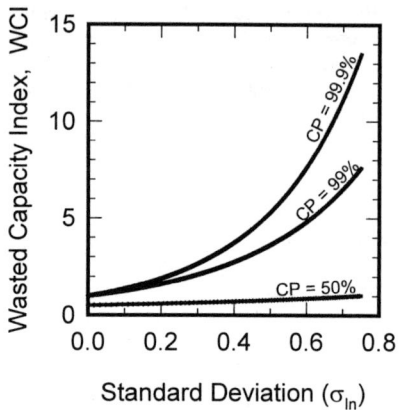

Figure 4 WCI versus σ_{ln}

PREDICTION OF CAPACITY FOR DRIVEN PILING

The remaining portion of this paper identifies the value of predicting capacity using dynamic formulae, wave equation analysis, and from dynamic monitoring without load tests. Accordingly, methods investigated herein are grouped into two categories: methods that base their predictions on field observations of penetration resistance, and methods that require measurements of the temporal variation of pile force and acceleration during driving.

The predictive methods investigated herein for estimating capacity based on field measurements of penetration resistance are: 1) EN (Engineering News) formula (Wellington, 1892), 2) Gates formula (Gates, 1957), and 3) WEAP (Wave Equation Analysis Program) (Goble and Rausche, 1986). The predictive methods investigated that require field measurements of pile acceleration and pile force with time are: 4) PDA (Pile Driving Analyzer) (Pile Dynamics, Inc., 1995), 5) Measured Energy (ME) approach (Paikowsky et al, 1994), 6) CAPWAP (Hannigan, 1990).

The first three predictive methods estimate pile capacity using simple field measurements of driving resistance and driving energy. The EN and Gates methods are simple dynamic formulae that require only the hammer energy and pile set (or blow count). WEAP requires numerical modeling of the pile hammer, the pile, and the soil, and develops a relationship between capacity and driving resistance. The remaining three methods (PDA, Measured Energy Method, CAPWAP) require detailed measurements of the temporal variation of pile force and velocity during driving.

Each method can use results of pile driving resistance at the end of driving (EOD) or beginning of restrike (BOR). Analysis of restrike is preferred since time effects can influence significantly the final capacity for a pile. Because of tight construction schedules, it may be common to restrike after 24 hours, but the capacity will continue to change after 24 hours

(Long et al., 1999). A detailed description for each predictive method is beyond the scope of this paper and may be found in the appropriate references cited for each method.

A collection of approximately one hundred load tests from an FHWA database (Rausche et al, 1996) was used in this study. Results from each load test could not be used for every method because some lacked information needed for a specific method.

Comparison of Prediction Methods

The ability of a method to predict axial capacity accurately is important to quantify for two purposes: to assess the effect of uncertainty associated with a specific method, and to allow comparison of the uncertainty associated with several different methods. Davisson's (1973) method is used to define the failure load from static load test results.

A plot of Q_p vs Q_m for the six predictive methods is shown (Fig. 5a-5f) to illustrate the agreement between predicted capacity and measured capacity. Data obtained at end of driving (EOD) were used to determine capacity for each method. All piles in the database from which predictions could be made were used in each plot. Predictions using the EN formula used weight of ram times drop as the measure of energy when the information was available, and rated energy when the information was not available or inappropriate. Prediction of capacity using the Gates method only considered tests in which the energy could be determined as weight of ram times drop.

The EN method is shown, in general, to overpredict measured capacity significantly (Fig. 5a) and, on the average, overpredicts the measured capacity by 2.6. The Q_p versus Q_m plot exhibits significant scatter. The performance of the Gates formula is shown in Fig. 5b. The method generally underpredicts capacity (average Q_p/Q_m =0.49). Capacities predicted using WEAP (Fig. 5c) generally underpredict capacity (average Q_p/Q_m=0.64), while the ME approach (Fig. 5d) exhibits good agreement between predicted and measured (average Q_p/Q_m = 0.93), and a small amount of scatter. The performance of PDA is shown in Fig. 5e. The method slightly underpredicts measured capacity (average Q_p/Q_m = 0.71) while for the CAPWAP method (Fig. 5f), considerable scatter is shown with the average ratio (Q_p/Q_m) for the CAPWAP method is near 0.58. Capacities for most of the methods are underpredicted, in part, because pile capacities usually increase with time after driving. The statistics for the EOD methods are given in Table 2.

TABLE 2. Statistics for predictive methods (piles)

Datatset	Method	EOD/BOR	n	μ	σ_{ln}	WCI
all	EN	EOD	123	2.60	0.675	5.7
all	GATES	EOD	51	0.49	0.323	2.1
all	WEAP	EOD	88	0.64	0.501	3.3
all	ME	EOD	73	0.93	0.462	3.0
all	PDA	EOD	77	0.71	0.454	2.9
all	CAPWAP	EOD	75	0.58	0.591	4.4
all	EN	BOR	75	4.03	0.491	3.3
all	GATES	BOR	40	0.63	0.328	2.1
all	WEAP	BOR	114	1.11	0.385	2.4
all	ME	BOR	92	1.41	0.363	2.3
all	PDA	BOR	85	0.91	0.319	2.1
all	CAPWAP	BOR	112	0.86	0.269	1.8

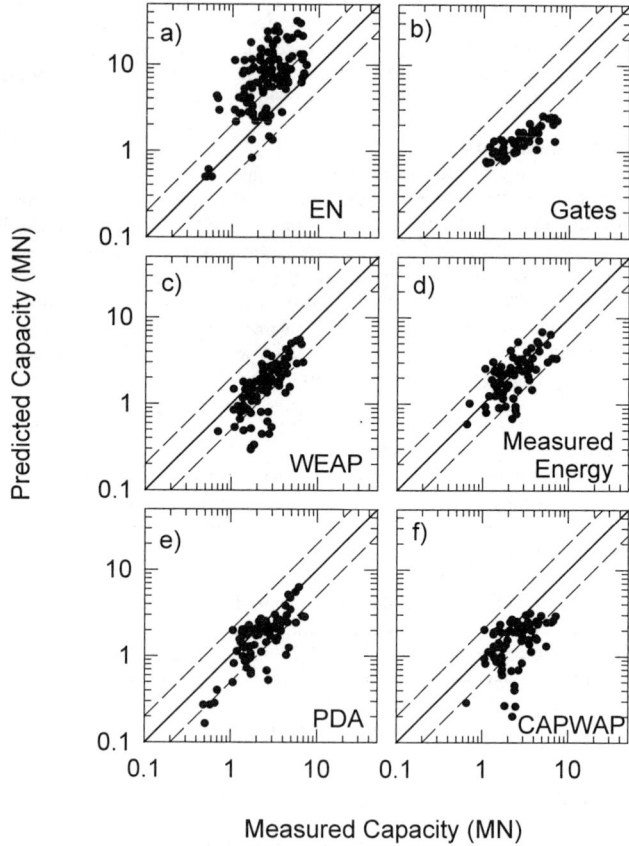

Figure 5 Predicted vs measured capacity for piles (EOD)

The six methods were also used to predict capacity for all available piles using data available at beginning of restrike (BOR). BOR driving information may provide superior estimates (to EOD information) because time effects on pile capacity can be observed. Predicted versus measured relationships for the six methods are shown in Fig. 6a-6f. Use of BOR data results in reduced scatter for every method except Gates. Furthermore, most predictions of capacity are greater and therefore closer to the measured capacity (average Q_p/Q_m closer to 1). The largest reduction in scatter is observed for predictions made with CAPWAP.

Values of the converted mean(μ) and standard deviation (σ_{ln}) are given for each of the predictive methods in Table 2. The associated WCI for each method for a reliability of 99 percent is also shown in Table 2. These tables can assist an engineer in deciding whether the greater level of effort to obtain PDA estimates justifies less waste of pile capacity. When EOD data are used, methods that require a greater level of effort to monitor and model pile

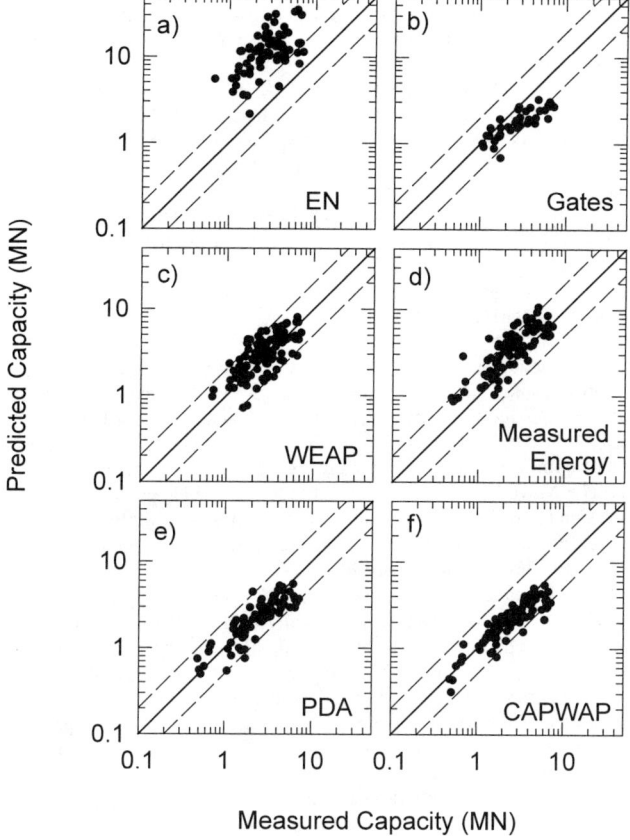

Figure 6 Predicted vs measured capacities for piles (BOR)

driving do not necessarily provide a lower WCI. However, when BOR data are used, there is generally a benefit (lower WCI) associated with dynamic monitoring and greater attention to modeling of pile driving.

LIMITATIONS

This study only includes load tests conducted to failure. A more complete database would include all load tests whether they failed or not. Including unfailed load tests would serve to decrease the mean and standard deviation; therefore, a database including only failed load tests is inherently conservative.

There are inevitable limitations due to specifics of the load tests, soil types, construction methods, and shafts geometry in this study. In addition, this study neglects uncertainties in the measured load and excludes specific details for each soil type and installation method. Finally, the database is limited in its ability to illustrate the effect of variation in soil parameters within a site. Some geologic profiles exhibit more variability than others. There are also inevitable limitations due to specific distributions of the load tests, soil types, and pile geometry in this study.

Some examples use reliabilities equal to 99 percent which require extrapolation. Over 100 load tests would be required to allow estimating a reliability of 99 percent without extrapolating. Thus, extrapolating to 99 percent assume that all scatter is random and extrapolation to 99 percent is representative of the extreme values. Furthermore, the method provides guidance for selecting the appropriate resistance factor for the foundation, but does not include uncertainties associated with the design load. The reliability of a foundation system depends on both the uncertainty in the ability to predict capacity and the ability to predict the design load.

SUMMARY AND CONCLUSIONS

Conducting load tests at a site provides the greatest accuracy for predicting the load carrying capability of a foundation system. Without conducting load tests, the LPC cone method provides the most accurate estimates for capacity of drilled shafts. Traditional methods, such as the FHWA88 method provide a reasonable degree of accuracy, but some modifications (modified FHWA88 method) seem to improve its accuracy. Some methods are best avoided. The Meyerhof cone method exhibits significant scatter. Beta methods (using index properties and correlations) for shafts in clay exhibit poor accuracy. No conclusion can be made for the β-method in which parameters are determined directly (e.g. K_s and ϕ') since the databases do not contain this information.

The ability of EN, Gates, and ME methods to predict pile capacity accurately benefits little from using BOR data. WEAP, PDA, and especially CAPWAP benefit from the use of BOR data. Pile capacity using EOD data is shown to predict with about the same precision with Gates, WEAP, ME, PDA, and CAPWAP. The ME method appears provides the most precise measurements when only EOD data are used. The database shows that use of CAPWAP with BOR data to predict capacity results in the greatest precision of all predictive methods investigated. For non-monitored methods, BOR results generally exhibited slightly less scatter than when using EOD data.

The "cost" for using a method is expressed as a wasted capacity index (WCI) which is a function of the reliability required for a foundation and the precision with which capacity can be determined. Thus, significant reduction in WCI can result from using methods that are more accurate. While WCI and cost are related, the two are not directly proportional. Site specific details will govern whether requirements for extra capacity will be relatively expensive or inexpensive.

REFERENCES

Alsamman, O.M. (1995), "The use of CPT for calculating axial capacity of drilled shafts," Ph.D. Thesis, Univ. of Illinois Urbana-Champaign, 300p.

American Society for Testing Materials, "Practice for Preparing Precision Statements for Test Methods for Construction Materials, C670-90a, Annual ASTM Book of Standards, Vol. 04.08, 1990.

Bustamante, M. and Gianeselli, L. (1982), "Pile bearing capacity prediction by means of static penetrometer CPT," *Proceedings*, 2nd European Symposium on Penetration Testing, Amsterdam.

Chandler, R.J. (1968), "The shaft friction of piles in cohesive soils in terns of effective stress," *Civil Engineering And Public Works Review*, v. 63, pp. 48-51.

Chen, Y.-J. and Kulhawy, F.H. (1994),"Case history evaluation of the behavior of drilled shafts under axial and lateral loading," EPRI Research Project 1493-04, Final Report.

Cornell, C. A. (1969), "A probability based structural code," *J. American Concrete Institute*, pp. 974-985.

Davisson, M. T. (1973), "High Capacity Piles," Innovations in Foundation Construction, Soil Mech. Div., Ill Section, ASCE, Chicago, pp. 81-112.

Gates, M. "Empirical Formula For Predicting Pile Bearing Capacity," Civil Engineering, Vol. 27, No.3, March 1957:65-66.

Goble, G. G. and F. Rausche, "Wave Equation Analysis of Pile Driving - WEAP86 Program," U.S. Department of Transportation, Federal Highway Administration, Implementation Division, McLean, Va., Volumes I-IV. 1986.

Hannigan, P. J., "Dynamic Monitoring and Analysis of Pile Foundation Installations," Deep Foundations Institute Short Course Text, First Edition, 1990, 69p.

Jelinek, R., Koreck, H.W., and M. Stocker (1977), "Load tests on five large diameter bored piles in clay," *Proceedings*, 9th ICSMFE, Vol. 1, Tokyo, pp. 571-576.

Long, J.H. and Shimel, S. (1989), "Drilled shafts - a database approach," *Proceedings*, ASCE Foundation Engineering Congress, Northwestern Univ., Evanston, IL. v. 2, pp. 1091-1108.

Long, J.H., J. Kerrigan, and D. Bozkurt (1999), "Measured Time Effects for Axial Capacity of Driven Piling," proceedings of 1999 TRB, January, Washington, D.C.

Meyerhof, G.G. (1976), "Bearing capacity and settlement of pile foundations," *Journal of Geotechnical Engineering*, ASCE, v. 102, pp. 197-228.

Paikowsky, S. G., J. E. Regan, and J. J. McDonnell, "A Simplified Field Method For Capacity Evaluation Of Driven Piles," Publication No. FHWA-RD-94-042 , U.S. Department of Transportation, Federal Highway Administration, McLean, Virginia, September 1994.

Pile Dynamics, Inc., "Pile Driving Analyzer Manual; Model PAK," Cleveland, Ohio.

Rausche, F., G. Thendean, H. Abou-matar, G.E. Likins, and G.G. Goble (1996), Determination of Pile Driveability and Capacity from Penetration Tests, Final Report, U.S. Department of Transportation, Federal Highway Administration Contract DTFH61-91-C-00047.

Reese, L.C. and M.W. O'Neill (1988), *Drilled Shafts: Construction Procedures and Design Methods*, ADSC-TL-4, Dallas, Texas, 564 p.

Tomlinson, M.J. (1957), "The adhesion of piles driven in clay soils," *Proceedings*, 4th ICSMFE, London, v. 2, pp. 66-71.

Watt, W.G., Kurfurst, P.J., and Z.P. Zehman (1969), "Comparison of Pile Load Test-Skin Friction Values and Laboratory Strength Tests," *Canadian Geotechnical Journal*, Vol. 6, No. 3, Aug. 1969, pp. 339-352.

Wellington, A. M. discussion of "The Iron Wharf at Fort Monroe, Va.," by J. B. Duncklee, Transactions, ASCE, Vol. 27, Paper No. 543, Aug. 1892, pp. 129-137.

Whitaker, T. and R.W. Cooke (1966), "An investigation of Shaft and Base Resistance of Large Bored Piles in London Clay," *Proceedings*, Symposium on Large Bored Piles, ICE, London, Feb., pp. 7-49.

Wysockey, M.H. (1999), "The axial capacity of drilled shafts," Ph.D. Thesis, Univ. of Illinois Urbana-Champaign (in preparation).

Some Effects of Construction on the Performance of Deep Foundations

Michael W. O'Neill[1]

Abstract

Several of many potential effects of construction details on the performance of deep foundations are described, and bounds on these effects are quantified from experimental and analytical studies. For these effects (drilling slurry, borehole roughness, time of construction and method of inserting driven piles) alteration in details of the installation process can produce resistances that differ by factors of 3 to 10.

Introduction

One of the consistent theses advocated by Professor Reese, whom we honor with this symposium, is that the way in which a foundation is constructed has a profound effect on its behavior. Not only are general construction practices important, but construction details specific to every site and every project likewise play a major role in foundation behavior. In a bored pile, for example, it is not sufficient just to know that a borehole in a soil with specific shear strength and deformation properties will be excavated and filled with concrete in order to predict accurately the resistance and deformations of the pile, but it is also necessary to forecast or otherwise rationally deal with such diverse factors as

- the roughness produced by the drilling tool on the sides of the borehole,
- creation of smear on the borehole walls during excavation,
- construction-time-dependent stress relief,
- Geometric errors in installing piles in groups,
- type of drilling fluid used (if any) and time of exposure to the soil,
- effects of construction of nearby bored piles, and
- cracking strength of concrete in tension and the influence of minor construction defects on tensile strength of the concrete if the pile is loaded laterally.

The method of installation can affect the behavior of driven piles, as well. Otherwise identical piles in identical soils can exhibit different behavior if one is installed, for example, by impact driving and one is installed by vibratory driving.

---

[1] Cullen Distinguished Professor of Civil Engineering, University of Houston, 4800 Calhoun Road, Houston, TX 77204-4791

The issue of the effects of foundation construction will become important in the current progression toward reliability-based design of foundations. While these effects cannot yet be predicted deterministically for any specific project, some understanding of their general magnitudes, and perhaps of the frequency of their occurrence, can be developed through careful experimentation and analysis. It is not possible to relate the consequences of all construction effects for even one type of deep foundation in the space available in this paper; however, several of the more important ones will be addressed. The effects that will be examined are:

- the details of drilling slurry on side resistance in bored piles,
- borehole roughness and smear in rock sockets,
- stress relief in the excavation of boreholes, and
- installation of driven piles in sand by impact vs. vibration.

Drilling Slurry

Several bored piles were constructed and load-tested to failure at the Texas A and M University National Geotechnical Experimentation Site (medium dense sand) to assess the effect of the details of drilling slurry on side resistance. In design usage, the average unit side resistance along a bored pile, $f(max)_{avg}$ in sand is often expressed through Equation (1).

$$f(max)_{avg} = \beta\, \sigma'_{v\,avg} \qquad (1)$$

β is a proportionality constant relating $f(max)_{avg}$ to $\sigma'_{v\,avg}$, the mean vertical effective stress along the length of the pile. β was measured for three full-sized piles with diameters of 0.914 m penetrating to the bottom of the sand stratum, which was about 11 m thick, as follows:

(1)   Using a light bentonite-water slurry in which the bentonite colloids had fully hydrated the mixing water and had reached a Marsh funnel viscosity of 33 sec/quart before the slurry was introduced. The borehole was drilled under the slurry and was allowed to remain exposed to the slurry for 2 hours after completion of drilling. The thickness of the mudcake that was produced was less than 1 mm. By developing a careful concreting curve the pile was found to be very slightly pear shaped.

(2)   Using a heavy bentonite-water slurry in which the bentonite colloids had not fully hydrated the mixing water before introduction into the borehole but which, *at the time of introduction*, had an acceptable Marsh funnel viscosity of 50 sec/quart. In order to achieve this viscosity at an early time, the slurry had to be dosed with more bentonite than for the pile in (1). The soil on the walls of the borehole was exposed to the slurry for 72 hours before concrete was placed. At the time of concreting, the mudcake was 10 mm thick.

(3)   Using a PHPA polymer slurry that was premixed at a dosage of 0.22 N (dry polymer) / L (potable water) and which developed a Marsh funnel viscosity of 44 sec./quart. The borehole walls were exposed to the slurry for about 1 hour prior to concreting. The borehole was calipered and found to be essentially cylindrical. No mudcake could be detected, as is expected with a PHPA slurry.

Figure 1 summarizes the results of compression loading tests in terms of the developed β factors, which are compared with the β factor that can be deduced from the design procedure of Reese and O'Neill (1988). The bentonite pile with the thin mudcake and the PHPA pile developed β factors exceeding 1, which is higher than recommended by Reese and O'Neill. However, the pile that was constructed by hastily placing the mixed (overdosed) bentonite-water slurry in the borehole, even though the viscosity of the slurry at the time of placement was adequate, and exposing the soil for three days, produced a value of β of less than 0.1. The general chemical makeup of the drilling slurry thus made very little difference; however, the way in which the slurry was handled had a profound effect on side resistance.

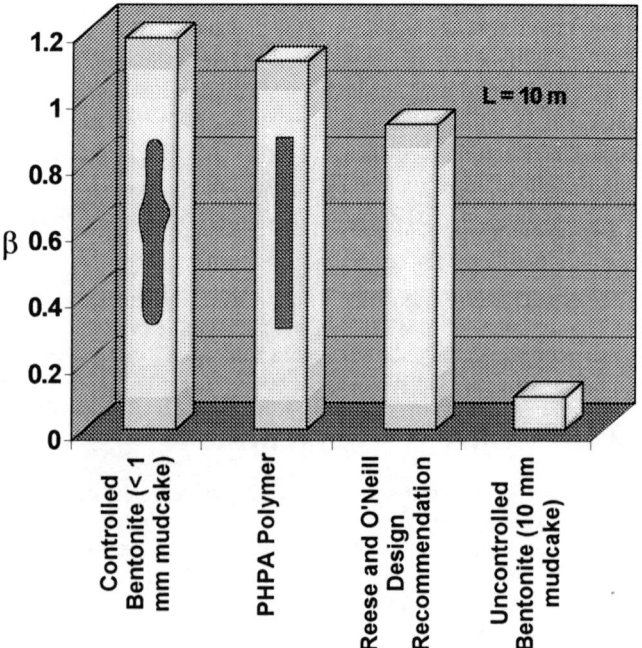

Figure 1. Comparison of β based on method of bored pile construction in sand (Ata and O'Neill, 1997a).

Further investigating the use of PHPA (partially hydrolyzed polyacrylamide) polymer slurry, a relatively new innovation in pile drilling, at the TAMU sand site, samples of soil were recovered and exposed to PHPA polymer in a direct shear box at the same dosage as above for one hour. Mortar was then cast on the interface, and the interface was sheared either 3 or 7 days after casting. After the mortar had set, the normal pressure was set equal to the vertical effective stress at the depth from which the sample was recovered. Companion samples were exposed only to water before casting the mortar. As can be seen in Figure 2, when the interfaces were sheared 3 days after being exposed to PHPA polymer slurry they exhibited lower shearing resistances than the samples exposed only to water. However, when shearing occurred 7 days after exposure of the interfaces to the polymer slurry or water, the interfaces that had been exposed to the polymer exhibited a much higher shearing resistance than the interfaces exposed to water.

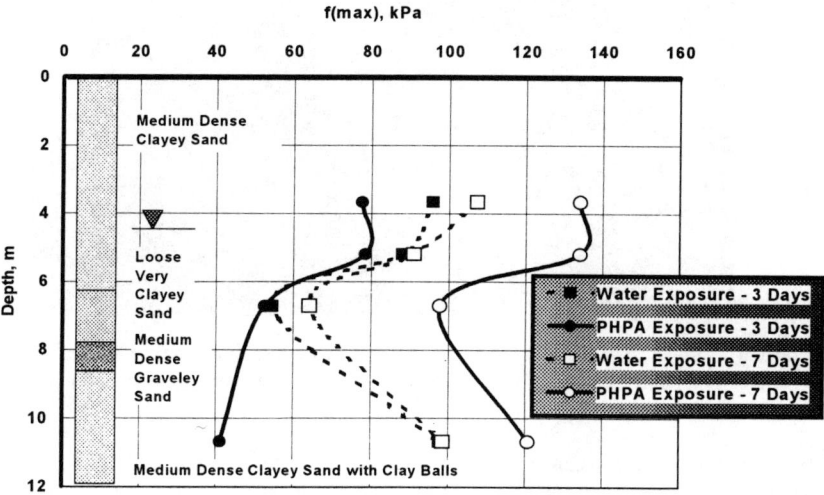

Figure 2. Interface shearing resistances for direct shear samples exposed to water and PHPA polymer before casting mortar with varying times before shearing (Ata and O'Neill, 1997b).

Figure 3 may provide a clue regarding the reason for this effect. It is a photograph of bored pile concrete cast against soil in which water had been used as the drilling fluid (left) and in which PHPA polymer slurry had been used as the drilling fluid (right). The rough surface is a combination of pitting and build-up of by-products of as-yet-unknown chemical reactions between the residual PHPA and the portland cement. There was clear evidence in scanning electron microscope studies of agglomeration of residual polymer chains near the interface caused by release of calcium ions from the cement, which may

have resulted in the initially low interface resistance (at 3 days) (Ata and O'Neill, 1997b).
It is speculated that the high concentration of residual polymer in time (7 days) drew
enough calcium out of the surface of the concrete to form the pits seen in the right image
of Figure 3, which in turn increased the shearing resistance of the bored pile at the
interface with the soil.

Figure 3. Comparative photographic images of bored pile-soil interface exposed only to
water (left) and exposed to PHPA polymer slurry (right) before casting concrete.

The developed side resistance of piles in clay soils also appears to depend upon the
manner in which the pile is constructed. Summarized in Figure 4 are the results of
measurements of average unit side shearing resistance in full-sized deep foundations in
the Pleistocene-aged Beaumont clay formation at the National Geotechnical
Experimentation Site at the University of Houston (Texas) and at a (geotechnically)
nearly identical site 5 km away. The soils are borderline CL-CH and have been
overconsolidated by desiccation. One test bored pile was constructed with PHPA
polymer, two were installed under light bentonite, five were installed dry (without
slurry), and two were driven steel pipe piles (closed toe with flush boot plate). Figure 4
shows the average unit side resistance along each group of piles through the use of the
proportionality factor α, which is defined by Equation (2).

$$f(max)_{avg} = \alpha \, s_{u \ avg} \qquad\qquad (2)$$

in which $s_{u \ avg}$ is the average undrained shear strength of the clay from UU triaxial
compression tests along the length of the pile (in the range of 80 - 120 kPa for the

Beaumont clay. The data in Figure 4 were derived from piles with diameters ranging from 273 to 914 mm and lengths ranging from 7.0 m to 24.4 m. The relative results were somewhat contrary to intuition. The piles installed under the PHPA and light bentonite slurry performed better than those drilled without slurry and the driven piles. Both the light bentonite and PHPA polymer appear to protect the soil from drying, and thus producing joints through which excess water from the concrete mix can penetrate the soil, to assist in maintaining lateral effective stresses, and to reduce interface disturbance due to drilling by lubricative action.

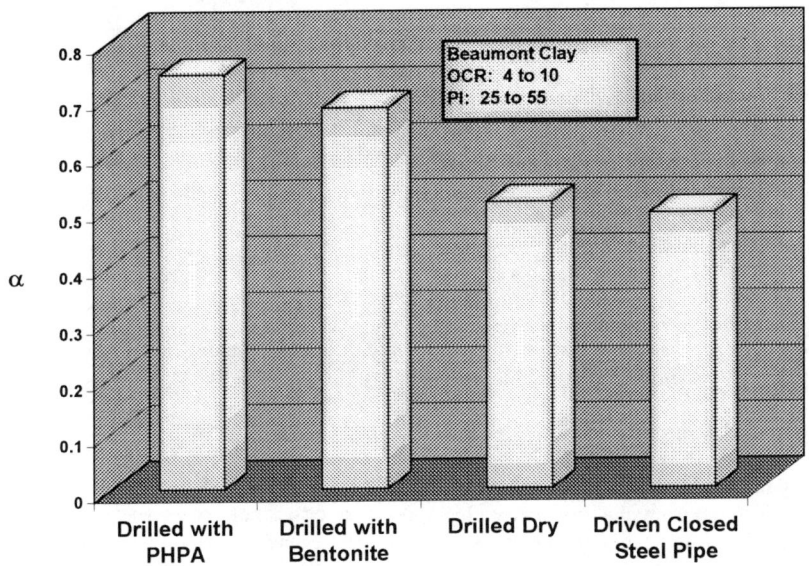

Figure 4. Comparative side load transfer in piles in overconsolidated clay.

Finally, Figure 5 shows the results of side resistance measurements in four bored piles in the Beaumont clay as a function of base configuration and method of construction. Here α is a local value that varies with depth. The curve on the left was obtained from tests on a pile that was installed with bentonite slurry, then cased, then concreted. The slurry from within the casing was removed before concreting, but some residual slurry remained in the overbreak zone. This slurry was not controlled and evidently had higher viscosity and sand content than was desirable, because some of it was trapped between the concrete and the soil after the concrete was placed and the temporary casing was removed. Therefore, the large amount of residual drilling slurry had a much more profound effect on side load transfer than any of the base conditions (underream, void or normal cylindrical base).

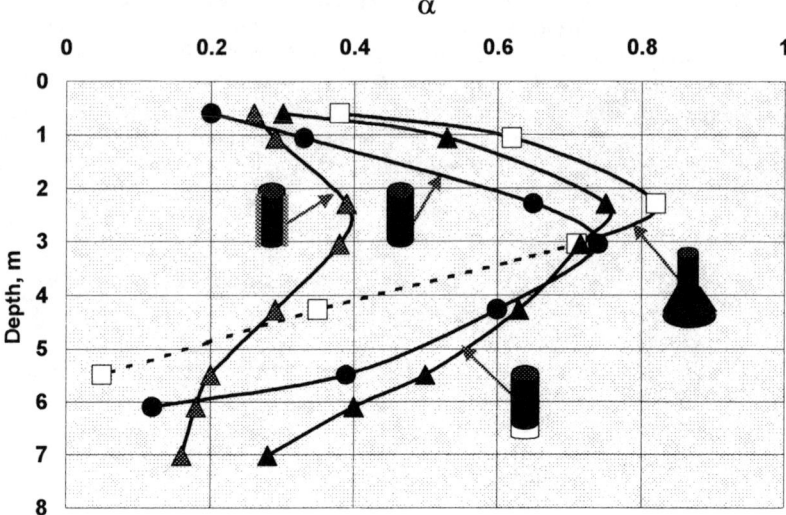

Figure 5. Depthwise variation in α in bored piles in Beaumont clay due to base
conditions and residual drilling slurry (O'Neill and Reese, 1970).

Borehole Roughness

It is generally understood that much of the side resistance developed by bored piles in
rock is derived from dilation between the concrete and rock at their interface (sides of the
bored pile). Some bored pile designers also appreciate the fact that if the rock is
degraded by drilling, mixed with water and smeared on the borehole wall by the drilling
tool, the normal stiffness at the interface will be reduced and the component of shearing
resistance due to dilation will be diminished. The smeared geomaterial may also become
part of the shear zone and further reduce load transfer. This behavior can be especially
troublesome in argillaceous rocks, such as clay shales and mudstones, where perched
water is not cut off by the contractor or where the contractor may actually add water to
the borehole to assist in picking up cuttings.

A long-wave length, sinusoidal interface pattern, approximating extreme observed
interface topography in clay shale formations in Texas that are excavated with augers
(without special rifling) for bored piles, is shown in Figure 6. The development of
shearing resistance along this interface was studied using axisymmetric FEM procedures
by Hassan (1994), who applied a simple constitutive model for the soft rock (Drucker-
Prager with non-associated flow); slide line elements along the interface to allow for
debonding and drained, frictional behavior; and elastic elements in the concrete pile. The
interface had provisions for a smear zone of soft, undrained material to approximate the

presence of smear. Similar analyses were made for smooth interfaces and for interfaces
without smear.

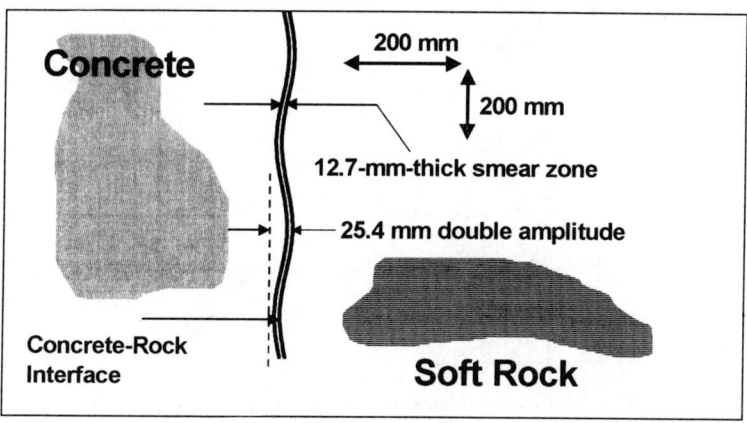

Figure 6. General conditions for modeling interface roughness and smear.

A "smooth" interface in a soft clay-shale is depicted in Figure 7, which shows the results
of measurements along one face of an extracted bored pile that was installed through clay
overburden into soft clay shale by augering. Along most of the sockdet, the double
amplitude of the radial dimension was less than 2 mm, usually less than 1 mm. The
rougher interface near the top of the socket was relatively short and presumably had little
effect on pile behavior. A wave pattern could not be detected. Therefore, "smooth" piles
were analyzed as if they were perfect cylinders.

Figure 8 shows the load-deformation behavior predicted by the FEM solution compared
with the measured behavior for the smooth-interface pile using the material parameters
shown. The mass modulus of the soft rock E_m was determined from initial pressuremeter
moduli, and the unconfined strength q_u was measured from compression tests on samples
recovered using double-walled core barrels. The factor ϕ represents the drained interface
friction angle for shearing on concrete on clay shale. The value of 25° is less than the
corresponding value measured in direct shear in the laboratory (30°), which suggests that
the interface shearing angle itself is dependent on the construction procedure used in the
field and cannot be duplicated with accuracy in the laboratory.

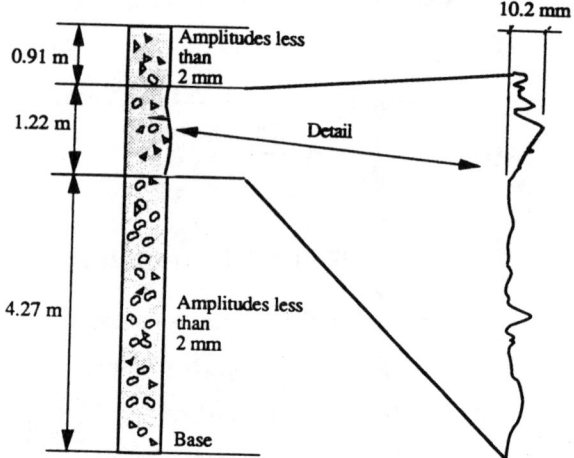

Figure 7. Interface roughness profile along auger-excavated bored pile
extracted from clay shale formation (Hassan, 1994).

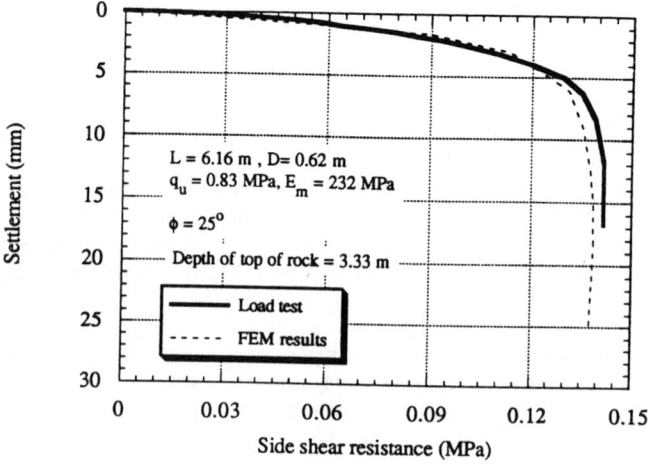

Figure 8. Comparison of measured and predicted side resistance-settlement curves for
smooth interface in soft clay shale.

By contrast, a comparison was made between measured and predicted behavior of a short socket in harder shale, in which circular grooves had been cut as shown in Figure 9. (Horvath et al., 1983). The spacing of the grooves along the axis of the socket was about 150 mm, so the grooving pattern represents a pattern generally similar to the extreme pattern shown in Figure 6 except for the 150 mm wave length. The same method of modeling the rock and the interface that was used for the pile depicted in Figure 8 was used for this pile. The comparison of load-settlement behavior is shown in Figure 10. By comparing Figures 8 and 10, the effect of interface roughness becomes apparent. No clear plunging load developed in the rough socket and the percentage of the compressive strength of the rock achieved in side shear was much higher than in the smooth socket. Clearly, then, the establishment of design values for side resistance must include some quantification of borehole roughness.

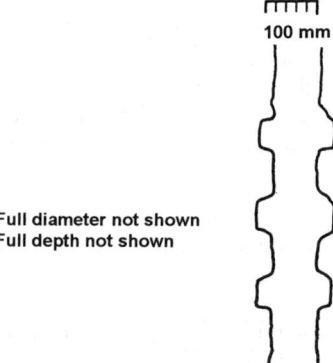

Figure 9. Borehole roughness pattern for rough borehole in shale (Horvath et al., 1983).

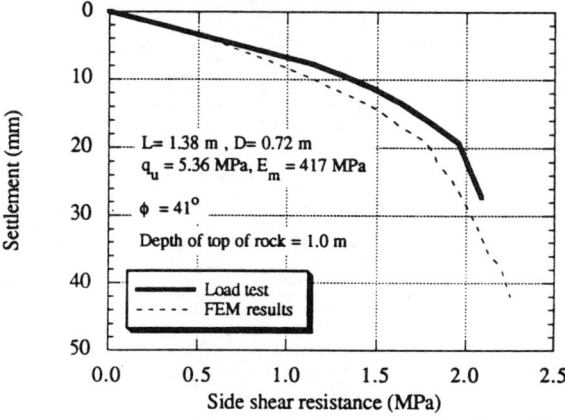

Figure 10. Comparison of measured and predicted side resistance-settlement relations for rough interface in clay-shale (measurements from Horvath et al., 1983).

The FEM appears to be a reasonable tool for investigating the general effects of borehole roughness in rocks, such as clay shales, that are ductile. Some results of a parametric finite element study that are relevant to the theme of this paper are shown in Figures 11 through 13. Figure 11 depicts the pattern of soft rock deformation along the interface for a case in which q_u = 1.2 MPa, E_m = 276 MPa, ϕ = 30° along the interface, the initial normal effective stress along the interface is 3.75 atmospheres (deep socket), and the rock mass itself is undrained. With the sinusoidal interface pattern, there is only a small amount of dilation; instead, the concrete mostly slides along the rock at the tip of the concrete asperity and begins to push against the rock asperity along its upper edge. The resulting pattern of shear stresses (τ_{rz}) in the rock mass are shown in Figure 12, which suggests a pattern of failure in which the rock asperities are gouged out along their roots. The relevant shearing resistance that a designer must use is therefore a combination of the interface shearing resistance between the rock and the concrete at the tips of the concrete asperities and that of the rock a few millimeters away from the interface beneath the roots of the asperities, with both components being affected by changes in normal effective stresses due to dilation. Hassan and O'Neill (1997) showed that this process eventually produced horizontal fractures in the rock that were self-limiting but which would presumably cause a reduction in shear strength in the rock near the interface with time if the load is not released.

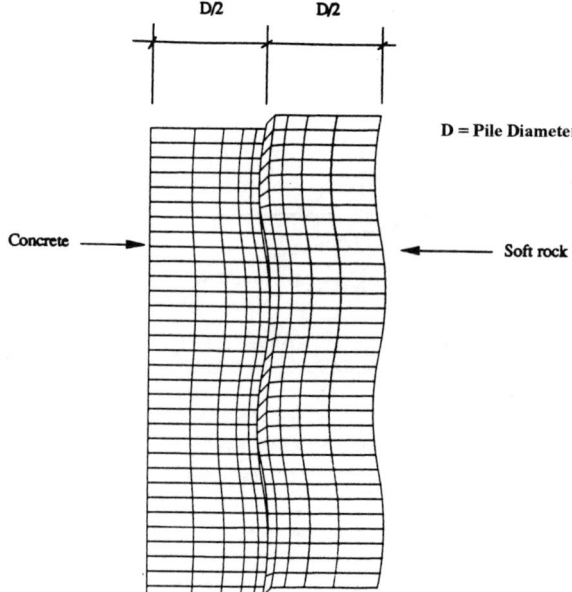

Figure 11. Pattern of deformation along an interface between a deep, rough bored pile and soft rock (Hassan, 1994)

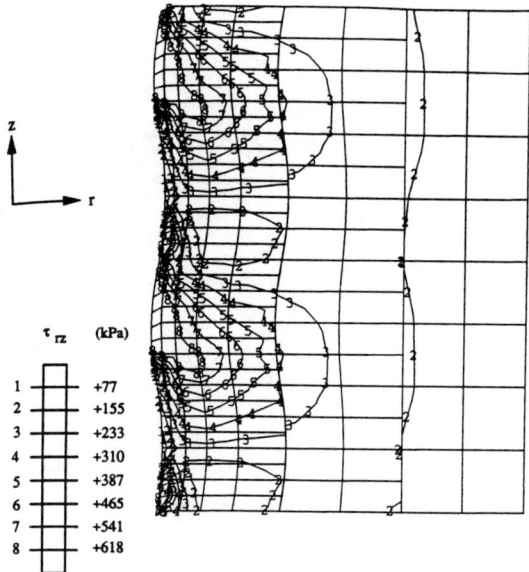

τ_{rz} (kPa)	
1	+77
2	+155
3	+233
4	+310
5	+387
6	+465
7	+541
8	+618

Figure 12. Contours of τ_{rz} along rough interface between concrete bored pile and soft rock (Hassan, 1994).

This model was applied to determine the effect of roughness, smear and rock strength/stiffness on the development of mean side shearing resistance along a 0.61-m-diameter bored pile 6.1 m long in rock with q_u = 2.4 MPa, E_m = 552 MPa, ϕ = 30°, initial normal interface stress of 1.25 atmospheres (shallow socket) and undrained behavior in the rock mass. The results, in terms of side resistance versus settlement, are shown in Figure 13. The curve denoted "stiff parent geomaterial, rough interface" presumes an interface pattern such as shown in Figure 6, with no smear zone. The curve denoted "stiff, smooth" presumes no roughness and no smear along the interface. The curve denoted "soft, rough" is for conditions identical to "stiff, rough," except that q_u is 0.48 MPa and E_m = 138 MPa. The rock around the pile represented by the curve marked "stiff parent geomaterial, rough, smeared interface" has the properties of the "stiff, rough" pile, except that the smear zone has the properties of the "soft" material described above, mimicking a rock that has been degraded by drilling and perhaps exposure to water.

It is of importance to observe that in the same rock the side resistance varied by a factor of three (comparing stiff, smooth with stiff, rough at large displacement). The resistance in the stiff, rough socket that had been smeared was almost identical to that in the stiff, smooth socket, although somewhat less displacement was required to produce plunging. If the rough, non-smeared pile results are compared for the soft and stiff rock, the limiting unit resistance is in direct proportion to the values of q_u for the two rocks. This behavior clearly confirms that it is important to know both the strength (and stiffness) of the rock and the character of the interface (roughness and smear), the latter being a product of construction details.

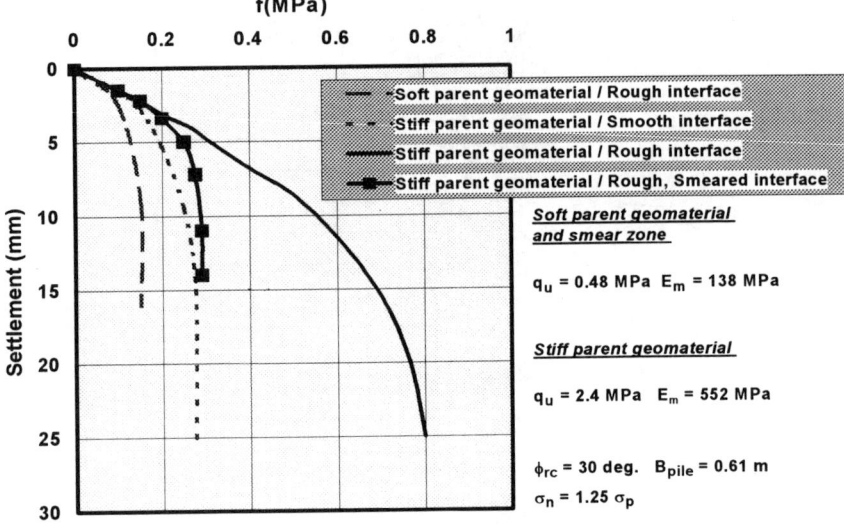

Figure 13. Comparative side shear behavior of bored pile rock sockets
(After Hassan and O'Neill, 1997)

These results can be generalized only qualitatively. Different drilling techniques can produce different roughness patterns in different types of rock, which can produce very different side resistances from those indicated here.

Stress Relief

The process of boring an excavation for a pile results in stress relief in the soil or rock laterally surrounding the borehole and beneath the bottom of the excavation. Such stress relief, which is time dependent, can result in reduction of shear strength of sands and overconsolidated clays. For example, O'Neill and Hassan (1994) report the results of tests performed by de Beer in which boreholes had been left open for about 2 and 350 hours, respectively, under a bentonite slurry, and the record of q_c (cone tip resistance) was markedly reduced within one meter of the latter borehole relative to the first.

In overconsolidated clay the process of drilling the borehole obviously remolds soil at the surface of the borehole -- the ultimate concrete-soil interface. However, the stress relief in the surrounding soil mass that is brought about by advancing the borehole apparently produces tension in the soil far from the borehole, which puts the pore water under negative stress (highest near the borehole), which with time reduces the effective stress in the soil as the negative pore water pressure dissipates. The magnitude of this effect is illustrated in Figure 14, which shows the low-strain shear modulus in the soil as a function of radial distance from the borehole interface around a 0.914-m-diameter borehole in saturated Beaumont clay, which was described briefly previously (Ata et al., 1998). The low-strain moduli were measured by Stokoe and Kalinski using SASW techniques. At the shallow depth of the excavation, 3.0 m, the *in situ* coefficient of earth pressure at rest is about 1.6 The borehole had been open for about three hours before the measurements were made. The low-strain modulus is less than the modulus in the free field for a distance of 2.5 borehole radii from the face of the borehole. At the face of the borehole the low-strain shear modulus is 73 per cent of the free-field value. If one assumes that the shear modulus varies with the square root of the undrained shear strength, s_u, at the surface of the borehole, s_u is about 53 per cent of its free-field value, which correlates reasonably well with the α factors reported for dry drilled shafts in this soil (Figure 4). When the borehole remained open and exposed to the air for a further five hours, the surface of the hole apparently became desiccated and the shear stiffness increased very close to the edge of the borehole.

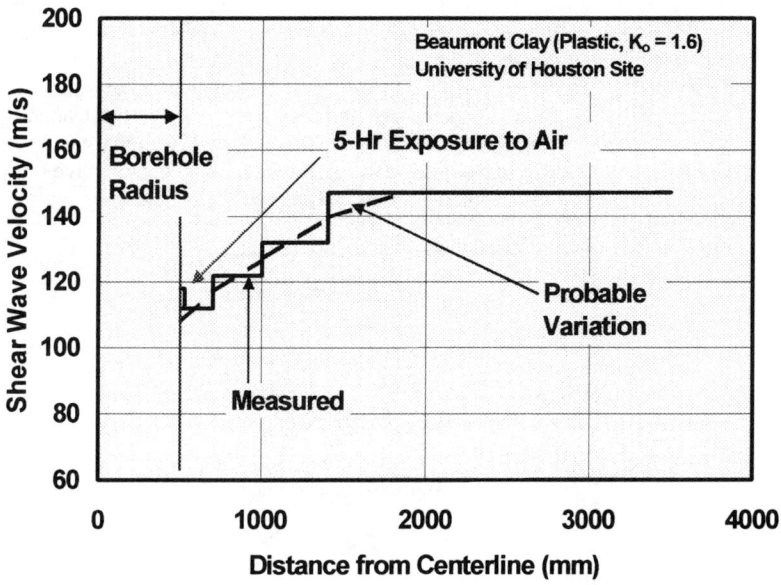

Figure 14. Radial variation of low-strain shear modulus around a drilled shaft.

If the mechanism of modulus reduction described here is correct, then one could speculate that the α factor depends, at least to a degree, upon the time the borehole is remains open before the stress is restored by placing concrete and by the permeability of the clay formation being penetrated. Prolonged exposure to air may dry the face of the borehole, but the suction that is created may tend to draw additional water out of the concrete mix and soften the soil back to the shear strength that it would have had if desiccation had not occurred. Further research into this phenomenon is warranted. It seems to be clear, however, that if elastic models are used to predict pile settlement (e. g., Randolph and Wroth, 1978) due consideration should be taken of the radial variation in shear modulus.

Stress relief in boreholes can also have a negative effect on p-y curves. Majano et al. (1998) describe a lateral socket test in friable sandstone performed by splitting a cylindrical drilled shaft socket vertically with an Osterberg cell. The socket was 53 m deep. For various reasons, the contractor left the test socket open for nine days before finally cleaning it, placing the Osterberg cell, and placing the concrete. The friable sandstone was very susceptible to stress relief because it had such little cementation that it was very difficult to sample. When the "split socket" test was run, the lateral pressure versus lateral displacement (averaged for both halves of the socket) were converted to p-y curves through finite element analysis. The results are compared in non-dimensional form in Figure 15 with p (rock resistance per unit length of pile) -y (local lateral pile deflection) curves inferred from Reese (1997) under more favorable circumstances. The normalizing factors are D, the pile diameter, and q_u, the unconfined compressive strength of the rock. It can be seen that the initial lateral stiffness of the p-y curve in the friable

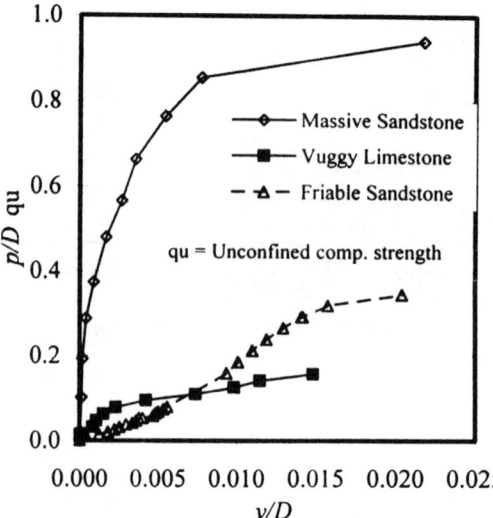

Figure 15. Comparison of p-y curves in sedimentary rock (Majano et al., 1998)

sandstone is several times smaller than that inferred from tests in other sedimentary rocks described by Reese. While this condition could have been caused by factors other than stress relief, it is clear from the concave shape of the p-y curve in the friable sandstone that stress relief played a major role in shaping the p-y curve.

Method of Driving Piles

Piles can be installed in many ways: by boring and concreting, by using continuous flight augers, by using screw elements, by oscillating casing into the ground and excavating the soil inside, by impact driving, and by vibratory driving, to name a few. Each of these construction methods can potentially produce different behavior. An example is given here showing the differences in behavior of piles installed by impact and by vibratory driving. A parametric study was conducted on the load-settlement behavior of 102-mm-diameter, closed-toe steel pipe piles driven 19 to 20 diameters into fine, saturated siliceous sand in a pressure chamber and then subjected to compression loading tests (O'Neill et al., 1990). Figure 16 summarizes the results of the static loading tests and shows the general boundary conditions in the test chamber. When the relative density of the sand was 65 per cent, the impact-driven pile produced about 20 per cent higher capacity than the pile that was driven by vibration, even when the pile that was vibrated into position was subjected to several restrike blows that advanced the toe about 25 per cent of the diameter of the pile.

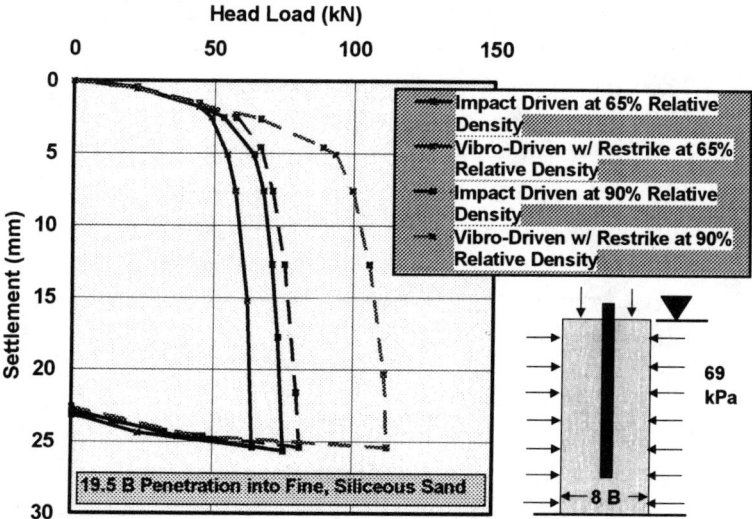

Figure 16. Comparative load-settlement behavior of piles driven by impact and vibration.

The superior performance of the pile driven by impact in the medium-dense sand would probably be anticipated by most foundation engineers. However, when the sand had a relative density of 90 per cent, the opposite effect occurred. The pile that was installed by vibration developed a much higher resistance than the pile installed by impact. It appears that driving the pile by vibration assisted in maintaining the high relative density in the soil around the pile, whereas driving by impact may have loosened the soil. There was no evidence of grain crushing and resulting arching in either case. The pile that was installed at 90 per cent relative density by vibration was inserted at a constant rate of penetration 10 mm / sec in the last 2 diameters of driving. The pile that was installed at 90 per cent relative density by impact was inserted at a rate of 8 mm per blow at 40 blows per 60 sec in the last 2 diameters of driving. Since the rates of penetration were generally comparable and since no visible grain crushing occurred, the suggested loosening effect may have been produced by the generation of body waves in the soil mass at the toe of the pile driven by impact.

Further details are shown in Figures 17 and 18 for identical piles driven into a coarser sand than the sand represented in Figure 16 and without restriking the vibrated piles. In the top half of the pile, the pile driven by vibration experienced an increasing side resistance as deflection increased, probably due to dilation of the sand at the interface with the pile, while no such behavior appears to have occurred around the pile driven by impact. Closer to the bottom of the pile, away from the soil surface, dilation does not appear to have occurred whether the pile was installed by vibration or by impact. The presence of dilation along the upper part of the shaft of the pile appears to confirm the postulation that vibratory driving tends to preserve the density of the initially dense sand.

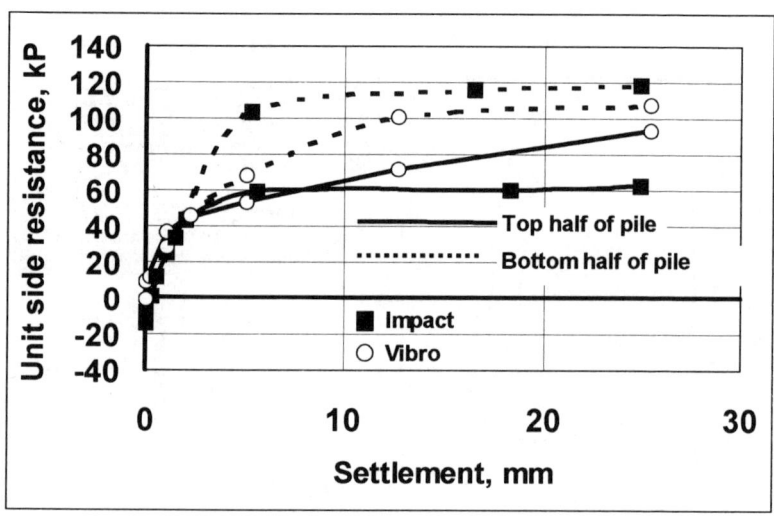

Figure 17. Unit side resistance vs. settlement for piles installed in saturated sand by impact and vibration.

In Figure 18 are shown the relationships of net unit toe resistance to toe settlement for the same piles that are depicted in Figure 17. At 90 per cent relative density, vibro-driving also produced higher toe resistance than impact driving. From a practical perspective, while one would not normally expend the energy to drive a pile though very dense sand with a vibratory driver, it is common practice to use a vibratory driver to install the pile through less dense sand and drive the toe into very dense sand to secure a high bearing capacity. Therefore, Figure 18 may have more practical significance than Figure 17.

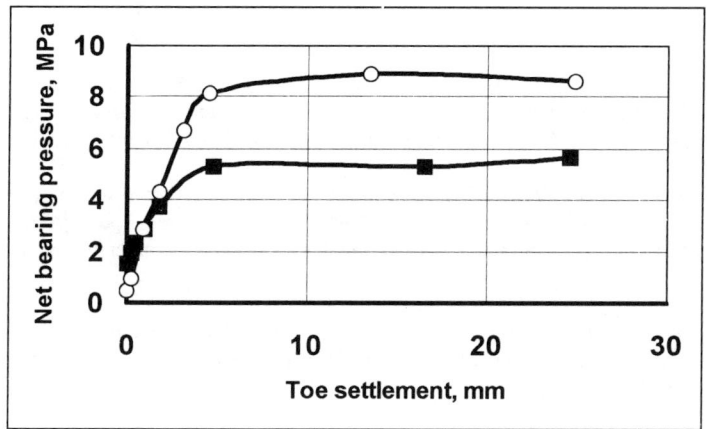

Figure 18. Toe resistance vs. settlement for piles driven by impact and vibration into very dense, saturated, coarse sand.

Not only did the pile driven by vibration develop a higher toe resistance, but the residual compression stress at the end of driving was lower in the pile driven by vibration. Similar comparisons made in very fine sand placed at a relative density of 90 per cent indicated the same trend but with smaller differences than those shown here.

Conclusions

The few examples given in this paper demonstrate some of the many potential effects of construction details on the performance of individual piles. When piles are installed in groups, these effects can be magnified. Factors such as the properties and handling of a drilling slurry that is used to advance a borehole and the natural roughness of a borehole in rock have a major effect on the resistance and deformation of bored piles, perhaps an even greater effect than the more traditional concerns of shear strength and deformational properties of the *in-situ* soil or rock or the method used to model the foundation system.

This is not to suggest that the best subsurface data and analytical techniques should not be used, but an accurate prediction of the real behavior of the foundation must consider the effects of construction factors as well. Since these effects are not now predictable without an appreciation of the details of the manner in which construction will be carried out, it is necessary to consider probable ranges in the effects when selecting factors of safety or resistance factors and, when such factors become economically unacceptable, to develop and enforce specifications that ensure that the construction practices envisioned by the designer will be carried out and that the effects of construction will remain within limits that are predictable and tolerable. Likewise, the salutary effects of construction details can be used to the foundation engineer's advantage.

References

Ata, A., and O'Neill, M. W. (1997a). *Characterization of the Effect of Polybore Polymer on the Construction and Performance of Drilled Shaft Foundations, Phase I: Field Tests*, Department of Civil and Environmental Engineering, University of Houston, May.

Ata, A., and O'Neill, M. W. (1997b). *Characterization of the Effect of Polybore Polymer on the Construction and Performance of Drilled Shaft Foundations, Phase I: Laboratory Tests*, Department of Civil and Environmental Engineering, University of Houston, October.

Hassan, K. H. (1994). *Analysis and Design of Drilled Shafts Socketed into Soft Rock*, Ph. D. dissertation, Department of Civil and Environmental Engineering, University of Houston, May, 1994, 264 pp.

Hassan, K. H., and O'Neill (1997). "Side Load-Transfer Mechanisms in Drilled Shafts in Soft Argillaceous Rock," *Journal of Geotechnical and Geoenvironmental Engineering*, ASCE, Vol. 123, No. 2, February, pp. 145 - 152.

Horvath, R. G., Kenney, T. C., and Kozicki, P. (1983). "Methods of Improving the Performance of Drilled Piers in Weak Rock," *Canadian Geotechnical Journal*, Vol. 20, pp. 758 - 772.

Majano, R. E., O'Neill, M. W., and Person, G. (1998). "Lateral Split Socket Test in Franconia Sandstone," in *Deep Foundations on Bored and Auger Piles*, W. F. van Impe, Editor, Balkema, Rotterdam, 1998, pp. 151 - 156.

O'Neill, M. W., and Hassan, K. H. (1994). "Drilled Shafts: Effects of Construction on Performance and Design Criteria," in Proceedings, International Conference on Design and Construction of Deep Foundations, U. S. Federal Highway Administration, pp. 137 - 187.

O'Neill, M. W., and Reese, L. C. (1970). "Behavior of Axially Loaded Drilled Shafts in Beaumont Clay," *Report No. 89-8*, Center for Highway Research, University of Texas at Austin, December.

O'Neill, M. W., Townsend, F. C., Hassan, K. M., Buller, A., and Chan, P. S. (1996). "Load Transfer for Drilled Shafts in Intermediate Geomaterials," *Report No. FHWA RD-95-172*, Federal Highway Administration, Office of Engineering and Highway Operatioins R and D, McLean, VA, November, 194 pp.

O'Neill, M. W., Vipulanandan, C. and Wong, D. (1990). "Laboratory Modeling of Vibro-Driven Piles," *Journal of Geotechnical Engineering*, ASCE, Vol. 116, No. 8, August, pp. 1190 - 1209.

Randolph, M. F., and Wroth, C. P. (1978). "Analysis of Deformation of Vertically Loaded Piles," *Journal of the Geotechnical Engineering Division*, ASCE, Vol. 104, No. GT12, December, pp. 1465 - 1488.

Reese, L. C. (1997). "Analysis of Laterally Loaded Piles in Weak Rock," *Journal of Geotechnical and Geoenvironmental Engineering*, ASCE, Vol. 123, No. 11, November.

Reese, L. C., and O'Neill, M. W. (1988). "Drilled Shafts: Construction Procedures and Design Methods," *Report No. FHWA-HI-88-042*, Federal Highway Adminstration, Washington, D. C., July.

Improvements in Pile Driving Simulation and Installation
Techniques for Deep Penetrating Casings and Piles in Deepwater

Rathindra N. Dutt[1], Frank Rausche[2], and James T. Collins[3]

Abstract

As the offshore oil and gas industry moves into the deepwater frontier there is a growing need to install the well casings and large diameter piles to deeper and deeper penetrations. Installations require casings and piles to be impact driven and therefore a prediction of driveability has to be made prior to installation in order to select an appropriate hammer. The current industry practice of single blow analysis, using the so-called wave equation, tends to over-predict the driving effort at deeper penetrations since it neglects the positive effect of energy stored in pile and soil between hammer blows. This difference can be particularly significant when piles are driven to a great penetration where sediments are stiffer by virtue of the high overburden. Holloway et al. (1975) and Hery (1983) who developed the so-called Residual Stress Analyses in the DUKEFOR and GRLWEAP wave equation programs respectively had already recognized this fact. Similarly, a recent study by Choe and Juvkam-Wold (1997), evaluating pile driving of long casing strings in deepwater concluded, that in order to properly simulate the driving behavior in stiff sediments at deep penetrations the interaction between consecutive hammer blows should be considered.

This paper discusses the shortcomings of available residual stress models and describes the development of a new model. The effectiveness of the new residual stress model is illustrated by comparing the actual and predicted behavior of a long casing string in deepwater. Also described in the paper are installation aides that have been effectively used in deepwater casing installation to reduce the driving resistance encountered by the casings during driving,

----

[1]Dutt & Associates, Inc., 6575 West Loop South, Suite 430, Bellaire, Texas 77401
[2]GRL & Associates, Inc., 4535 Emery Industrial Parkway, Cleveland, Ohio 44128
[3]Marathon Oil Company, P. O. Box 3128, Houston, Texas 77253-3128.

Introduction

The current industry practice to evaluate pile driving behavior is to perform wave equation analyses for an idealized hammer-pile-soil system for one hammer blow at various pile penetrations. The most commonly used program is GRLWEAP (Goble Rausche Likins and Associates, Inc., 1997). Like other similar programs (e.g. the well known TTI program, Lowery et al. 1969, Edwards, 1969), GRLWEAP is based on mathematical solutions developed by Smith (1960) where the soil, pile, and the hammer are idealized as discrete elements and the impact pile driving is represented by one-dimensional wave propagation theory. In the traditional analysis, referred here to as the Single Blow Analysis (SBA), Smith incorporated three simplifications:

1. At the beginning of analysis the forces in the pile and the soil are zero.
2. The analysis is finished when the pile starts to rebound. However, the full rebound is not analyzed and the program assumes that the pile rebounds to a stressless state.
3. Analysis is performed for one particular depth and a bearing graph is constructed relating peak stresses and pile bearing capacity (SRD) to blowcount (the inverse of pile penetration per blow).

In most cases assumptions 1 and 2 are satisfactory for the majority of pile installations. In fact, most of the correlations and case histories described in the literature are based on these simplified assumptions. In reality, a pile does not rebound completely after the hammer blow is complete and residual stresses remain in the pile and soil between blows. Holloway et al. (1975) recognized the importance of residual stresses remaining in a pile from prior blows and developed software for the calculation of residual stresses and an improved penetration per blow prediction in an iterative manner. This required repetitive analyses of several impacts (blows) to achieve convergence of the residual stress pattern and penetration per blow.

Hery (1983) adopted Holloway's Residual Stress Analysis (RSA) model for the GRLWEAP program. He allowed up to ten consecutive analyses; however, for deep pile penetrations, convergence was not always achieved. Furthermore, preliminary investigations indicated that for long piles repetitive "blows" would not necessarily develop repetitive stress patterns.

In a recent study conducted at Texas A & M University, Choe and Juvkam-Wold (1997) developed a Multiple Blow Analysis (MBA) model to evaluate driving of casings to deep penetrations in the deepwater environment of the Gulf of Mexico. They reported that MBA, which also takes into account the effect of residual stresses in the pile, predicts much lower penetrations per blow over the whole depth range compared to SBA.

Instead of only producing a bearing graph as an output, GRLWEAP offers a "driveability analysis" option, which calculates driving resistance and peak dynamic pile driving stresses as a function of depth considering changes in driving parameters and pile length. Important also for the analysis of piles with deep penetrations is the automatic reduction of the soil's static resistance values to the "Resistance to Driving" which is based on a setup factor defined as the ratio of long term capacity divided by minimum capacity during driving. The option recognizes differences in loss of setup from layer to layer and the regain of strength during driving interruptions. The driveability analysis can also be performed for piles with varying lengths. GRLWEAP's driveability option has been extensively used in the preparation of the examples presented in this paper.

This paper describes the considerations necessary for RSA or MBA for piles with very deep penetrations. These considerations address the difficulties with convergence, i.e. how many consecutive blows have to be analyzed before the residual stress pattern is reliably determined. A case history is presented to illustrate how residual stress analysis provides a better blowcount prediction when compared to single blow analysis for very long piles driven to deep penetrations.

Also, in order to reduce the pile installation effort, the industry has developed several pile installation aides to improve pile penetration. This paper describes one type and use of pile installation aide.

Wave Equation Analysis and Residual Stresses

GRLWEAP's pile model of a wave equation analysis consists of springs and masses, normally representing 1m long pile segments. The soil resistance is represented by both an elasto-plastic spring and a dashpot on each shaft segment and at the toe. The elasto-plastic spring represents the static soil resistance and includes a quake, which is that deformation at which the soil spring reaches its ultimate resistance. For pile displacements increasing beyond the quake the soil spring response is no longer elasto-plastic. The analysis of a pile under the impact of a hammer therefore yields pile segment displacements and velocities, which at first show how the pile penetrates into the ground and activates soil resistance. The analysis is normally carried out until the pile starts to rebound, which occurs after the soil resistance has become higher than the downward directed pile driving forces. The final permanent pile set is then calculated as the difference between the maximum tip displacement and the weighted average quake value. In effect, this calculation assumes that the pile rebounds to a state of zero stresses and neglects the effect of stresses remaining in the pile between hammer blows, as explained in the following.

When a long or slender pile is impact driven by a hammer, it experiences relatively large displacements near grade or seabed and much smaller displacements near the toe when pile driving becomes hard. Thus, at the end of the impact event the upper pile section rebounds strongly while the pile bottom moves only a short distance upwards.   The strong rebound in the upper zone first reduces shaft resistance values to zero and then, during continued rebound, produces negative, i.e. downward directed resistance forces.   In contrast, near the pile toe where the rebound is small, the resistance forces remain positive (upward directed).   As a result, the pile remains in a compressed state after the hammer blow.   When the next hammer impact is applied, the upper soil layers actually help with their downward directed resistance forces to overcome additional resistance near the pile bottom.   In effect, both pile and soil store energy.   Hence, slender piles with high shaft resistance encounter refusal at soil resistance values that are much higher than predicted from SBA.   On the other hand, pure end bearing piles do not benefit from energies stored in the piles.   In piles with high shaft resistance, driving stresses can be much higher than predicted from SBA, since the residual stresses are additive to those that the hammer imparts during an impact.

While elastic pile deformations of slender piles very frequently cause high residual stresses, a friction pile with significant end bearing and a large toe quake is another example.   The quake is the elastic deformation of the soil prior to reaching ultimate resistance.   In this case, the toe tends to push the pile upwards a relatively long distance.   As the shaft elements of the pile move upward, their skin resistance first decreases to zero and then becomes negative until equilibrium is reached between the positive toe resistance and the negative skin friction.

The above explanation is satisfactory for relatively short piles.   However, when pile penetrations are very deep and pile driving is hard, the pile may only successively overcome the soil resistance along its length.   In fact it may take several blows in the analysis before the pile toe starts to experience any impact force at all.   Therefore, the stress pattern remaining in pile and soil may be different from blow to blow.   Convergence checks can then only be conducted by considering a long series of blows.

The Residual Stress Model

To calculate residual stresses and realistic driving resistance values, Holloway et al. (1975) programmed the following steps in the DUKEFOR program.

Before the first dynamic analysis is performed, the weights of hammer and pile are applied to the pile-soil model and in a static analysis, initial pile deformations and soil resistance forces are calculated that balance these static weights.   A dynamic analysis is then performed until all forces in the pile have decreased below a certain threshold (4.45 kN).   Static equilibrium is then again assured by a static pile analysis.   The final pile deformation is calculated as the difference between the displacements of pile top and pile toe.   The process is repeated until the deformation between consecutive blows converges.   The pile set is then the difference between the pile displacements calculated from two consecutive blows. The blowcount is the inverse of the pile set.

In his adaptation of Holloway's model to GRLWEAP, Hery (1983) added a static analysis at the end of the dynamic analysis. GRLWEAP's approximate initial static balancing of hammer and pile weights and the duration of the dynamic analyses remained unchanged. In contrast to Holloway's analysis, which considered soil resistance forces in their elastic state, Hery's static analysis allowed for soil segments moving from the elastic to the plastic state and vice-versa. Hery checked for convergence of the pile compression (pile top displacement minus pile bottom displacement) between blows analyzed and allowed for up to 10 analyses. For long piles convergence did not necessarily occur within 10 analyses.

In their analysis Choe and Juvkam-Wold performed up to five consecutive analyses, to calculate the residual stresses. They did not indicate that they performed either a static analysis between dynamic analyses or a convergence check.

For the analyses of long piles with deep penetrations the following changes were made to the GRLWEAP code. First the approximate initial analysis was replaced by a complete static analysis. The maximum number of analyses was made user input; based on trial analyses with long piles its default was set to 40. Convergence checks were based on the calculated sets, i.e. the difference of pile top displacements from the static analysis of two consecutive blows. If consecutive blows did not indicate convergence then a check was made on potential patterns of convergence between consecutive groups of two or more blows. There were also changes made in the duration of each dynamic trial analysis. Obviously, this is important for considerations of computation time when both the number of pile segments (e.g.330 for a 330m pile) and the number of trial analyses (sometimes up to 40) is large and when an analysis is run at up to 99 different pile tip penetrations. Rather than relying on pile rebound as is common for traditional analyses, the dynamic analysis is now finished once all pile segment velocities have decreased below 5 percent of the impact velocity.

For a very long pile (330m long, 1219mm x 38mm pipe) with 244 m of penetration and driven by an IHC S200 hammer, calculated sets per blow from consecutive analyses are shown in Figure 1. One curve is for a uniform shaft resistance with a high total ultimate resistance of 56 MN and the second result is for a layered soil with 30 MN ultimate resistance. SBA indicated absolute refusal in both cases. The high resistance resulted in a decreasing trend of calculated pile top sets that indicated (by extrapolation) refusal after the 13th trial analysis. The second analysis with the lower resistance indicated a convergence to a set of 0.54mm after the 21st analysis or the equivalent of 468 blows per 250mm. Had the calculated set of the fifth blow (1.17mm) been used to calculate the driving resistance, the result would have been 214 blows per 250mm. The reason for the large difference between the early and late blows analyzed is the successively increasing residual pile compression of ultimately about 15mm, which is large compared to the permanent set. In these friction pile examples, calculated peak compressive stresses occurred at the pile top and therefore were the same for both SBA and RSA (no residual stresses exist at the pile top). Even though they were not critical, calculated compressive stresses from SBA and RSA at locations other than the pile top are worth comparing. For example, approximately at pile mid-length SBA only indicated 55 MPa. RSA, however, calculated 124 MPa with the difference being the stresses locked into the pile between hammer blows.

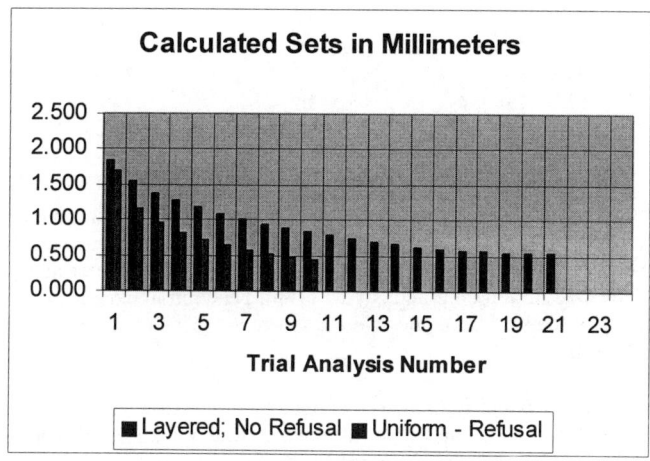

Figure 1. Pile Top Sets for Both a Very High Uniform and a Medium High Non-Uniform Shaft Resistance from Consecutive Trial Analysis.

It should be added that the GRLWEAP manual does not recommend the use of RSA without correlation testing because the recommended soil resistance parameters were based on correlation with SBA results. This paper provides information necessary for an improved confidence in the use of RSA for long pipe piles in marine clays.

Assessment of the New Residual Stress Model

Marathon Oil Company provided information for this evaluation. The water depth at the platform location is 236m. Twelve 2134-mm-diameter skirt piles, three piles per leg, support the four corner legs of the platform. The piles are 143m. long and penetrate 119m into the seabed.

The pile wall schedule for the skirt piles are summarized below:

Distance from Pile Top, m.		Wall Thickness mm	Steel Grade MPa
0.0	0.6	83	345
0.6	27.1	54	345
27.1	43.9	44	345
43.9	51.5	44	248
51.5	88.1	38	248
88.1	94.2	44	248
94.2	109.4	44	345
109.4	112.5	38	345
112.5	139.9	38	248
139.9	143.0	51	248

Piles were stabbed as a single piece and ran to an average penetration of 34m. (range 32m to 40m). The piles were driven with a Menck MHU-1700 underwater hydraulic hammer in the slender mode on the same batter as the jacket legs. Each pile was driven continuously to 119m. penetration except for stops due to mechanical problems. All 12 piles were stabbed and driven to design penetration in a 13-day period.

Installation of 660mm-diameter conductors was also evaluated. These are slightly deviated conductors and have a constant wall thickness of 19mm. The conductors have a 1.5m long by 32mm thick drive shoe. Each conductor was made up of 14 sections whose lengths varied from 64m to 13m. The 660mm-diameter conductors free fell to an average penetration of 35m and these conductors were driven with a Menck MHU-195 hammer. At deep penetrations below about 110m, a Vulcan 060 hammer was used to drive some of the conductors. Skin friction reducers were installed in the bottom section of 14 of the 16 conductors in an effort to increase conductor penetration. The 14 conductors with friction reducers were driven to their design penetrations. The two conductors without reducers encountered refusal some 12m to 24m short of design penetration.

Site conditions were investigated by drilling two soil borings at the platform location. Pertinent soil properties from the geotechnical investigation are summarized in Figures 2 and 3. The soils at the site are highly plastic clays for the full depth of the boring. The average plasticity index (PI) in the top 35m is about 40 and the average PI decreases to about 30 in the underlying clays. Except for the very top surface sediments, the natural moisture content decreases with increasing penetration in a characteristic fashion of a normally consolidated deposit. The clays are normally consolidated for the most part as evidenced in the consolidation data presented in Figure 3. However the top 35m of the sediment are in a state of over-consolidation.   High resolution geophysical data confirm the erosion of approximately 12m of sediments from the surface and also indicates that the

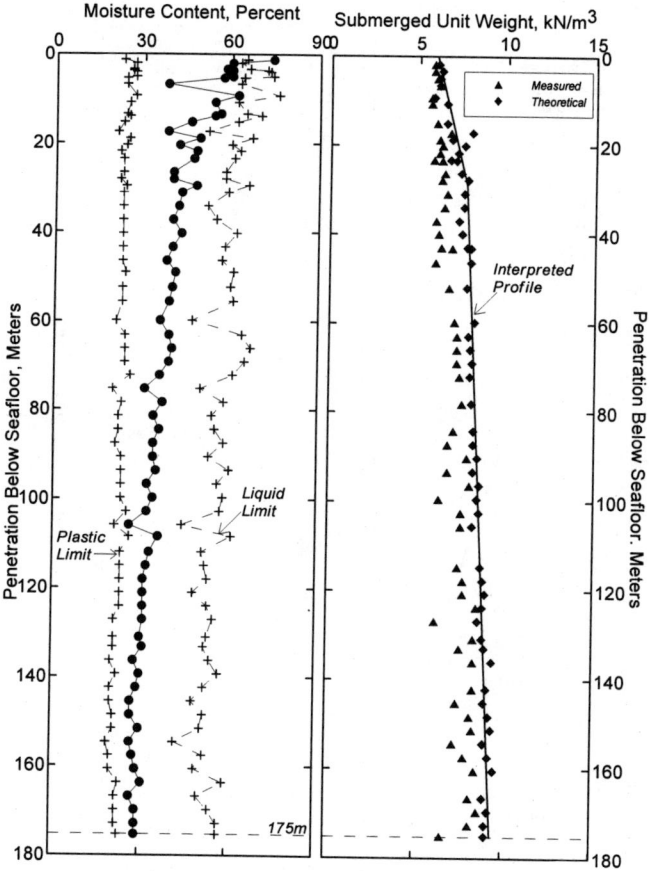

Figure 2. Soil Properties: Plasticity & Unit Weight

sediments at the site have been uplifted. The consolidation test data show evidence of under consolidation at depth with the degree of under consolidation increasing with penetration. In our opinion, this is more a reflection of sample quality than the true state of consolidation of the sediments. In such a deepwater environment soil samples from deep penetrations typically show evidence of some sample disturbance due to stress relief during sample retrieval. Remolded miniature

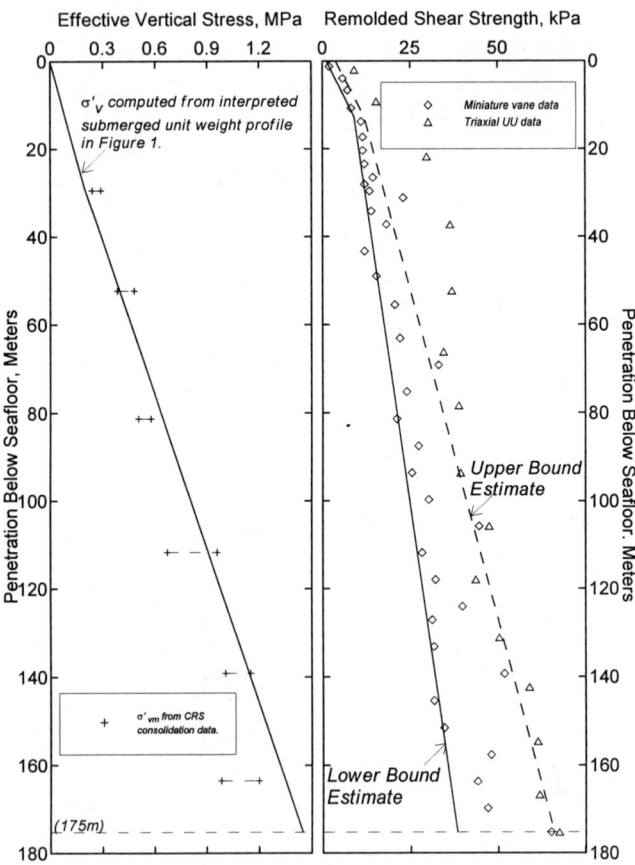

Figure 3. Soil Properties: Stress History & Strength

vane and triaxial UU strength data are summarized in Figure 3. Lower and upper bound remolded strength profiles have been interpreted to account for the variation in the remolded strength data. The lower bound and the upper bound profiles correspond to approximately 0.13 and 0.25 of the interpreted undrained shear strength, respectively.

Driving records for two of the piles and one conductor without skin friction reducers are presented in Figures 4, 5, and 6. Also presented on these plates are the measured hammer energies during driving of the 2134mm-diameter skirt piles with the Menck MHU-1700 hammer. As can be seen, pile SP-4 had higher blowcounts (cumulative 2,591 blows) than pile SP11 (cumulative 1,573 blows). Pile SP-4 was

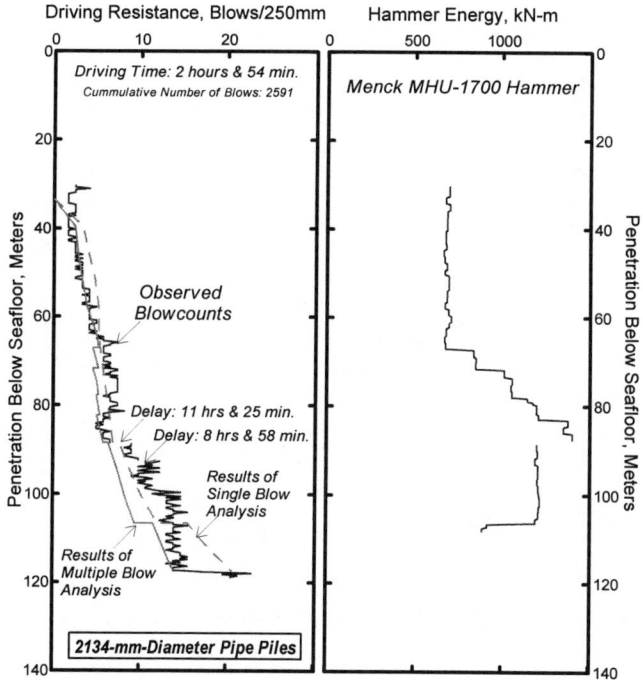

Figure 4. Predicted vs. Observed Blowcounts, Pile SP-4

driven over a 23-hour period with two significant driving interruptions because of mechanical problems. Also, the measured ram energies were significantly lower when driving SP-4. It is interesting to note that pile SP-7 (not presented here) stopped at a penetration of 117m. for 93 hours due to mechanical problems.

Restarting the pile was not a problem with blowcounts being 15 blows per 250mm before the stop and 37 blows per 250mm after restart. Based on wave equation analysis the soil-pile set-up was about 24 percent.

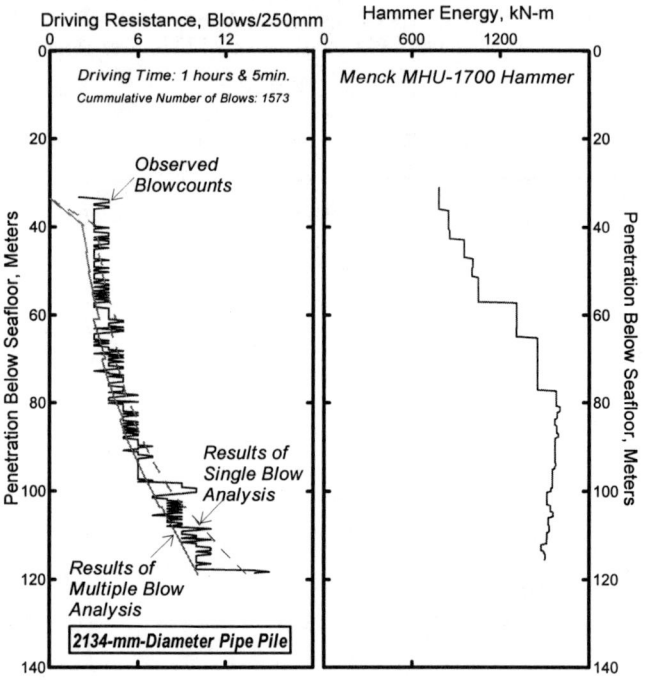

Figure 5. Predicted vs. Observed Blowcounts, Pile SP-11

Wave equation analyses were performed using the as-built pile characteristics and the measured ram energies. Both single and multiple blow analyses were performed for comparison with the actual blowcounts. Soil resistance to driving (SRD) required for blowcount prediction was obtained using lower and upper bound remolded shear strength profiles. The procedure recommended by Dutt et al. (1995) was utilized to estimate soil resistance to driving. In addition, their recommended soil quake and damping values for deepwater normally consolidated sediments were utilized in the wave equation analyses. Only the predicted blowcounts using the upper bound remolded shear strength data have only been presented here because the other analyses with lower bound remolded shear strength predicted very low blowcounts. Results of the hindcast wave equation analyses are also presented in Figures 4, 5, and 6 for comparison.

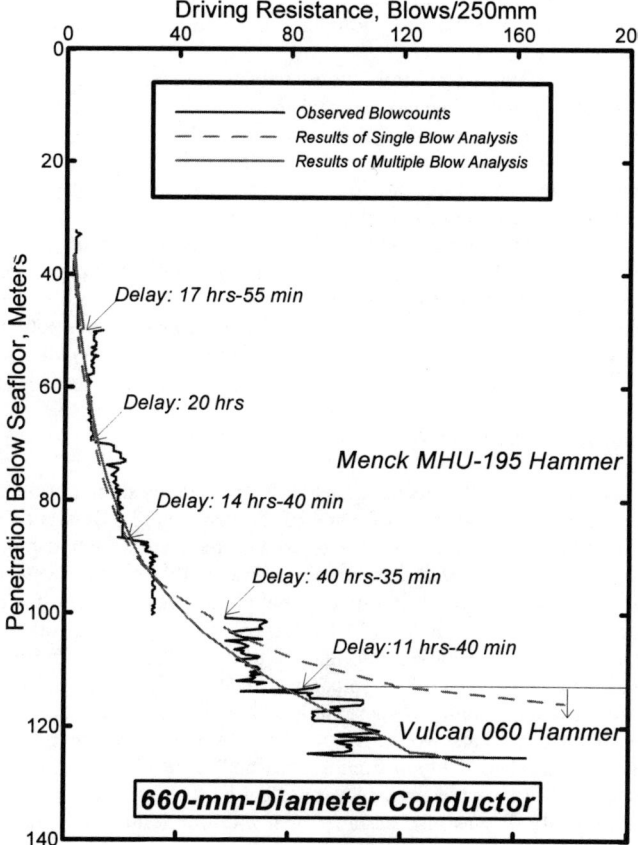

Figure 6. Predicted vs. Observed Blowcounts, Conductor CC with no Skin Friction Reducers.

The following conclusions are drawn from the comparison of the actual and predicted blowcounts presented in Figures 4, 5, and 6:

1. Use of upper bound remolded shear strength estimate at 0.25 times the interpreted undrained shear strength results in a better match with the observed blowcounts.
2. Both single blow and multiple blow analysis do equally well in predicting blowcounts at shallow penetrations where the sediments are relatively soft.
3. When a pile or a conductor is driven continuously (no driving interruption), multiple blow analysis provides a better prediction of the blowcounts
4. Multiple blow analysis provides a better estimate of the blowcounts irrespective of driving interruptions at deep penetrations.
5. The continuous driving cases also indicate that the use of a lower side damping value (lower than 0.6 sec./m) is warranted at deep penetrations where the clays are very stiff and hard.
6. Pile restart at deep penetrations even after significant delays is not a problem at this all clay site.

Use of Skin Friction Reducers to Improve Driveabilty

To achieve deeper conductor penetration by driving skin friction reducers were mounted on the bottom 61m of some of the conductors. Details of the skin friction reducers are shown in Figure 7. The friction reducers were half-pipes rolled to an ID to match the OD of the conductors. The profile of the conductor with reducers was similar to a conductor with external upset connections, so the reducers did not create operational difficulties. The purpose of the reducers was to expand the soil laterally as the ring passed through. Skin friction reduction during driving was assumed to be caused by either temporarily reducing the horizontal total stress acting on the conductor shaft, or increasing the radius of disturbance in the soil around the conductor. This would increase the time required for pore pressure dissipation and thus retard the skin friction "set-up." It was assumed that the skin friction reducers would not have a long-term effect on the ultimate skin friction that could be developed. It was also assumed that the long-term horizontal effective stresses would be unaffected by the reducers.

The three friction reducers were notionally grouped on the first section of each conductor, because the first (deepest) section would pass through the strongest soil. Also, the three reducers passed through the deepest of the conductor guides while the conductor was penetrating under self-weight. There was no interference or interaction between the reducers and the conductor guides during driving.

Observed blowcounts for two of the conductors with skin friction reducers are presented in Figures 8 and 9. Comparison of the observed blowcounts in

Figures 6, 8, and 9 shows that that the use of skin friction reducers resulted in 12m to 24m deeper penetrations for the conductors.

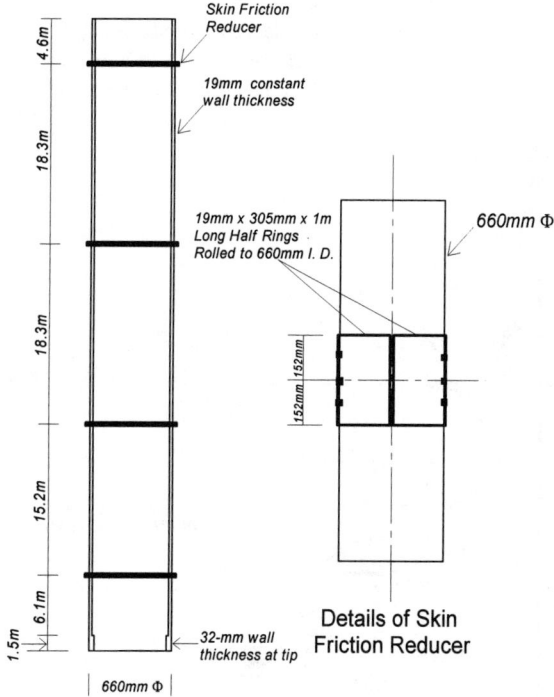

Figure 7. Details of the Bottom Section with Skin Friction Reducers

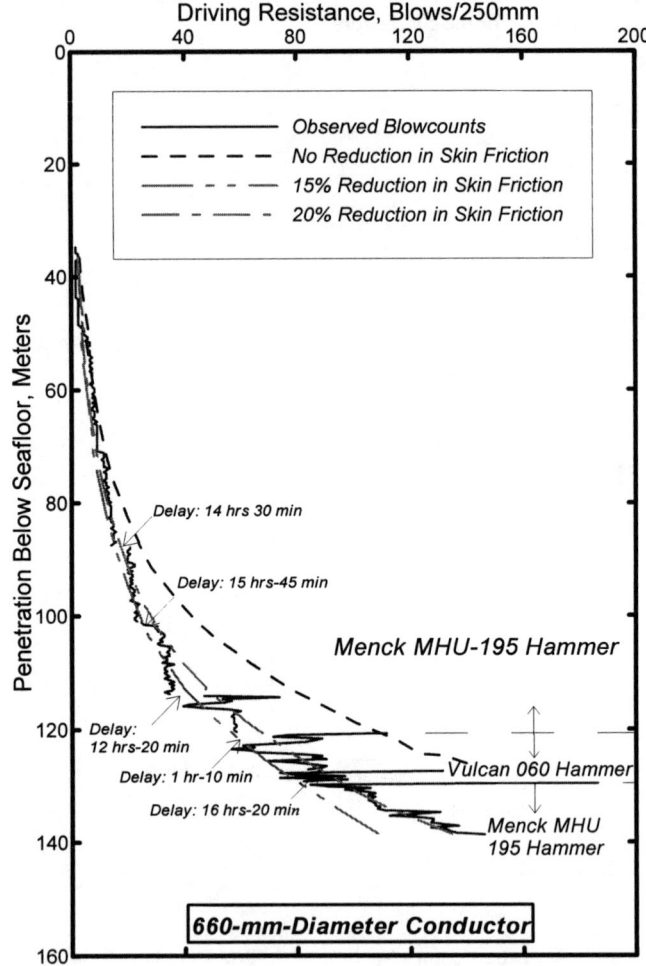

Figure 8. Predicted vs. Observed Blowcounts, Conductor D with Skin Friction Reducers.

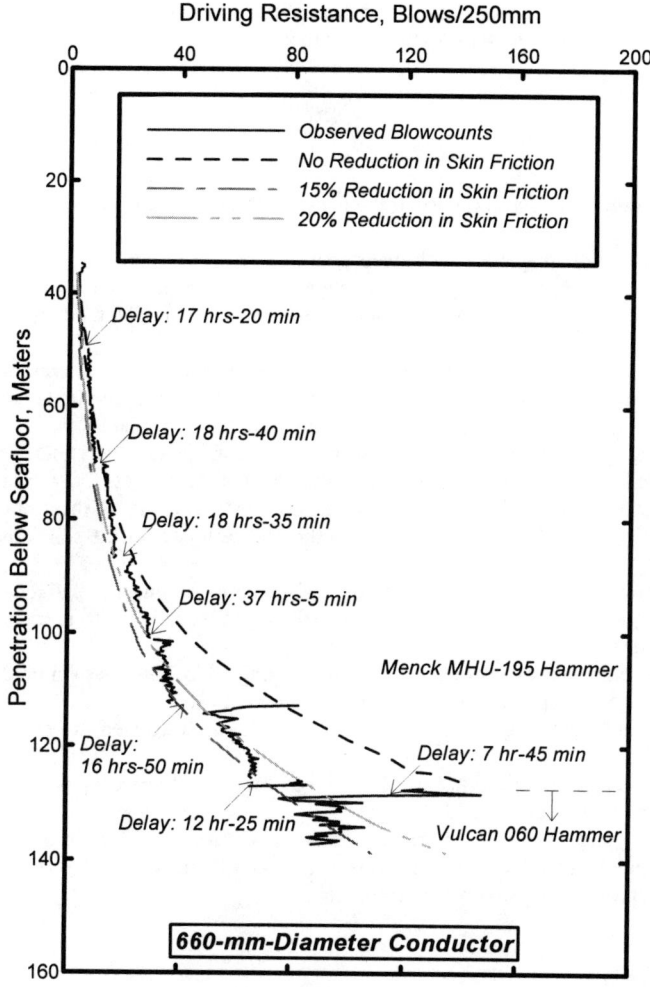

Figure 9. Predicted vs. Observed Blowcounts. Conductor GG with Skin Friction Reducers

A series of multiple blow wave equation analyses were performed to evaluate percent reduction in the skin friction resistance due the skin friction reducers. Results of the analyses are also presented in Figures 8 and 9. Analyses show that the 10 percent increased penetration of the conductors resulted from 15 to 20 percent reduction in the skin friction resistance.

Conclusions

The following is a summary of the major findings and conclusions:

1. The analysis of long piles with deep penetrations has to consider the effect of residual stresses on calculated blowcounts. This requires both a dynamic analysis and equilibrium check by static analysis for each one of a series of trial analyses.
2. Because of potentially large differences in the residual stress patterns in piles between early and late blows it is necessary to perform a large number of trial analyses before the driving resistance can be accurately calculated.
3. For long piles with deep penetrations, the combination of GRLWEAP's driveability option with its residual stress analysis provides for a relatively simple procedure considering the computation intensive nature of the problem.
4. Use of remolded shear strength provides the best estimate for the soil resistance to driving.
5. At shallow penetrations where the soil resistance to driving is relatively small both single blow and multiple blow analysis do equally well in predicting blowcounts.
6. Multiple blow analysis is better (compared to single blow analysis) in predicting blowcounts at deep penetrations in very stiff to hard clays.
7. A lower side damping value (less than 0.6 sec/m) is warranted in very stiff to hard clays.
8. The soil resistance to driving can be reduced by 15 to 20 percent by the use of skin friction reducers.
9. Restart of piles at deep penetration (between 107m and 122m) in a normally consolidated formation is not a problem at this all clay site. The set-up after about 4 days was approximately 24 percent.

Acknowledgment

The authors wish to thank the management of Marathon Oil Company for the release of the soil and pile installation data and their permission to publish this paper.

References

Choe, J, and Juvkam-Wold, H. C., " Advanced Pile Driving Analysis with User-Interactive Programming," SPE 3909, Paper submitted to the Society of Petroleum Engineers for publication, 1997.

Dutt, R. N., Doyle, E. H., Collins, J. T., and Ganguly, P., "A Simple Model to Predict Soil Resistance to Driving for Long Piles in Deepwater Normally Consolidated Clays," Proceedings, 27th Offshore Technology Conference, Houston, Paper No. 7668, Vol. 1, pp 257-269, May 1995.

Edwards, T. C., "Pile Analysis Wave Equation Program Manual," Research Report No. 33-11, Texas A & M University, College Station, 1967.

Gibson, G. C., and Coyle H. M., "Soil Damping Constants Related to Common Soil Properties In Sands and clays," Research Report No. 125-1, Texas A. & M. University, College Station, September 1968.

Goble Rausche Likins and Associates, Inc., "GRLWEAP Manual, 1997-2 Update," Cleveland, September 1997.

Hery, P., (1983), Residual stress analysis in WEAP, Master's thesis, Department of Civil, Environmental, and Architecture, University of Colorado, Boulder, CO.

Holloway, D.M., Cough, G.W., and Vesic, A., S., (1975), The mechanics of pile-soil interaction in cohesionless soils, Soil Mechanics Series No. 39, School of Engineering, Duke University, Durham, N.C.

Lowery, L. L. Jr., Hirsch, T. J., and Samson, C. H. Jr., "Pile Driving Analysis – Simulation of Hammers, Cushions, Piles, and Soils," Research Report No. 33-9, Texas A. & M. University, College Station, September 1967.

Lowery, L. L. Jr., Hirsch, T. J., Edwards, T. J., Coyle H. M., and Samson, C. H. Jr., "Pile Driving Analysis – State of Art," Research Report No. 33-13, Texas A. & M. University, College Station, 1969.

Smith, E. A. L., "Pile Driving Analysis by the Wave Equation," Journal of the Soil Mechanics and Foundation Division, ASCE, Vol. 86, No. SM4, Proc. Paper 2574, August 1960.

STUDIES OF THE BEHAVIOR OF A SUCTION PILE
WITH A SIMPLIFIED NUMERICAL MODEL

by S. T. Wang[1] and José A Arréllaga[2]

ABSTRACT

Suction piles are sometimes used to support offshore structures where the soil at the base of the pile has sufficient permeability. A complicated finite-element solution can be used to study the behavior of a suction pile. However, in the particular case reported in this paper there were tight constrains in time and a relatively simple model was thus utilized. The results obtained from the simplified model were considered to be reasonable.

The simplified model of this study employed well-accepted, empirical, soil-resistance curves; p-y curves for lateral resistance and t-z curves for axial resistance. The numerical techniques presented herein are suitable for suction piles with any combination of vertical and lateral loading. The results of the analysis of suction piles with the simplified method are compared with results from a more involved finite-element solution. Assumptions and limitations of the simplified method are also addressed.

INTRODUCTION

A suction pile typically has a large diameter with a relatively short penetration (one or two pile diameters). The top of the pile has a stiffened cap for connection to the superstructure and the bottom of the pile is open-ended for ease of penetration, as illustrated in Fig. 1. Suction piles are sometimes used to support

[1] 'Project Manager, 'Ensoft, Inc./Lymon C. Reese & Associates, P.O. Box 180348, Austin, Texas 78723, ensoft@ensoftinc.com.

[2] Project Manager, Ensoft, Inc./Lymon C. Reese & Associates, P.O. Box 180348, Austin, Texas 78723, ensoft@ensoftinc.com.

offshore structures where the soil at the base of the pile has sufficient permeability. This type of piles has been used to support well platforms in the North Sea. A complicated finite-element solution can be used to study the behavior of a suction pile. However, in the particular case reported in this paper there were tight constrains in time and a relatively simple model was thus utilized. The results from the simplified model were considered to be reasonable.

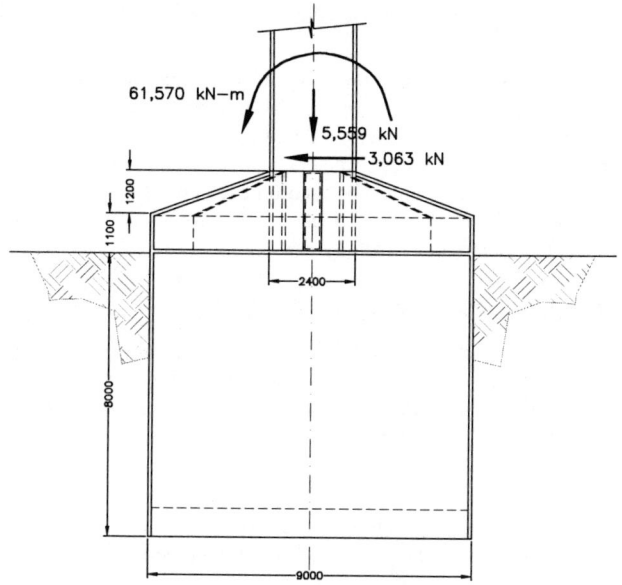

Figure 1. Sketch of a Suction Pile

The simplified model employed well-accepted, empirical, soil-resistance curves; p-y curves for lateral resistance and t-z curves for axial resistance. Typically, an external moment applied to a rigid cap at the top of the suction pile will cause the pile to rotate. Rotation is restrained by axial resistance consisting of both side friction and end bearing. Lateral deflection is restrained by lateral soil resistance.

The unique feature of the simplified model was the division of the circumference of the large-diameter suction pile into a number of small strips. Each strip was assumed to represent a *unit pile,* and the suction pile was modeled as a pile group consisting of unit piles placed contiguously. Each unit pile had the same geometry and stiffness. The total axial resistance was the sum of the axial resistance from each unit pile, with skin friction determined from t-z curves and end bearing from q-w curves. Similarly, the total lateral resistance was the resistance of the composite pile determined from p-y curves.

STUDY OF STABILITY PROBLEM OF SUCTION PILES

One critical element in the design of suction piles at different environments is to analyze the penetration needed for the suction piles to sustain wave loads created by the maximum storm. Two methods were used for this study. One method employed three-dimensional, finite-elements, which can provide detailed distribution of stress-strain in the vicinity of the pile. The other used a simplified pile-group model, which employs well-accepted, empirical, soil-resistance embodied p-y and t-z curves.

The soil profile selected for this study is a typical subsurface condition encountered at offshore locations. In general, the top layer is fine to medium, silty-silica sand, dense to very dense, with a thickness of 5 meters. The next stratum is a stiff to hard clay with a thickness 3 meters. The stiff clay is underlain by 1 meter of very dense, fine to medium sand. The total unit weight of the sand in the third layer is 21 kN/m^3 and the internal friction angle is 39 degrees. Another layer of stiff clay is extended from 9 meters to 12 meters.

LINEAR-ELASTIC, FINITE-ELEMENT ANALYSIS

The first step of the study is to build a linear-elastic model for the suction pile embedded into the surrounding soil in a three-dimensional space. The steel and the soils were treated as linear-elastic materials for the preliminary study in order to check the stress distribution and for verification of the solution.

Numerical Model

A three-dimensional, finite-element model, shown in Figure 2, was used in the preliminary analyses. The characteristics of the numerical model are:

- The pile is modeled using quadratic, 8-noded, shell elements.
- The soil is modeled using quadratic, 20-noded and 15-noded, solid elements.
- The soil is extended vertically about 2 times the diameter of the pile, and radially about 3 times the diameter of the pile.
- The model has 1408 elements and 5796 nodes.
- The boundary conditions are: at the bottom, displacements restrictions in all the components and at the vertical ends, both horizontal displacements components are restricted. The forces and moments are applied at level 1.5 over the top of the pile, and are transmitted to the pile through rigid elements.

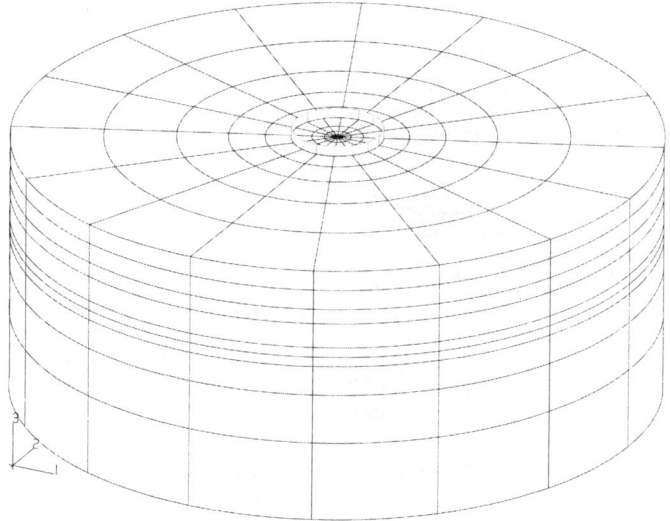

Figure 2. Three-dimensional Model

Displacements

The suction pile with a diameter of 9 meters and a wall thickness of 30 mm, which is similar to a steel can, was represented by shell elements. The applied forces and moments are: Fy=-3063 kN, Fz=5559 kN, Mx=93736 kN-m. The total penetrations of the steel can for the analyses were taken as 8 meters and 9 meters. The deformed shape shown in Fig. 3 resulted from the use of linear-elastic elements and appears to be reasonable.

It is clearly noted that the applied moment has a significant effect on the rotation of the short pile. Calling the left edge of the pile (at x=-R=-4.5m) LEFT, and the right edge of the pile (at x=R=4.5m) RIGHT, the variation of the horizontal displacement along the height of the pile is shown in Fig. 4.

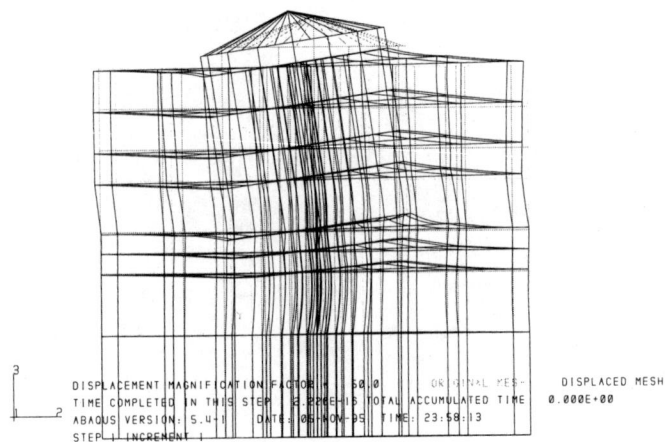

Figure 3. Deflected Shape for H = 8 m

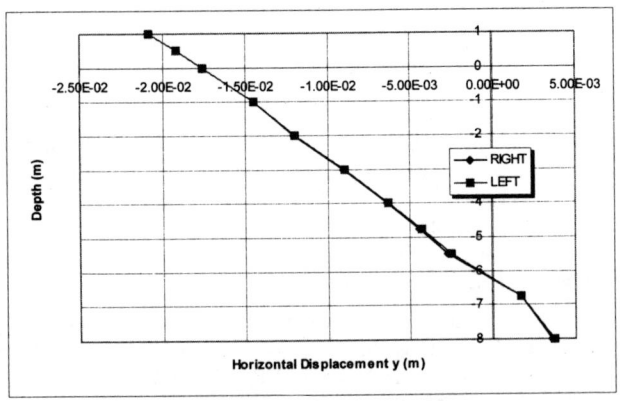

Figure 4. Horizontal Displacements for H = 8 m

For the case of H=9m, the horizontal displacements are shown in Fig. 5. The displacements are similar to the first case with 8-meter penetration. Figure 5 shows that the pile in the second case with 9-meter penetration is stiffer than the first one. In addition, the effect of the sand layer at the depth of 9 meters, which is stiffer with respect to the clay above it, exhibits a change in the rotation of the shell. It indicates that the dense sand provides a rotation restrain at the tip of the pile.

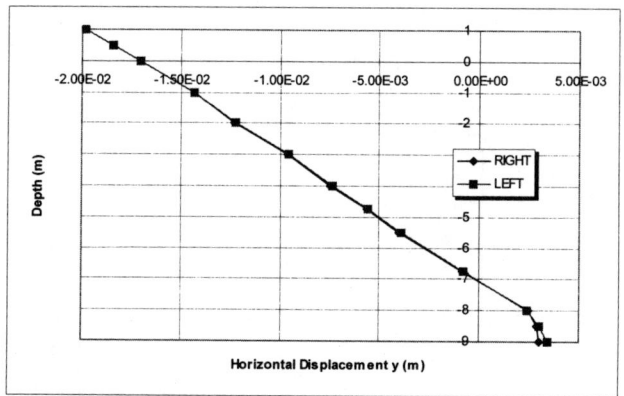

Figure 5. Horizontal Displacements for H = 9 m

PILE-GROUP MODEL

Description of Numerical Model

By assuming that the steel cap on the top of the suction pile is rigid, the suction pile can be simulated as a closely-spaced pile group as illustrated in Fig. 6. The unique feature of the analysis was the division of the circumference of the large-diameter suction pile into a number of small strips. Each strip was assumed to represent a *unit pile,* and the suction pile was modeled as a pile group consisting of unit piles placed contiguously. Each unit pile had the same geometry and stiffness. The total axial resistance was the sum of the axial resistance from each unit pile, with skin friction determined from t-z curves and end bearing from q-w curves. Similarly, the total lateral resistance was the resistance of the whole pile determined from p-y curves.

The method of pile-group analysis used herein is based on the concept presented by Reese and Matlock (1966a), but the problem of superimposing nonlinear systems was resolved by means of a successive-displacement-correction method for foundations with a rigid cap (Awoshika and Reese, 1971).

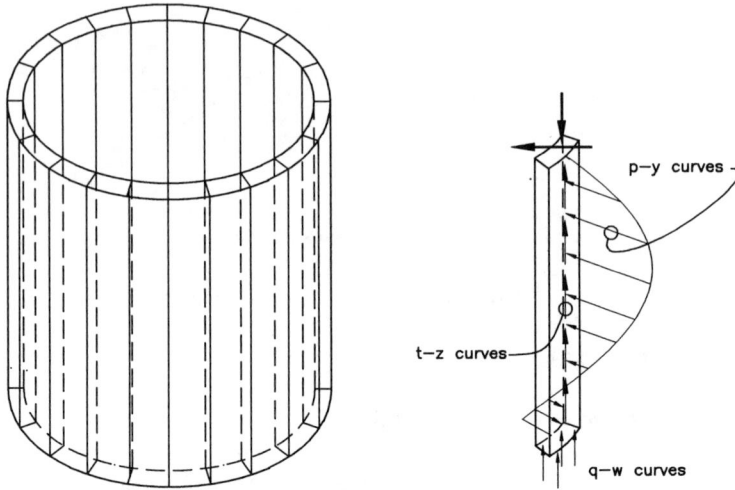

Figure 6. Pile Group Model

In the computer model, the pile cap (Fig. 7) is assumed to be non-deformable. Thus, the relative positions of the pile tops remain the same for any pile-cap displacement. The solution yields the movements of the pile cap, as shown in Fig. 7a, and the equilibrium of the cap, as shown in Fig. 7b. As shown in Fig. 7c, the pile heads may be pinned, fixed against rotation, or partially fixed against rotation.

The computer solution can be readily extended to the case of a deformable pile cap, which probably would require successive iterations between the superstructure and the foundation. The usual pile caps are much stiffer than the individual piles and pile-cap deformation can usually be neglected without introducing serious errors.

The coordinate systems for the group and for an individual pile are shown in Fig. 8a. The components of the inclined and eccentric loading, axial force P_v, the lateral force P_h, and the moment M, are computed and made to act at the origin of the global coordinate system, a-b. The x-y coordinate system is used for each pile in the group, as shown in Fig. 8, with the position of the pile head defined by appropriate distances in the global coordinate system and with the batter angle θ.

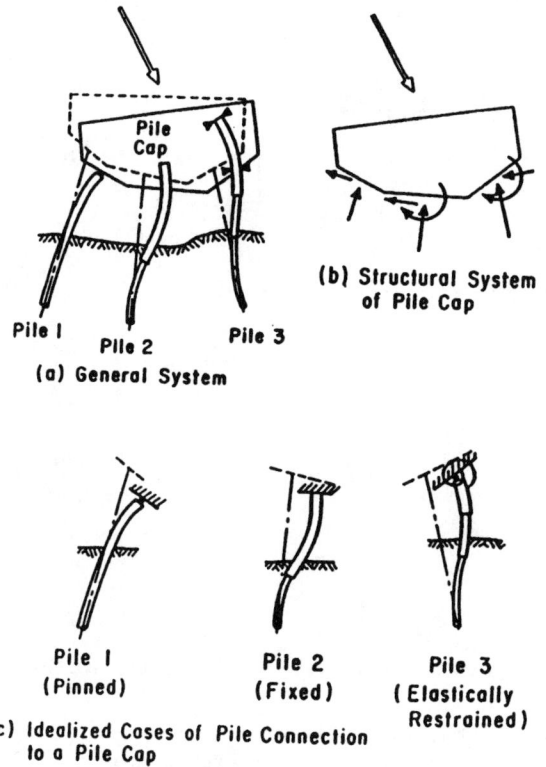

(b) Structural System
    of Pile Cap

Pile I   Pile 2   Pile 3

(a) General System

Pile I   Pile 2   Pile 3
(Pinned)   (Fixed)   (Elastically
                            Restrained)

(c) Idealized Cases of Pile Connection
    to a Pile Cap

Figure 7. Basic Structural System

The origin of the global coordinate system (pile cap) is given two displacements, Δv and Δh, and a rotation α. The corresponding movements of the pile head can be computed from the given geometry and are represented in the sketch. The resulting pile-head forces, in both of the coordinate systems, are depicted in Fig. 8b.

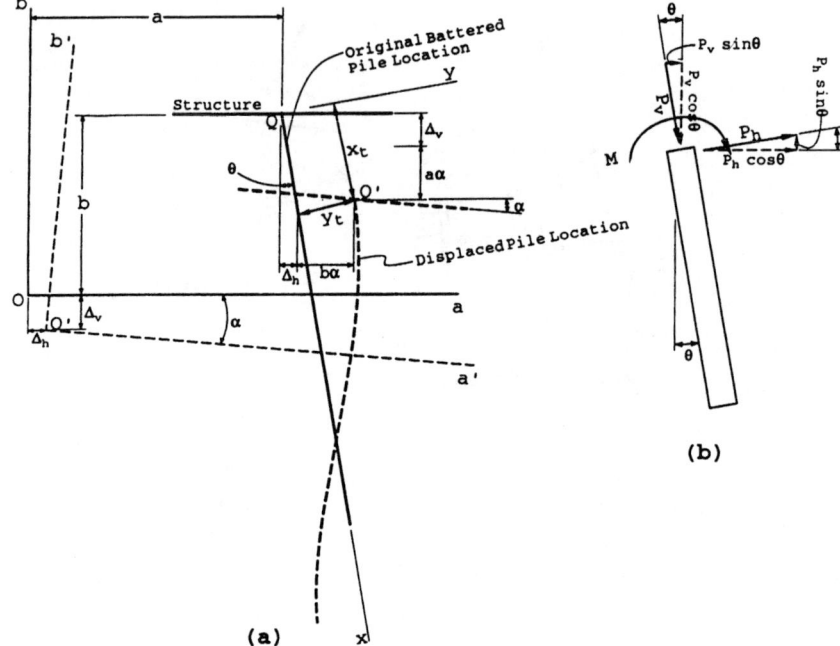

Figure 8. System for the Analysis of a Two-Dimensional Group of Piles

The pile-head forces may be computed by multiplying pile head movements by stiffness factors J_x, J_y, and J_m, for computing axial load, lateral load, and moment, respectively. These stiffness factors are nonlinear functions of the pile-head movements, with the nonlinear functions computed by use of the procedures for isolated piles. With the first iteration, the forces in the global coordinate system may be computed as shown in Fig. 8b. The equilibrium of the pile cap can be computed, as follows.

$$P_v + \sum F_{v_i} = 0 \tag{1}$$

$$P_h + \sum F_{h_i} = 0 \tag{2}$$

$$M + \sum M_{s_i} + \sum a_i F_{v_i} + \sum b_i F_{h_i} = 0 \tag{3}$$

Equilibrium is not achieved unless the nonlinear J-values had been correctly estimated. Therefore, the equilibrium is sought by the successive correction of pile-cap displacement. After each correction of the displacement, the difference between the load and the pile reaction is computed and the next correction is obtained through the calculation of a new stiffness value at the previous pile-cap position. The stiffnesses are obtained by giving a small increment to each component of displacement, one at a time. The proper magnitude of the increment may be set at 1 x 10^{-5} times a unit displacement to attain acceptable accuracy.

Soil Resistance on Suction Piles

The soil resistance on the suction pile consists of the lateral soil resistance, friction on the interior and exterior surfaces of the steel pipe, and the tip resistance at the metal area of the pile tip. As noted in the finite-element result, the steel pipe will rotate significantly due to the applied moment. The axial resistance, including the friction and the end bearing, will restrain the rotation. The group-model has the capability to compute the behavior of the suction pile by taking into account the nonlinear effects in the lateral soil resistance as well as the soil friction and tip resistance.

The skin friction and tip resistance for each member strip were computed, based on methods presented by the American Petroleum Institute (RP2A). The group model has the capability of developing load-transfer curves internally, showing resistance in skin friction (t-z curves) and in end bearing (q-w curves). A computer program was prepared for the group model employing those curves while computing the movements of the individual members. The resulting forces satisfy the conditions of equilibrium.

The lateral resistance on the suction pile can be computed, based on the p-y method. The soil resistance (p-y curves) applied on each member is equivalent to the soil resistance derived based on the full size (9-m in diameter) divided by the number of members. Twenty member strips were used for this particular application, thus resulting in strips with diameters of 0.45 m. The same loads used for the finite-element analysis were applied to the pile cap. The behavior of the suction pile for penetration of 7 m, 8 m, 9 m, and 10 m was studied by the simplified pile-group model. As examples, p-y curves at a depth of 1.75 m for a 9-m-diameter pile and a 0.45-m-diameter strip member are shown in Fig. 9.

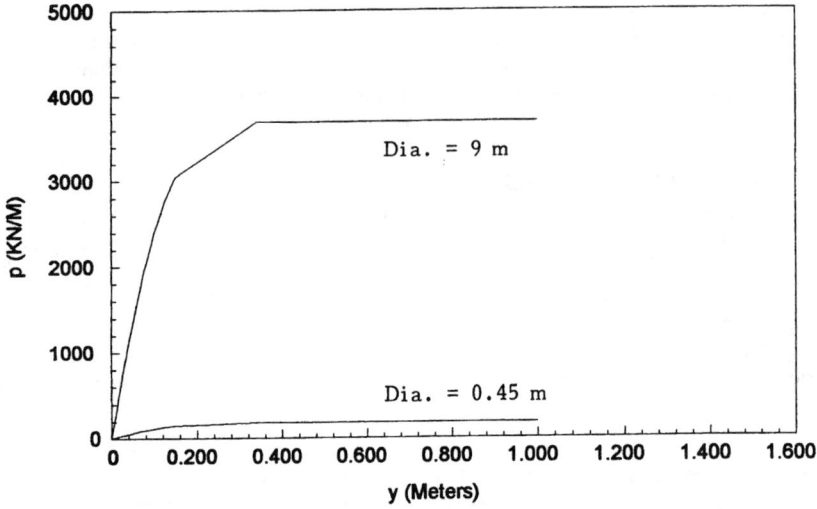

Figure 9. p-y Curves at a Depth of 1.75 m for a 9-m-diameter
Pile and for a 0.45-m-diameter Member

Deflection for All Penetrations

As shown in Fig. 10, the computed deflection at the pile cap for a penetration of 8 meters is about 55 mm, which is significantly greater than the deflections for cases with penetration equal to 9 and 10 meters. The case with penetrations equal to 7 meters will cause failure in bearing as indicated by the computer result. It is apparent that the suction pile becomes unstable when the penetration decreases to 8 meters or less for the soil conditions at this site. The pile tip will be embedded into the clay layer if the penetration is equal to 10 meters. The deflection at the top for the case with the penetration equal to 10 meters is slightly greater than that for 9-meter penetration because the tip resistance decreases in the last clay layer.

The deflection at the pile head computed by the linear-elastic finite element method is about 21mm for the case with 8-m penetration and 20 mm for the case with 9-m penetration. For a 9-m penetration, the suction pile is in a stable condition as discussed earlier, and the deflection at the top has a good agreement between the two methods. However, the linear-elastic solution is not able to simulate the wedge failure of the soil if the soil is overstressed into plastic state. Observing the results from this example, the advantages provided by the nonlinear group model is thus obvious for the design of suction piles.

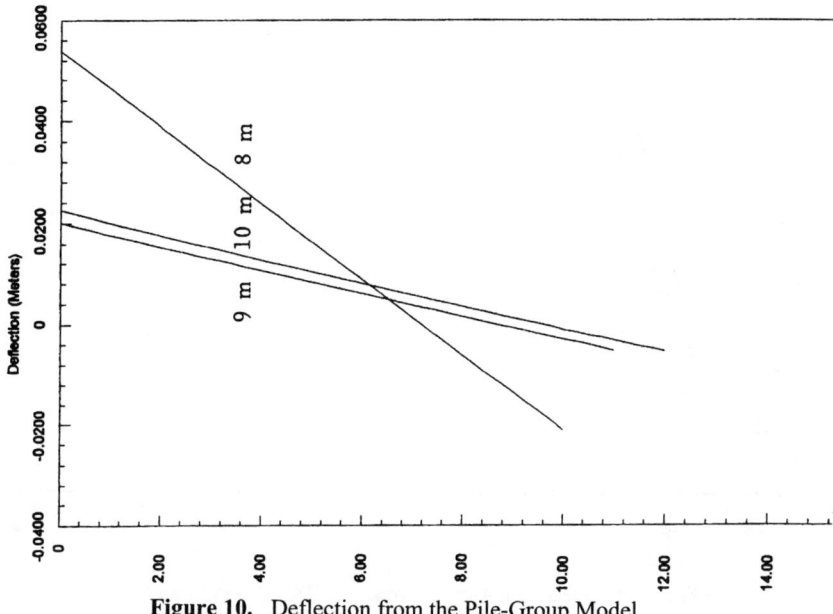

Figure 10. Deflection from the Pile-Group Model
as a Function of Depth for Various Penetration

Bending Moment for All Penetrations

The bending moment as a function of depth for the member at the front edge is shown in Fig. 11. The cross section of each member is about 450 mm by 30 mm. However, it is believed that the actual bending moments may be less than those obtained from the group model because the hoop stresses between moments were not considered in the simplified model. In any event, a structural member design using the simplified model would be conservative in flexure.

CONCLUSION

The method presented in this paper provides engineers a suitable model to study the stability of suction piles. With this method engineers can easily investigate alternate geometries, can study the influence of the various parameters on a solution, and can produce results that are instructive and valuable. Continued improvements of the simplified group model for the design of suction piles depend strongly on the availability of additional measured data of the response of suction piles to loading.

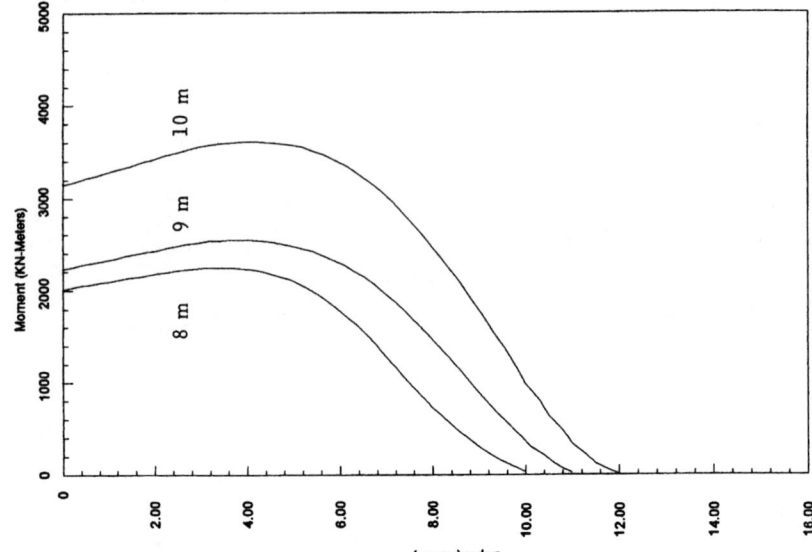

Figure 11. Bending Moment from the Pile-Group Model
as a Function of Depth for Various Penetrations

ACKNOWLEDGEMENTS

The finite-element model was conducted by a team led by Dr. José Roesset at the Offshore Technology Research Center of the University of Texas at Austin. We also appreciate the useful recommendations provided by Dr. Lymon C. Reese.

REFERENCES

American Petroleum Institute, "Recommended Practice for Planning, Designing and Constructing Fixed Offshore Platform, API Recommended Practice 2A (RP 2A)," Nineteen Edition, 1993.

Awoshika, K. And L. C. Reese, "Analysis of Foundation with Widely-Spaced Batter Piles," Research Report 117-3F, Center for Highway Research, The University of Texas at Austin, February, 1971.

Reese, L.C. and H. Matlock, "Behavior of a Two-Dimensional Pile Group Under Inclined and Eccentric Loading," Proceedings, Offshore Exploration Conference, Long Beach, California, February, 1966.

Total and Effective Stress Analysis of Suction Caissons for Gulf of Mexico
Conditions

Beena Sukumaran[1], Associate Member, ASCE and Bill McCarron[2], Member, ASCE

Abstract

This paper presents numerical analyses of suction caissons for Gulf of
Mexico soil conditions. Deepwater Gulf of Mexico surface sediments are
predominantly normally consolidated with shear strengths negligible at the mud line
and increasing linearly with depth. Total and effective stress analyses are presented
using commonly available material models. The effects of loading direction, load-
point application, and caisson length to diameter ratio are investigated. The
numerical analyses include plane strain and Fourier interpolation element
simulations of three-dimensional response. Considerations are given to the effect of
the strength model used to describe the soil mass. Practical observations on the
importance of various modelling assumptions are given.

Introduction

Suction caissons are ideal foundations for laterally moored production
structures in the deepwater Gulf of Mexico. Laterally moored systems such as
SPARs, semi-submersibles and tanker-shaped structures have global performance
characteristics that are in many respects less demanding on foundations than those
for vertically anchored tension leg platforms. The principal foundation loading
characteristic that makes suction installed caissons attractive for laterally moored
systems is that the predominant direction of foundation loading is horizontal, so that
lateral foundation capacity is the controlling issue. This is opposite to the
requirements of vertically moored structures.

Sukumaran et al. (1999) have demonstrated the successful application of
finite element techniques to the analysis of suction caissons under elastic perfectly-

[1] Assistant Professor, Rowan University, 201 Mullica Hill Road, Glassboro, NJ 08028
[2] Project Engineer, BP Amoco p.l.c., Houston, TX

plastic material assumptions and total stress conditions. In this paper, the application will be extended to effective stress analysis. Stress paths experienced in the soil mass will be examined for total and effective stress analyses. The effects of caisson length-to-diameter ratio L/D will be examined with respect to limiting lateral pressures.

The present work includes use of von Mises, Tresca and a capped-plasticity models. The latter model is used in effective stress analyses, and the remaining are used in total stress analyses. The soil models parameters are selected to simulate a cohesive material with strength and stiffness increasing with depth. These conditions are typical of Gulf of Mexico deepwater conditions. The finite element analyses are performed with the program ABAQUS (HKS, 1997).

Background

Sukumaran et al. (1999) have demonstrated the successful application of finite elements to suction caisson analysis. Those authors considered a number of two- and three-dimensional verification problems for which classical solutions or experimental results were available to confirm details of the numeric modelling. Both displacement and hybrid formulation finite elements were found to produce well-defined limit loads that compare reasonably with analytical and experimental results. Since the analyses were performed under a total stress assumption, suction induced pore pressures were not explicitly considered. The following are key conclusions or observations made those authors.

- The maximum anchor capacity is obtained when the load attachment point forces the caisson to have a translational mode of failure rather than a rotational mode of failure.
- Inclined loads applied at the face of the caisson tend to reduce caisson rotation, resulting in greater lateral capacity.
- A pseudo three-dimensional model available in ABAQUS, utilizing the use of Fourier interpolation elements to define approximate three-dimensional conditions, can be used instead of an actual three-dimensional model for analyzing the capacity of a suction caisson with considerable computational timesavings.
- Second-order reduced integration elements are sufficient for numerical analysis of three-dimensional caisson geometries.

Modelling of Foundation Soil

The objective here is to simulate the response of caissons in deepwater Gulf of Mexico cohesive sediments, which are typically normally consolidated or lightly overconsolidated near the surface. The analyses presented here incorporate two slightly different strength profiles. This is largely due to a desire to retain the previous total stress analysis work for demonstration purposes, and the need for more

precise definitions of material parameters for effective stress analyses. The strength
profile used for total stress investigations is

$$S_u^{DSS} = 1.41 \cdot z \ (kPa) \tag{1}$$

where z is the depth below seabed in meters and S_u^{DSS} is the undrained static direct
simple shear strength.

The model employed for the effective stress analyses is the "clay-plasticity"
model available in ABAQUS (HKS, 1997). This model falls within the critical state
type of strain-hardening soil plasticity models. The specific formulation is similar to
the Modified Cam-clay model (Roscoe and Burland, 1968). The clay-plasticity
model allows for stress path dependent strength by virtue of a non-circular trace of
the yield surface in the deviatoric plane. For the effective stress analyses, the triaxial
compression (TC) strength profile is interpolated linearly between the values in
Eq. 2.

$$
\begin{array}{llll}
S_u^{TC} = & 5.5 \ \text{kPa} & \text{Depth} = & 0 \ \text{m} \\
& 6.0 \ \text{kPa} & & 2.6 \ \text{m} \\
& 22.6 \ \text{kPa} & & 12.0 \ \text{m} \\
& 67.0 \ \text{kPa} & & 36.6 \ \text{m}
\end{array} \tag{2}
$$

Figure 1 shows the deviatoric cross sections of the three different strength
models used in the present work. These models are the von Mises, Tresca and cam-
clay type models available in ABAQUS. The various models are plotted so that the
so-called simple shear (SS) strength is the same for each model. The term 'simple

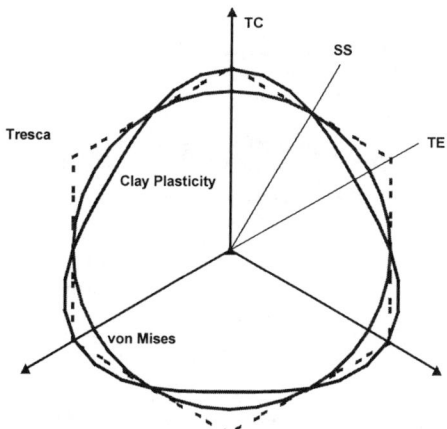

**Figure 1. Deviatoric Cross-Sections for Strength
Models.**

shear' is not associated with the geotechnical test procedure. Shown in Fig. 1 are the relative positions of triaxial compression (TC), triaxial extension (TE) and simple shear stress paths.

Natural deepwater sediments show considerable variation in physical and mechanical properties in the top few meters. For the purpose of demonstrating numerical analyses, the values in Table 1 are considered to be representative of a typical deepwater Gulf of Mexico clay to a depth of 12 m below the mud line. These values are based on the authors' experience and published data (e.g., Jeanjean et al, 1998), and represent a simplification of natural conditions.

Table 1: Characteristic Soil Properties at Shallow Depths.

$S_u^{TC}/S_u^{DSS} = 1.25$ (OCR = 1) $S_u^{TE}/S_v^{DSS} = 0.85$ (OCR = 1) $S_u^{DSS}/\sigma_v = 0.26$ (OCR = 1)	$\gamma' = 6.3$ kN/m^3 $\varphi = 27°$ $w = 0.6$ $v = 0.3$	$e = 1.66$ $G_s = 2.77$ $\kappa = 0.031$ $\lambda = 0.194$

Results From Plain Strain Analysis of Pile Cross-Section Resistance

The topic of ultimate lateral soil resistance to circular pile cross-sections is briefly addressed in order to point out some important modeling assumptions and establish a frame of reference for later discussions of normalized lateral capacities of caissons and general responses of caisson foundations. Three modelling issues are addressed: pile-soil interface conditions, selection of material model, and the use of initial stress conditions. Figure 2 shows the responses for four analyses performed with permutations of the modelling parameters mentioned. Three of the analyses use

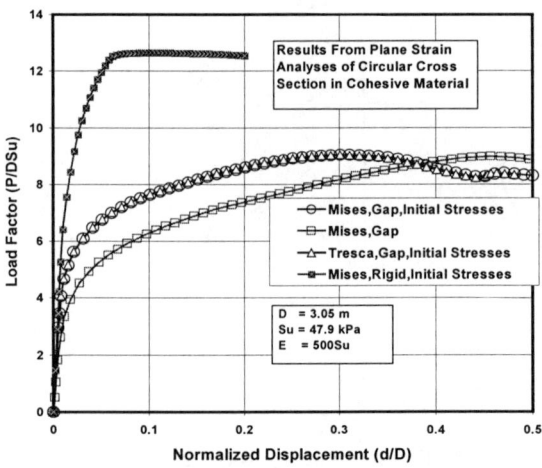

Figure 2. Load-Displacement Response for Plane Strain Analyses.

the von Mises strength model, but have different interface conditions or initial stress conditions. Under the assumption of rigid/perfect contact between the pile-soil interface, the bearing factor at failure $N = P_{ult}/DS_u$ is 12.6, but when separation/gapping is allowed at the interface the load factor is reduced to 9. Both of these cases included initial hydrostatic stresses equal to 1.2 times the undrained shear strength S_u. A third analysis allows gapping, but does not include initial stresses. The bearing factor for this case is also 9, but occurs at a larger displacement, indicating the importance of an initial stress in preventing separation of interfaces. The final analysis uses the Tresca strength model, has gap elements and initial stresses. The Tresca model strength is matched with the von Mises strength model at the simple shear SS location (Fig. 1). Softening responses shown for analyses with gap elements are due to changing boundary conditions at the soil-pile interface when separation occurs. This behavior can occur at shallow to intermediate depths for real pile condition, but is unlikely at large depths because the soil cannot support unlimited overburden pressures in the presence of a free surface.

Figure 3 shows the extent of the plastic zone in the finite element mesh for the von Mises strength model and gapping allowed. The modeled diameters of the pile and exterior boundary are 3.05m and 54.8m, respectively. Plastic zones exist for a distance of six diameters from the pile section center. Figure 4 shows the stress paths for various angular locations at a distance of 1.5 diameters from the pile center for analyses with von Mises and Tresca strength models. The points at are located at angles of 0°, 45°, 90° and 180° around the model. The 0° and 180° locations are on the passive and active pressure sides of the model, respectively. Stress paths in the forward or passive side of the pile center are directed toward the SS meridian in the deviatoric plane. The models were calibrated so that their strengths at the SS meridian matched. Figures 2 and 4 demonstrate that the same results are obtained when the model parameters are so selected. The fact that this matching condition gives identical capacities for frictionless materials in plane strain conditions has been demonstrated analytically by Drucker and Prager (1952).

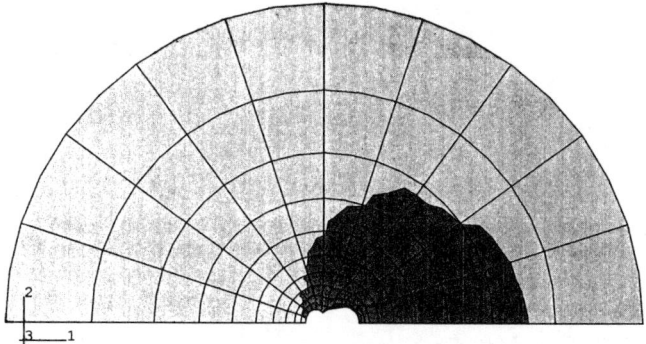

Figure 3. Zone of Plastic Strains for Plane Strain Problem.

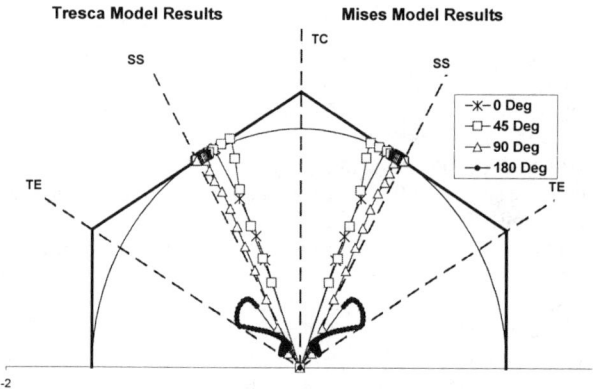

Figure 4. Deviatoric Stress Path for Plane Strain Analyses.

3-D Caisson Lateral Response With Total Stress Analyses

The results of three-dimensional caisson response with a total stress assumption for the foundation material are presented here to demonstrate the significant features affecting caisson capacity. The analyses make use of Fourier interpolation elements available in ABAQUS. Reduced integration second-order displacement-based elements are used in the present work. Sukumaran et al. (1999) provide a more detailed discussion and validation of the selection of these elements. Figure 5 shows the finite element mesh used. The depth of the caisson is 40m, and diameters of 3.05 and 6.1m are analyzed. Initial stresses are specified in the model with a K_o value of 0.8. The effects of load direction, material strength model, point of load application and length-to-depth ratio L/D on the capacity are illustrated in Fig. 6. The loads are applied on the face of the caisson at the top or mid-height, or rigid horizontal displacements are prescribed.

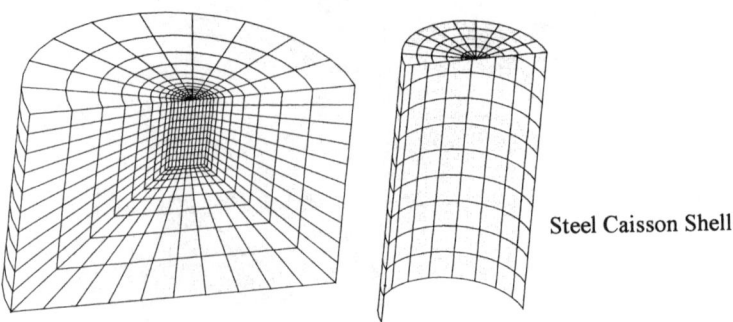

Steel Caisson Shell

Figure 5. Fourier Interpolation Finite Element Model for Three-Dimensional Caisson Geometry.

The following observations are made with respect to Fig. 6: 1) the minimum capacity results when the load is applied horizontally at the top of the caisson, 2) inclined loads result in increased capacity compared to identical conditions with purely horizontal loading, 3) larger L/D ratios result in reduced normalized capacities, 4) maximum capacities (at least for the cases investigated) are obtained for horizontal translation without rotation, 5) Tresca and von Mises strength models give approximately the same limit loads. The fact that analyses with inclined loads result in larger capacities compared to only horizontal loading is explained by the fact that the vertical load component creates a moment on the caisson that reduces the tendency of the caisson to rotate.

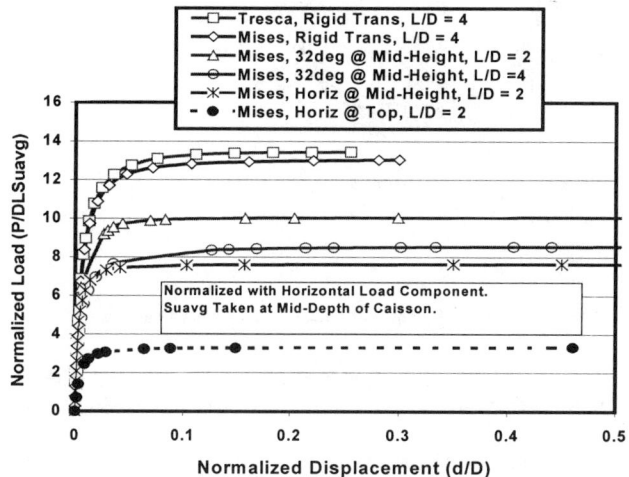

Figure 6. Load-Displacement Response for Total Stress Caisson Analysis.

The limit loads shown in Fig. 6 are well defined. The load factor 7.6 found for horizontal loading on a caisson L/D=2 compares favorably with the factor 8.4 determined experimentally by Randolph et al. (1998) for similar conditions.

Figure 7 contains polar coordinate plots of radial pressures on the caisson surface for various depths on its surface. The results are for a prescribed horizontal displacement of the caisson in the 0° direction without caisson rotation. The caisson pressures are normalized by the undrained strength at the same depth. The left side of Fig.7 shows results for the Mises model and the right side has the Tressca model results. The normalized pressures σ_r/S_u increase from 8 to 12 for 25% and 95% of the caisson depth, respectively. There are only minor differences between the von Mises and Tresca strength model results. Because the strength models allow tensile stresses, tensile stresses do develop in the analyses for coordinate angles greater than 120°. The magnitudes, however, are small. The stress patterns shown in Fig. 7 can be used to perform structural design of suction caissons. These stresses compare well with experimental and analytical limit results discussed by Murff and

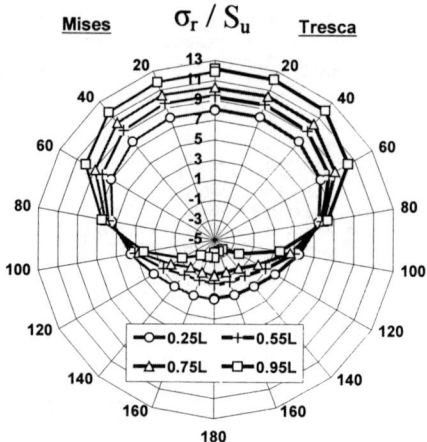

Figure 7. Normalized Radial Caisson Stresses; Total Stress Analysis, L/D = 4.

Hamilton (1993). For analyses with prescribed horizontal translation and without rotation, the resultant force from soil pressures is located at ¾ of the caisson depth.

3-D Caisson Lateral Response With Effective Stress Analyses

The finite element mesh in Fig. 5 is used for the effective stress analyses of a caisson with diameter and length of 6.1 m and 12.2 m, respectively. Initial stresses are specified with a K_o value of 0.7. The material model used is the 'clay plasticity' model available in ABAQUS. A complete description of the model is found in the ABAQUS Users Manuals (HKS, 1997). The model was calibrated to match as closely as possible the strength profile described in Eq. (2) and the properties contained in Table 2. The ratio S_u^{TC}/σ_v for $K_o = 0.6$ consolidation is 0.27 for the selected model parameters, which matches closely laboratory data. Because of limitations of the model, the minimum ratio of the TE/TC strength ratio is 0.78. This is higher than that implied in Table 2, but is nevertheless a reasonable value for this application. Permeability is described as being inversely proportional to depth with $k_v = 0.3/z$ m/yr.

Figure 8 shows the load-displacement (horizontal components in each case) and suction pressure response for three analyses. In each analysis, the load is applied at mid-depth on the face of the caisson at an inclination of 32° from the horizontal. The first analysis represents a monotonic loading until the analysis fails to converge at a normalized load level of 9.45. The time period of loading is two days. At the peak load achieved the suction pressure in the plug accounts for 70% of the resistance to the vertical load component. Excess pore pressures were found to be extremely uniform in the caisson soil plug. The second analysis is similar to the first,

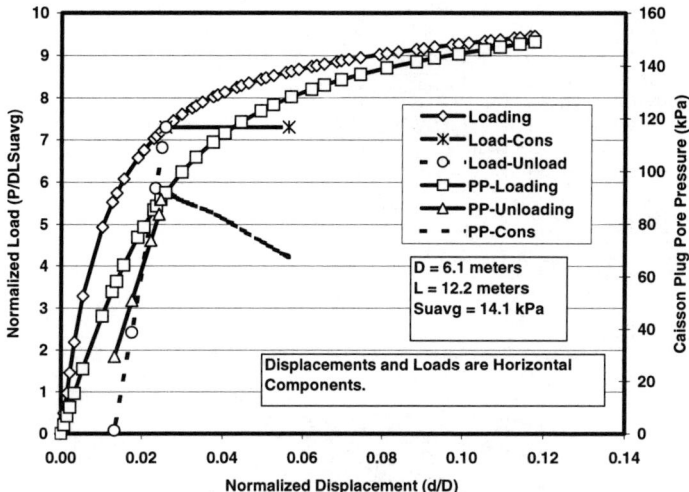

Figure 8. Load- and Pore Pressure-Displacement Response for Effective Stress Analysis.

except that the load is held constant after achieving a value of 7.3 and pore pressures are allowed to dissipate in the model for a period of 352 days. The final case represents the response to a load-unload cycle over a period of two days. The three analyses allowed pore fluid drainage at the free surface of the soil only.

Referring to Fig. 8, a number of observations can be made: 1) the increase in pore pressure is roughly proportional to the applied load, 2) consolidation results in a large increase in displacement, 3) consolidation for long periods of time may not reduce the plug pore pressures appreciably when the external load is still applied, 4) unloading leaves a residual displacement that is large compared to the loaded value, and 5) unloading results in a large reduction in plug pore pressures. Limiting loads the analysis is achieved at displacements of approximately 10 to15% of the diameter. The displacement reported in Fig.8 is that at the point of loading. Larger displacements occur at the top of the caisson.

Contour plots of excess pore pressures and plastic strain magnitudes are shown in Fig. 9. Note that the sign convention for the pore pressure in Fig. 9a is the geotechnical sign convention rather than the mechanics convention. From these figures, the following are noted: 1) plastic strains are limited to the soil within one caisson diameter of the caisson surface, 2) positive (geotechnical convention) excess

pore pressures occur on the passive pressure side of the caisson, 3) negative
(geotechnical convention) pore pressures occur in the soil plug (not shown for
clarity), below the caisson and on the active pressure side of the caisson.

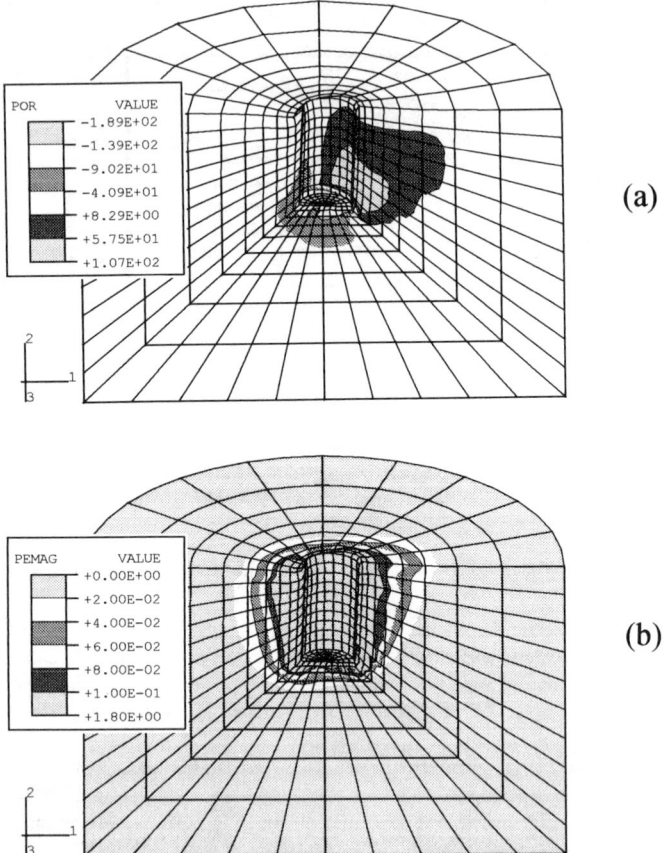

**Figure 9. Results from Effective Stress Analysis of 3-D Caisson for Peak Load
Condition; (a) Excess Pore Pressure, (b) Plastic Strain Magnitudes.**

Examples of stress paths on the active and passive side of the caisson are
shown in Fig. 10. In Fig. 10a, the stress path in the p-q plane is shown. Fig. 10b
shows the Lode angle response. The stress paths are taken from points approximately
3 m from the caisson face. Fig. 10a shows that the effective mean stresses are being
reduced on both sides of the caisson, and that the deviatoric stresses are increasing at
each of the locations selected. The Lode angle response shown in Fig. 10b illustrates
that on the passive side of the caisson, the stress path moves away from the initial TC

state and then returns. The Lode angle is the angle measured from the TC meridian to the stress state in the deviatoric plane shown in Fig. 1. On the active pressure side, the responses generally move monotonically from TC to SS and TE states. The depth of the various points in the model can be determined by noting that the mean stress increases approximately 5 kPa per meter depth. Thus, points with initial mean stresses greater than 60 kPa are below the base elevation of the caisson.

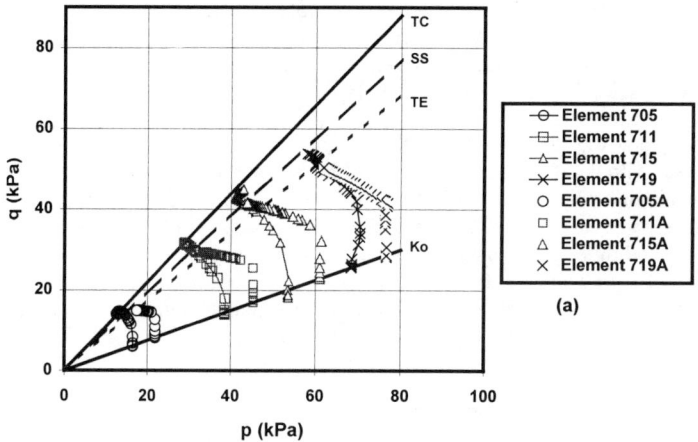

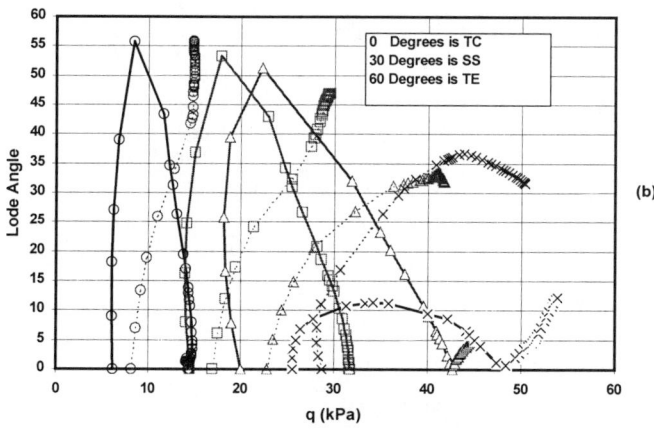

Figure 10. Effective Stress Paths for 3-D Caisson Analysis; (a) p-q Plane, (b) Lode Angle.

The measure of shear stress shown in Fig. 10 is the equivalent deviatoric stress q. In a triaxial test, q is equal to the difference in maximum and minimum principal stresses. Element numbers annotated with a following 'A' in the legend are on the active or trailing side of the caisson.

Figure 11 shows the variation of horizontal effective stresses, pore pressures, and total horizontal stresses inside and outside of the wall on the leading edge of the caisson. Also plotted on this figure is the ratio of the total horizontal stress to undrained shear strength (Sig_{tot} / S_u^{TC}). The total stress is the sum of the effective stress and pore pressure. The data are plotted only to a depth of 10 m because localized stress oscillations occur below that depth.

Referring to Fig. 11, it is evident that the pore pressures inside the caisson are nearly constant with depth. Additionally, the effective horizontal stresses on both sides of the caisson wall are nearly identical, and have changed only nominally from the initial at rest values specified. The majority of the passive pressure generated due to the applied loading is in the pore pressure component of stresses. The ratio Sig_{tot}/S_u^{TC} is between 7 and 8 below 3 meters depth (25% of the caisson depth). These values compare well with those reported by Sukumaran et al. (1999) for total stress analyses of the same geometry with inclined loading.

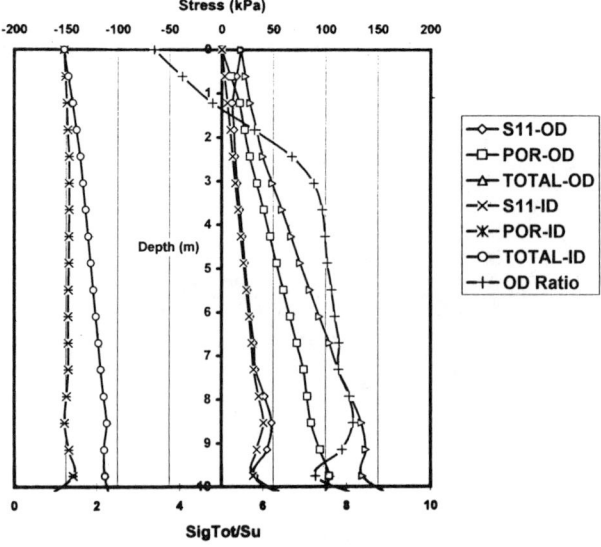

Figure 11. Horizontal Components of Total and Effective Stresses on the Leading Edge of the Caisson. (S11 = Horizontal Effective Stress, POR = Pore Pressure, TOTAL = S11 + POR)

Summary

Total and effective stress analyses have been performed to provide insight into the performance of laterally loaded suction caissons for normally consolidated soil profiles typical of deepwater Gulf of Mexico conditions. Total stress analyses produced well-defined limit conditions. Although not elaborated on here, the limiting radial soil bearing pressures compare well with experimental results reported by others. Plane strain and three-dimensional total analyses indicate that either von Mises or Tresca strength models can be used to calculate limit loads if they are matched at the SS meridian in the deviatoric stress plane.

Total stress analyses with inclined loads give maximum bearing factors $N = P_{ult}/DLS_{uavg}$ between 8.5 and 10, depending on the L/D ratio for the caisson. An effective stress analysis ended with a bearing factor of 9.45, but the limit load was not fully defined. These calculations are based on the horizontal load component. Plane strain and three-dimensional total stress analyses indicate limiting radial bearing pressures σ_r/S_u between 8 and 12 on caissons horizontally translating caissons without rotation. Previous work by Sukumaran et al. (1999) indicate values below $\sigma_r/S_u = 8$ are likely for horizontal or inclined loads accompanied by rotation of the caisson. These results combined with the stress distributions provided herein can be used to perform preliminary designs of suction caissons in normally consolidated cohesive soils.

Plane strain analyses demonstrated that specified initial horizontal stresses influence the predicted finite element response when gap elements are used at the soil-pile interface. Separation of pile-soil interface can lead to reduced capacities. This behavior has been observed in experiments (Randolph et al., 1998). Effective stress analyses demand that the initial stresses be specified in order that the correct initialization of material model is obtained. Without such initialization, stress-path dependent responses will be affected.

References

Drucker, D.C., and W. Prager, 1952. Soil Mechanics and Plastic Analysis of Limit Design. *Quarterly of Applied Mathematics*, Vol. 10, No.2, pp. 157-165.

HKS, 1997. ABAQUS Version 5.7 User's Manuals. Hibbett, Karlson and Sorensen Inc.

Jeanjean, P., K.H. Andersen, and B. Kalsnes, 1998. Soil Parameters for Design of Suction Caissons for Gulf of Mexico Deepwater Clays. *Proceedings of the Offshore Technology Conference*. Houston. Paper No. 8830, pp. 505-519.

Murff J.D, and Hamilton J.M., 1993. P-Ultimate for Undrained Analysis of Laterally Loaded Piles. *ASCE Journal of Geotechnical Engineering*. Vol. 119, No. 1. pp. 91-107.

Roscoe, K.H. and J.B. Burland, 1968. On the Generalized Stress-Strain Behavior of 'Wet' Clay. In *Engineering Plasticity*, J. Heyman and F.A. Leckie (eds.), Cambride University Press. pp. 535-609.

Randolph, M.F., M.P O'Neill, D.P. Stewart, and E. Erbrich., 1998. Performance of Suction Anchors in Fine-Grained Calcareous Soils. *Proceedings of the Offshore Technology Conference*. Paper No. 8831, pp. 521-529.

Sukumaran, B., W.O. McCarron, P. Jeanjean, and H. Abouseeda, 1999. Efficient Finite Element Techniques for Limit Analysis of Suction Caissons Under Lateral Loads. *Computers and Geotechnics*. Accepted for publication, to appear in 1999.

Definition of Terms

D	Diameter
DSS	Direct simple shear
e	Void Ratio
E	Young's elastic modulus
G_s	Specific gravity
K_v	Vertical Permeability
K_0	Ratio of horizontal to vertical stress
L	Length or depth of embedment
OCR	Overconsolidation ratio
p	Mean effective stress
P_u	Ultimate load
q	Mises equivalent deviatoric stress = $(3/2\ s_{ij}\ s_{ij})^{1/2}$
s_{ij}	Deviatoric stress tensor
S_u	Undrained strength
SS	Simple shear
TC	Triaxial compression
TE	Triaxial extension
w	Water content
z	Vertical coordinate
γ'	Submerged unit weight
κ	Slope of e-*ln* p virgin compression curve
λ	Slope of e-*ln* p rebound curve
ν	Poisson's ratio
σ_r	Radial pressure on caisson
σ_v	Vertical effective stress
φ	Friction angle for soil strength

The behaviour of distinctive pile type under the same building

W.F. Van Impe[1]

Abstract

During the construction of a new office building in Waddinxveen - The Netherlands, an elaborated research project on the pile foundation was set up. The aim of the research was to clarify the influence of various installation techniques of distinctive pile type on the corresponding pile capacities and to study the interaction of the pile foundation and the concrete office building structure. The pile foundation consists of 55 piles of 6 different type ; prefabricated concrete piles, Vibro piles, Vibrex piles, CFA piles, steel H-beam piles and concrete sheet piles all carrying top load cells from the start of the building construction. Static load tests on 23 piles were performed between 1997 and the first months of 1998. During the load testing, the own weight of the building was acting as the necessary reaction force. In this paper the main results of the pile load tests and their relation to the overall design predictions will be discussed.

Introduction

IFCO, a Dutch foundation engineering consultant, has used the pile foundation below their new 3 floors-office building, as pile research facility. The total building weight of approximately 24 MN is transferred by 55 piles, of 6 distinctive type (fig. 1), to a firm sand stratum, at 7 m below the soft clay and peat deposits. IFCO applied 15 slender precast piles, 10 steel H-beams, 9 Vibro-piles, 10 Vibrex-piles, 8 CFA-piles (fig. 2) and finally 3 prestressed concrete sheet piles. The initial site investigation consisted of 7 CPT, (fig. 1), while another 14 CPT have been carried out after pile installation, in order to check for changes in soil strength as a consequence of the installation of the various type of piles. After installation, pile

[1]Prof., Director Laboratory of Soil Mechanics, Ghent University, Belgium

Prof., Soil Mechanics, Catholic University Leuven, Belgium

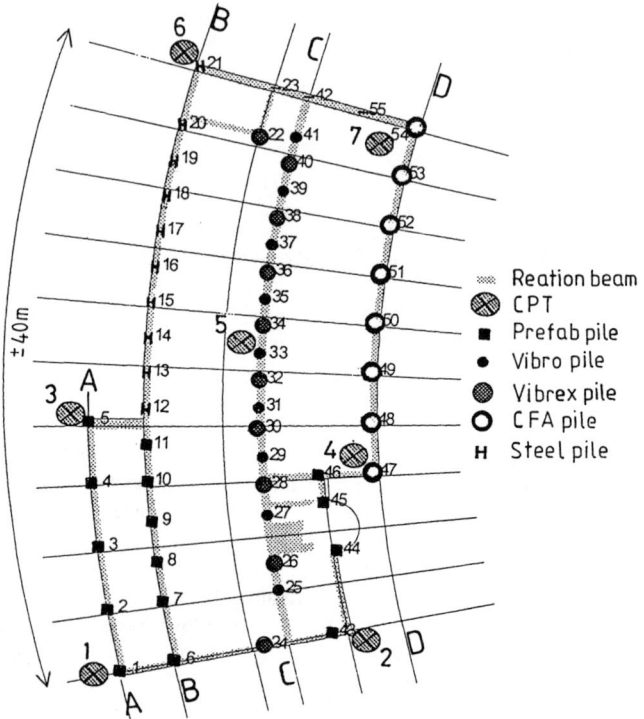

Figure 1 : IFCO office - pile lay out

integrity tests were performed as well as pseudo-static pile loading tests (PSPLT) on every pile. Pile Driving Analyses were performed on all driven piles. A load cell was installed in between each pile head and the building floor foundation beam. The 55 load cells were interconnected and monitored in a PC network. The recorded pile loads were stored every 15 minutes, during the entire building construction period and during the first full year after its completion. Once the building completed, in cooperation with the Ghent University Soil Mechanics Laboratory - Belgium, 23 static load tests were performed by means of 2 hydraulic jacks, placed between each pile top and the building foundation. During testing the actual pile load was increased to the piles' predicted failure load, with the aim to achieve a pile settlement of about 35 mm (10 % of pile diameter $\varnothing$).

It should be explicitly mentioned that all of the (8) typical (small stem) CFA piles, were (re)driven after the auger-type installation and concrete hardening up to an additional pile head penetration of about 0.3 m. This obviously has remarkable consequences on the overall load-settlement behaviour of those CFA piles.

VIBRO PILE

VIBREX PILE

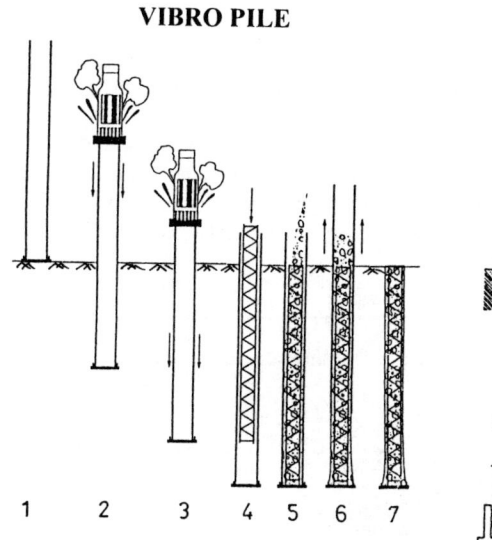

Figure 2a : Vibro pile type

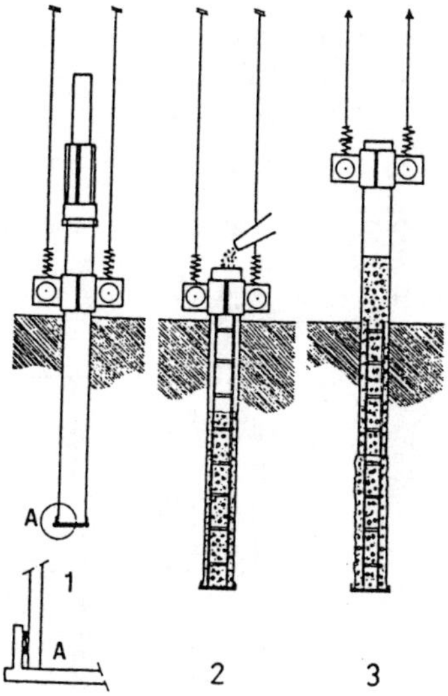

Figure 2b : Vibrex pile type

CFA PILE

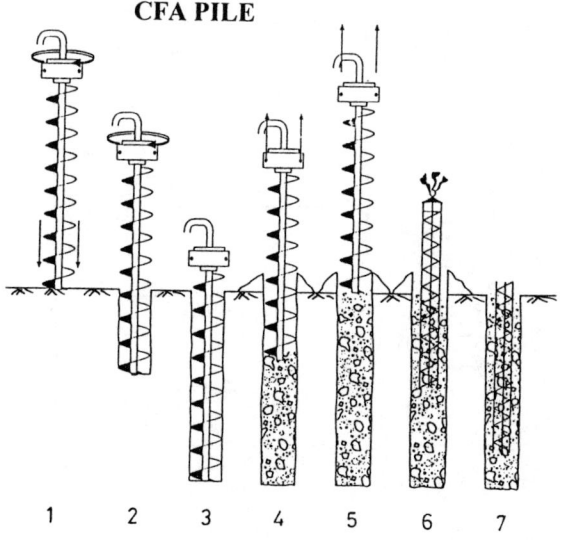

Figure 2c : CFA pile type

Soil layer characteristics

The natural ground level can be situated at about 5.3 m underneath the mean sea level (NAP), while the open phreatic level in the clay remains about 1.5 m under ground level. The tertiary clean very thick sand layer starts at about 7 m below ground level (fig. 3a/b ; level - 12). On top of this sand layer there can only be very soft clay with peat lenses in the -10 to -12 zone on fig. 3a/b. The pile tip level for the CFA-piles was fixed at -16.5, fig. 3a/b, for the Vibro piles and for the Vibrex piles at -16 (fig. 4a,b), for the prefab concrete piles varying from -13.4 to -16.9 and for the steel H-piles at -17.9 (fig. 4c).

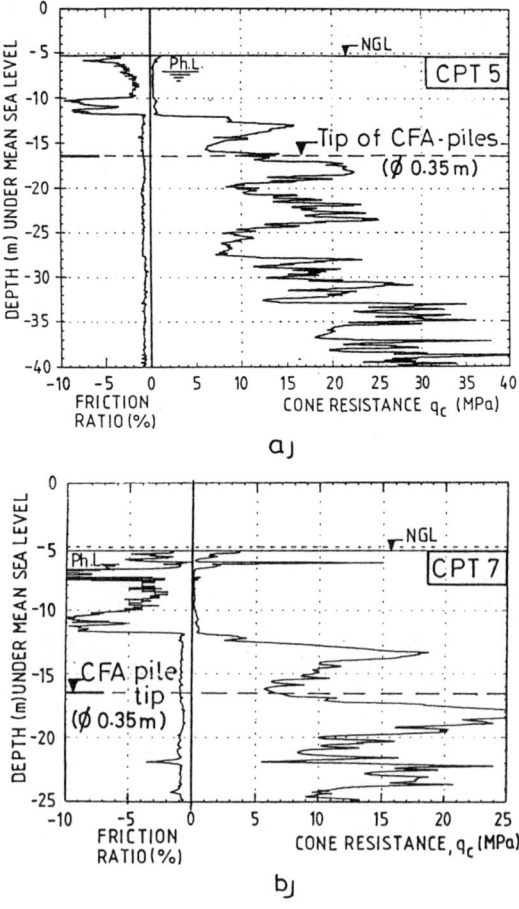

Figure 3 : CPT results at the building site

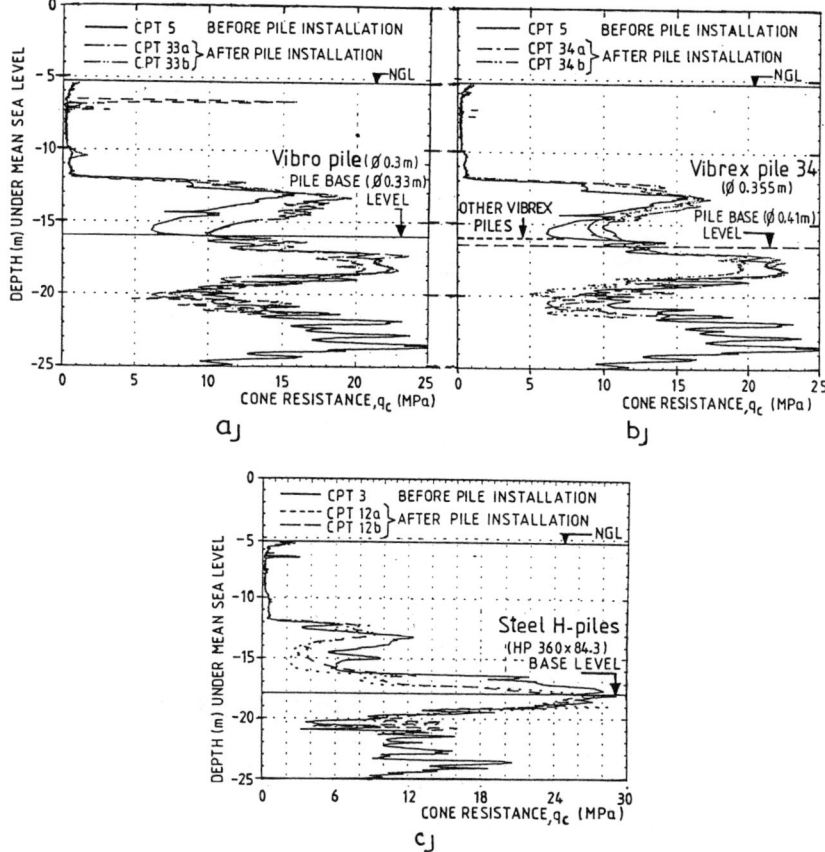

Figure 4: Pile installation - CPT results

In the sand layer a slightly perched water level (up to 0.2 m above the open phreatic level) could be detected. The pile installation quality was controlled by means of CPT before and after pile installation. The after-installation CPT measurements were performed at 0.5 m out of the pile axis. In case of one of the CFA piles (pile n° 51), the installation clearly caused a severe sand decompaction in an area up to 1.5 m around the pile tip level, cfr. fig. 5.

Pile Testing

In total, 8 prefab concrete piles, 5 Vibro piles, 3 Vibrex piles, 1 steel H-pile and 6 CFA-piles, were tested, jacking against the reinforced concrete beams of the ground

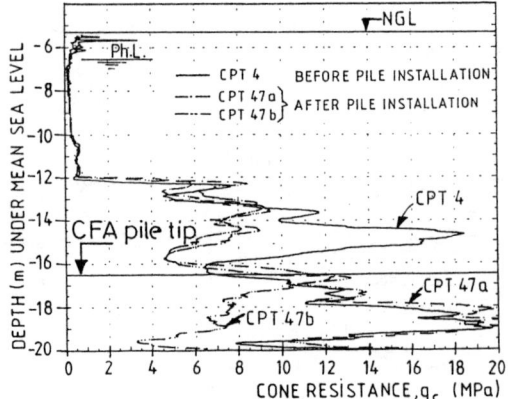

Figure 5 : CFA-installation - CPT results

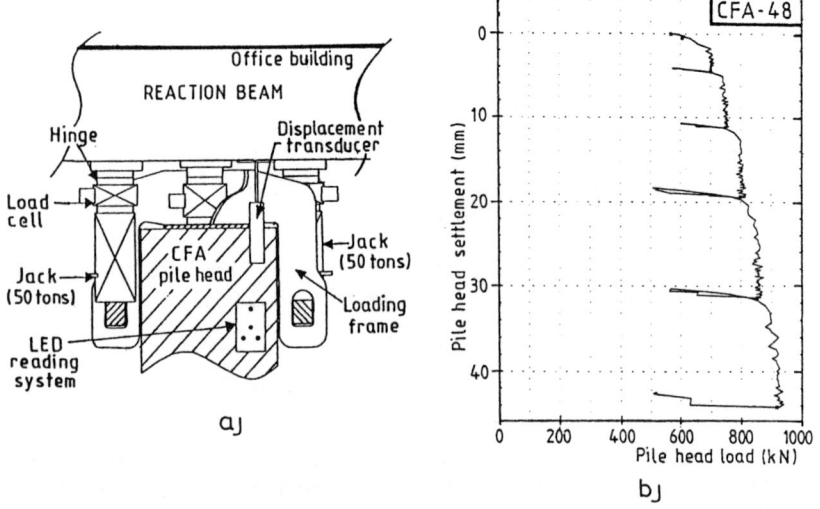

Figure 6 :   a : Test set-up for the static load test
             b : Example of pile loading cycles

level floor (fig. 6). The pile head was cut and separated from the floor beam, in order to prepare the pile load test. The principle of the load testing device, shown in fig. 6a, allows for a total maximum load of 1000 kN through two jacks on an inversed U-shaped loading frame, pushing on the pile head. The two jacks are generating this pile head load through the reaction forces on the very stiff building floor concrete beam.

The pile (vertical and horizontal) head displacements were measured optically each 30 sec up to a 0.05 mm accuracy. The pile load was controlled and when required adjusted each 30 sec. Starting from an initial loading of about 400 kN, the load was increased in steps of about 50 kN, (fig. 6b).

Each loading step was kept constant for one hour. In a full loading-up-to-failure program, three unloading-reloading steps were included. Each minute the load on the pile head of the test pile as well as the load on the immediate neighbouring piles was controlled. The bending of the reaction beam was evaluated. A load test was stopped at a well determined pile head settlement of 40 mm or more (up to 80 mm for some exceptional CFA pile load tests, such as CFA n° 53).

Single pile load settlement analysis

Single pile capacity prediction

Pile design methods, generally accepted in Europe, are characterised by the semi-empirical, yet direct transformation of soil bearing parameters out of in situ testing. Mostly used in Belgium are design methods based on the CPT diagram. Methods have been validated and calibrated through an extensive experience spanning over more than 30 years of full-scale co-operative research.

Because the basic CPT-based design method provides the ultimate base and shaft resistance of jacked or driven compression displacement piles, installation coefficients have to be introduced to account for the installation effects of pile type differing from the displacement pile type used to validate the basic design method.

The pile ultimate base resistance R_{bu} is deduced from the CPT data by :

$$R_{bu} = \beta \cdot q_{ub} \cdot \Omega_b = \beta \cdot \alpha_b \cdot \varepsilon_b \cdot q_{bu}^{(m)} \cdot \Omega_b \qquad (1)$$

with :

$\beta =$ a shape factor introduced for non-circular nor square-shaped bases ; for example $\beta = \dfrac{1 + 0.3B / L}{1.3}$ with B = width and L = length of rectangular base

$\alpha_b =$ an empirical factor taking into account the method of installation of the pile and the soil type ;

$\varepsilon_b =$ a parameter referring to the scale dependant soil shear strength characteristics (e.g. in case of fissured clay) in relation to the pile base dimensions

$q_{bu}^{(m)} =$ ultimate unit pile base resistance derived from the CPT results in the natural ground conditions, according to methods proposed by De Beer-1972 and Van Impe -1988, fig. 7.

$\Omega_b =$ the nominal pile base cross-sectional area.

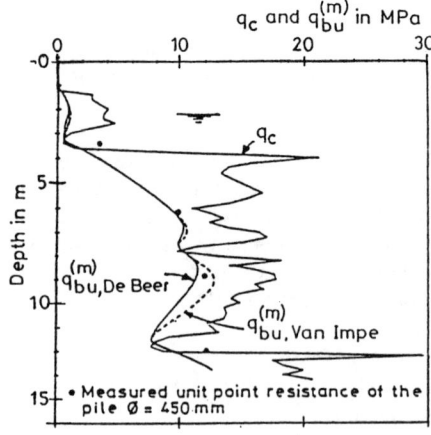

Figure 7: Comparison of calculated $q_{bu}^{(m)}$ with experimental data

Estimation of the ultimate shaft friction R_{su} is based on one of the three following CPT values : the total CPT shaft friction Q_{st} (easiest and most common method) ; the cone resistance q_c and/or the local unit CPT shaft friction f_s - Van Impe 1988, 1991. The total pile shaft resistance R_{su} can be directly evaluated from :

$$R_{su} = \frac{X_s}{\pi \varnothing_c} \cdot \xi_f \cdot \Delta Q_{st} \qquad \text{or} \quad = \frac{X_s}{\pi \varnothing_c} \cdot \Sigma \xi_{fi} \cdot \Delta Q_{st} \qquad (2)$$

with :

$\xi_f =$ an overall empirical factor ($= \alpha_s \cdot \beta_s \cdot \varepsilon_s$) introducing the effects of pile installation method (α_S), of the nature of the shaft's material and roughness (β_S) and soil structure scale effects (ε_s) ;

$X_s =$ the perimeter of the pile shaft ;

$\varnothing_c=$ diameter of the CPT rods.

The pile shaft friction can also be evaluated from a semi-empirical correlation between the ultimate unit shaft friction q_{su} and the cone resistance values q_c :

$$q_{su} = \eta_p \cdot q_c \quad \text{or further detailed as :} \quad q_{su} = \xi_f \cdot \eta_{pi}^* \cdot q_c \qquad (3a,b)$$

and consequently summarizing in each layer H_i :

$$R_{su} = X_s \cdot \Sigma H_i \cdot \eta_{pi} \cdot q_{ci} = X_s \cdot \Sigma H_i \cdot \xi_{fi} \cdot \eta_{pi}^* \cdot q_{ci} \qquad (4)$$

where η_p = an overall empirical factor depending on both soil and pile type. The correlation $\eta_p = q_{su}/q_c$ can be split into a pure soil parameter η_{pi}^* equal to the ratio of q_c to the average unit side friction $q_{su}^{(m)}$, and a pile/soil dependant empirical factor ξ_f as already defined in equation (2). Values for $q_{su}^{(m)}$, or corresponding η_{pi}^*, have been suggested for Belgian practice and are summarized in table 1.

Table 1 : $q_{su}^{(m)}$ values commonly used in Belgian Practice

		0.075	0.2	0.5	1.0	1.5	2.0	2.5	3.0	≥ 3.0
	q_c (MPa)									
CLAY	$q_{su}^{(m)}$ (kPa)	5	10	18	31	44	58	70	82	$q_c/36.6$
SAND	q_c	≤ 10 MPa			$10 < q_c < 20$ MPa					> 20 MPa
	$q_{su}^{(m)}$	$q_c/150$			Linear interpolation between $q_c/200$ and $q_c/150$					$q_c/200$

A third method relates the pile shaft friction to the directly measured local unit side friction f_s by

$$q_{su} = \alpha_{fs} \cdot f_s \tag{5}$$

Again, one can expect that α_{fs} depends on pile and soil type, and should hence be defined by calibration with static load tests.

The α_b, ξ_f and η_p factors in eq. (1), (2) and (4) have been deduced from static pile load tests over many years in various research projects. Tables 2 and 3 are giving an overview of values commonly used in the Belgian design practice for sand and stiff

Table 2 : Commonly used α_b factors for various pile types

Pile type	α_b factor for	
	sand	stiff OC clay
Group I - High soil displacement		
Impact driven piles	0.8-1.15[1]	0.8-1.0[1]
Screwed piles-(Van Impe, 1991)	0.8-1.0	1.0
Jacked piles (smooth)	1.0	1.0
Group II - Low soil displacement or low relaxation		
Drilled with special previsions for casing	0.6-0.8	0.8
Group III - Soil excavation		
Cast-in-situ bored piles (large diameter and CFA)	0.33-0.67	0.8
[1] : highest value for expanded base with semi-dry concrete only ; for cast-in-situ with plastic concrete, function of the diameter of the bottom plate relative to diameter of driving tube ;		

Table 3 : Commonly used ξ_f factors for various pile types and shaft material

Pile type	ξ_f factor for	
	sand	stiff OC clay
Group I - High soil displacement		
Shaft in compacted semi-dry concrete	1.6	1.15
Shaft in plastic concrete or prefabricated concrete	0.8-1.0[1]	0.65-1.0
Screwed piles - plastic concrete (Van Impe, 1991)	0.8-1.25[2]	0.8-1.25[2]
Steel shaft	0.6	0.45-0.65
Group II - Low soil displacement or low relaxation		
Drilled with special previsions-wet concrete	0.6-0.8	0.65-0.85
Group III - Soil excavation		
Cast-in-situ bored piles (large diameter and CFA)	0.4-0.6	0.5
[1] : for cast-in-situ with plastic concrete, function of diameter of the bottom plate relative to diameter of driving tube ; [2] : highest values for auger flange shaped shaft		

OC Clay ; intermediate values are adopted for intermediate soil types. For certain pile type, installation factors that have not yet been calibrated, based on an objective conducted full scale load test program, require the input of engineering judgement. For the η_p factors, Van Impe-1991 has summarized the values resulting from such Belgian research work.

Interpreting the load settlement results

In the analysis of the theoretical single pile load settlement curve, one can go out from several methods. Chin assumes a hyperbolic function relating time and settlement :

$$\frac{t}{s} = a \ . \ t + b \tag{6}$$

which means that time to settlement is a linear function (b is a constant). As t goes to infinity, the settlement "s" reaches the value 1/a. This value can be easily computed, but is strongly depending on the number of accurately measured settlement data.

As it appears, at large values for time t, some deviation appears between the measured values and the single hyperbolic law according to Chin. Therefore, Fleming et al. proposed to use a double hyperbolic function, separating the time settlement behaviour caused by the tip load and shaft load :

$$s = \frac{W_s \ . \ t}{T_s + t} + \frac{W_b \ . \ t}{T_b + t} \tag{7}$$

in which

W_s	=	asymptotic value for shaft related deformation
W_b	=	asymptotic value for base related deformation
T_s	=	mobilisation time for half of the shaft related deformation
T_b	=	mobilisation time for half of the base related deformation
t	=	is time elapsed from application of the pile head load
s	=	relative pile head settlement at time t

As the real values of the parameters W_s, W_b, T_s and T_b are not known, they are estimated using a least square approximation using the data of the measured time settlement curve. The final settlement can be found by adding W_s and W_b.

Some examples

Examples of the pile load test results are shown in the figures 8 and 9a,b,c,d,e. In figure 10 all measured load-settlement curves were gathered. The pile CFA n° 51, as mentioned already before, because of the severe soil decompaction around the pile tip, becomes much more flexible than the surrounding piles and shows a much worse load-settlement behaviour. The stiffness of the building however redistributes the loads very appro- priately among the surrounding piles in order not to overload pile n° 51, as will be illustrated here lateron in fig. 12. The building floor reaction beam nevertheless showed a 2 mm additional differential sag settlement at the location of the CFA pile n° 51 pile head.

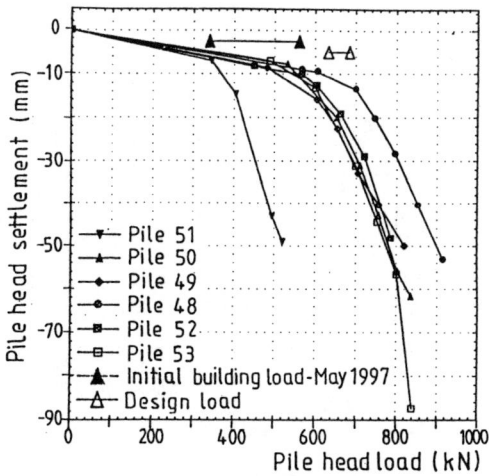

Figure 8 : Pile test results - CFA piles

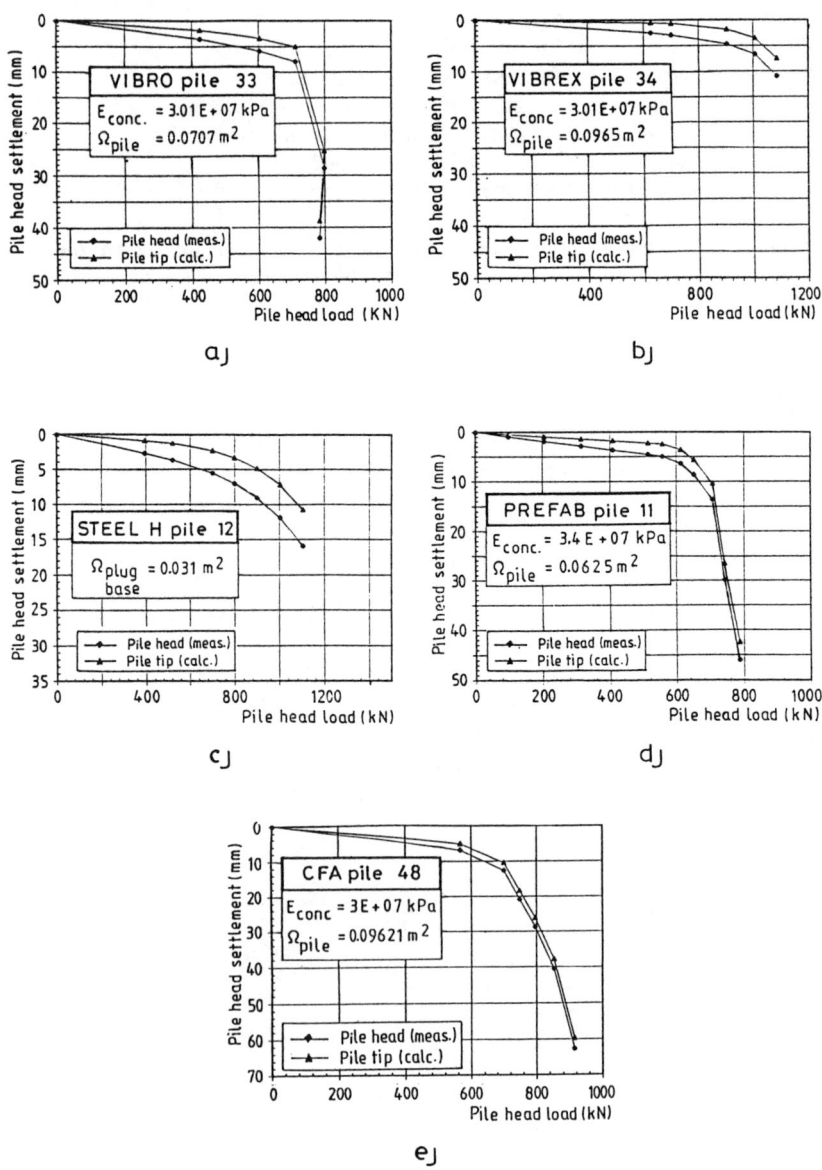

Figure 9 : Pile test results

The pile integrity tests on CFA n° 51 indicated no particular damage for this pile; the pile shaft is clearly intact but the pile tip soil stiffness decreased remarkably (fig. 8) due to the destructive augering technique in the clean sand layer.

Prediction versus measurements of single pile behaviour

The "predicted" unit pile tip capacity $q_{u,b} = \alpha_b \cdot q_{bu}^{(m)}$ for each pile was evaluated, going out from the CPT-results before pile installation ; the unity shaft capacity $q_{s,u}$ was similarly evaluated from the CPT cone resistance q_c (eq 3) :

- $q_{s,u} = 18$ kPa in the very soft cohesive upper layers and for all pile type (except the CFA).
- $q_{s,u} = \xi_f \cdot (\eta_p^* \cdot q_c)$ which $\eta_p^* = 1/150$ in the Pleistocene sands around the lower end of all piles (except CFA)

Assuming $\alpha_b = \xi_f = 1$ means for example that one looks at ideally installed piles not influencing the soil stress conditions in any way. In case of CFA piles and for the lower end pile length in the Pleistocene sand $\eta_p^* \approx \dfrac{1}{200}$; while $q_{s,u}$ in the very soft upper layers for this soil excavation pile type is fully neglected. The corresponding mean predicted total pile capacity is summarized in table 4.

Table 4 : **Predicted** mean ultimate total pile capacity **at $s_b = 10\%$ $\varnothing_{pile}$**
from CPT results **assuming $\alpha_b = \xi_f = 1$**

Pile type and n°	Ω_{pile}^{base} -0.0625m² prefab concrete (kN)		Ω_{pile}^{base} =0.0707m² Vibro (kN)		Ω_{pile}^{base} =0.0962m² Vibrex (kN)		Ω_{pile}^{base} =0.031m² Steel H (plugged)	Ω_{pile}^{base} =0.0921m² CFA (kN)	
	n°6	n°7	n°31	n°35	n°34	n°36	n°12	n°47	n°54
-14.9	695								
-16.9		881							
-16.0			1034			1407			
-16.5				1240	1684			663	591
-17.9							1317		

Using the measured load settlement curves from fig. (8 and 9) and the common "failure" criteria, the estimated ultimate total pile load, from the measurments, is indicated in tables 5a&b.

Probably a more clear way to look at the pile load test curves, is to plot every load settlement curve in a non-dimensional diagram (fig. 10), making use of the

Table 5a: Ultimate total load -**measured** - CFA piles
($\Omega_{\text{pile}}^{\text{base}} = 0.9621\ m^2$)

	Ultimate total pile load (kN) ($Q_{u,tot}^{meas}$)					
At failure (S_b)	CFA 48	CFA 49	CFA 50	CFA 51	CFA 52	CFA 53
2.5 % ∅	658	519	545	350	573	542
5 % ∅	742	625	604	406	655	629
25 % ∅ (convent.)	-	-	-	-	-	834

Table 5b: Ultimate total load -**measured** - for other pile type
at s_b = 10 % ∅

Pile n°	6	7	34	36	31	35	12
Pile type	Ultimate total load at 10 % ∅ = s_b,in kN						
Prefab	672	888					
Vibrex			1225	1195			
Vibro					714	915	
Steel-H							1315

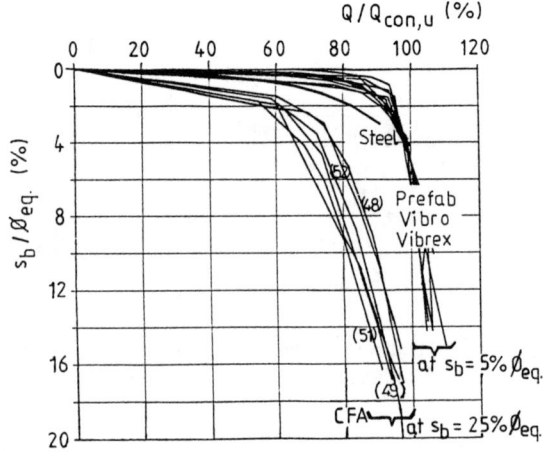

Figure 10 : Dimensioneless graphs of pile test results

"conventional" ultimate pile capacity corresponding to each of the piles. It's obvious from such graph that the redriven CFA-piles underneath this existing building, even at a relative pile head settlement of 25 % of the pile diameter, still did not fully develop their ultimate pile tip capacity. The CFA's in this case although were additionally driven in over about 0.3 m after CFA-screwing and concrete hardening. The initial pile stiffness however was still not up to the same level as the displacement type of pile, (as seen from the comparison with the other type of piles).

Installation factors

Comparing the splitting of the pile load test results by the Cemset analysis, into pile tip and pile shaft capacity values and compare them to the predicted values, it's possible to define "installation quality" parameters, such as : α_b and ξ_f, defined already earlier. The results of such analysis are summarized in table 6.

Table 6 : Installation quality parameters

Pile type & n°	Prefab concrete		Vibro			Vibrex		Steel H	CFA		
	n°6	n°7	n°35	n°37	n°39	n°36	n°38	n°12	n°48	n°51	n°53
α_b pile tip (calc)	0.83	0.84	0.56	0.90	0.72	0.71	0.57	1.05	0.92	0.48	0.80
$\xi_{f,clay}$ over pile shaft length in upper solft layer (ass)	1	1	-	-	-	-	-	1	1	1	1
$\xi_{f,sand}$ over upper pile shaft length in Pleistocene sand (calc)	1.34	1.33	-	-	-	-	-	0.75	0.92	0.53	0.81
Comments	α_b<1 because of slcnderness of pile and driving through soft upper layer ; horizontally vibrating pile		α_b remains very low, probably because of too low concrete fluidity at pile tip level						α_b remains rather high (except n° 51) because of redriving of CFA piles		

Comparing the ultimate total load estimations, listed in tables 5a&b, with the common total pile capacity prediction methods from of CPT results at the pile site, one could on the other hand also define a practitioners' overall quality installation coefficient, "κ^*" as the ratio :

$$\kappa^* = \frac{Q_{u,tot}^{meas}, \text{out of load test results}}{Q_{u,tot}^{pred}} \quad \text{at the same relative displacement} \quad (8)$$

Some estimations of this κ^*-coefficient are gathered in the fig. 11a,b,c,d. An overall installation coefficient for piles lower than $\kappa^* = 1$ suggests a non-displacement pile behaviour as compared to the indicated CPT-capacity prediction method with both α_b and $\xi_f = 1$.

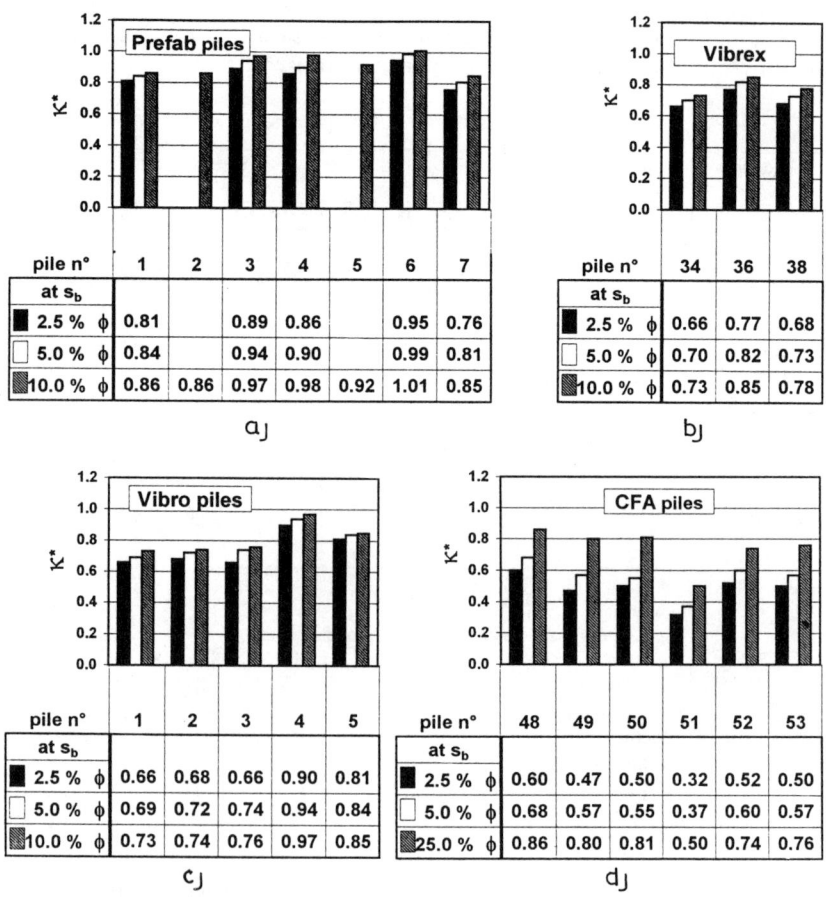

Figure 11 : κ^*-values related to the various "failure" criteria

Pile group - office building structure interaction

Along the building sections A-A/B-B/C-C/D-D of the foundation pile rows (fig. 1), the measured pile head loads at full completion of the building (May 1997) are shown in fig. 12.

Carefully implementing here the various pile type systems (out of fig. 2, 8 and 9), some considerations from the measured pile head load distributions are quite obvious:

- the relative overall building stiffness (as compared to the Pleistocene sand layer in which the pile group is embedded) remains quite low. The pile loads of the centre piles are indeed always somewhat higher than at the border piles. In this case of a clearly end bearing pile group, the upper soil layers around the pile shaft don't practically contribute to this interaction problem.
- the prefab and steel (plugged) H-piles do show very similar stiffnesses along building section B-B ; which will also be compatible with their installation quality parameter values at small strain levels ($\varepsilon \ll 1$ % $\varnothing$)
- it's very striking that the Vibro pile type at service loads (so at small strain levels $\varepsilon \ll 1$ % $\varnothing$) apparently (cfr. section C-C) do show a much more flexible interaction behaviour than the more stiff Vibrex piles (fig. 9b and 10). This suggests, at building service loads and corresponding strain levels, that the Vibrex pile type are much closer to displacement pile type behaviour (better and stiffer pile-soil interaction) than the Vibro pile type
- however, as soon as the mobilized shaft and pile tip capacity increases drastically up to 2.5 % $\varnothing$, 5 % $\varnothing$ and 10 % $\varnothing$ of strain level - the figures (11) confirm that the quality of installation and its corresponding κ^*-coefficient is very comparable for both pile type. At such testing strain levels, the installation quality level is not better for Vibrex as compared to Vibro. The much stiffer interaction behaviour of the Vibrex pile-type, at service load strain level, therefore is probably entirely due to the larger pile shaft diameter (350 mm versus 300 mm for the Vibropile) and larger pile tip dimensions (410 mm versus 330 mm for the Vibro piles).
- the comparable stiffness and interaction behaviour of the prefab piles and CFA piles, at service load strain level, along section D-D can only be explained because of the important redriving (over 300 mm) after CFA-auger installation and concrete hardening. Only pile n° 51 remained of very poor installation quality, because of much more important decompaction of the sand around the pile tip.

Using a 3-D-simulation of the 3 floors-office building-pile group interaction, a pile head load distribution pattern was computed going out from the equivalent spring stiffness for each of the piles as derived from the pile load tests.

Two different simulation models were used :

- an interaction model in which the office building structure was in direct contact with the pile heads

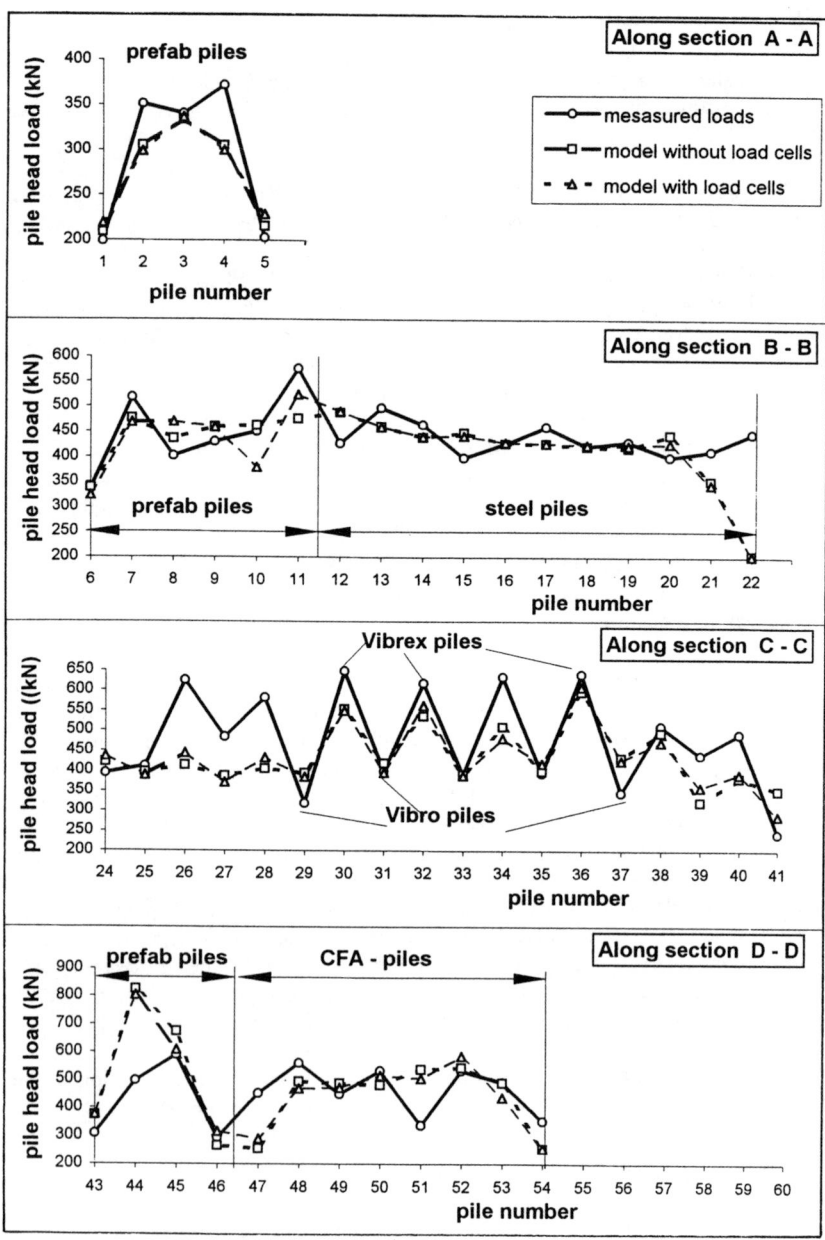

Figure 12 : Measured and modeled pile head service loads

- a slightly different model in which also each pile head load cell was modeled as a structural "column" and, as a hinge element, was acting with a stiffness of K_{LC} = 2463 MN/m.

The outcome of both simulations is very similar (cfr. fig.12) and as a conclusion it can be stated that such a pile head load prediction stating from the 3 D-office building modelling seems to match quite well the mearsured data. The most important deviation can be seen only in the zone of the Vibro-Vibrex piles n° 26 till 28. This is exactly corresponding to the area of the office building elevator shafts for which the changes in structural stiffness are important and more complex to model.

Concluding Comments

The load testing of the foundation piles using the own weight of the IFCO office building has shown that, as compared to the driven displacement piles below the same office building, some type of piles showed a much more flexible load-settlement behaviour. The overall installation coefficient indeed remained many times below κ^* = 1 ; for example even at 25 % of relative pile head settlement, the CFA piles still did not fully develop their ultimate tip capacity.

Acknowledgement

The author is grateful to the IFCO company and to Prof. A. Van Weele, owners of the building, for allowing the Ghent University - Soil Mechanics Department to test the various pile type underneath the building and allow for the use of the detained pile data. available. Especially the research students Mr. K. Verstraeten, Mr. Wittouck and Mr. Inghelbrecht are acknowledged for their appreciated help during the pile testing and the many pile data interpretation discussion sessions.

References

De Beer E., 1963, The scale effect in the transposition of the results of deep-sounding tests on the ultimate bearing capacity of piles and caisson foundation, *Géotechnique*, vol. 13, n° 1, p39-75, London.

De Beer E., 1971-1972, Méthodes de déduction de la capacité portante d'un pieu à partir des résultats des essais de pénétratrion, *Annales des Traveaux Publics de Belgique*, n° 4 (p191-268), 5 (p321-353) & 6 (p351-405), Brussels.

De Beer E., 1988, Different behaviour of bored and driven piles, Proceeding *BAP I*, p 47-82, Ghent.

De Cock F., Legrand Ch.,1997, Design of axially loaded piles - European practice, Balkema Rotterdam.

Fleming W.G.K., 1997, Execution and interpretation of pile loading tests, Conf. Di Geotechnica di Torino, XVI Ciclo.

Van Impe W.F., De Beer, E., Lousberg, E., 1988, Prediction of the single pile bearing capacity in granular soils out of CPT results, *ISOPT I*, p. 1-34, Orlando, Florida.

Van Impe W.F., Thooft, K., Van Den Broeck M., 1988, End and shaft bearing capacity evaluated separately out of static pile loading tests, *BAP I*, p. 489-498, Ghent.

Van Impe W.F., 1988, Considerations on the auger pile design, *BAP I*, p. 193-218, Ghent.

Van Impe W.F., 1991, Developments in pile design, Proceeding *DFI*, Stresa

Van Impe W.F., 1991, Deformations of deep foundations, *General report X ECSMFE*, Florence

Van Weele A.F., 1979, Some considerations with regard to the bearing capacity of foundation piles, *Geologie en mijnbouw*, 58, p. 405-416

Van Weele A.F., 1982, Wich load is allowable on a given foundation pile, Amici et alumni em. prof. Dr ir E. De Beer, p. 287-296

Van Weele A.F., 1993, Quality assessment foundation piles after installation, *BAP II*, p. 459-468, Ghent

Van Weele A.F., 1997, Pile foundation failure, Expensive but instructive, Singapore

Verstraeten, K., 1998, Vergelijkend last-zakkingsgedrag van vijf paaltypes onder een bestaand gebouw, Final Degree work, KUL.

Suction Caisson Foundations in the Gulf of Mexico

Sherif El-Gharbawy[1], Ph.D., Roy Olson[2], Ph.D., P.E.
Texas, USA

ABSTRACT

A suction caisson is a deep foundation consisting of a steel tube with an open bottom and a closed top that is inserted into the ground by applying suction to the inside to produce a pressure differential on the top. In the deep offshore environment, a sufficient pressure difference can be created across the closed top that high capacity, deep foundations can be installed.

Operations in the deep marine environment are expensive and new foundation types are unlikely to receive serious consideration unless they have been studied both experimentally and theoretically. We have been investigating the behavior of suction caissons in both sand and clay for about six years. The current manuscript deals with tests on 2-in. (50-mm) to 4-in. (100-mm) diameter suction caissons in clay. The caissons were subjected to tensile loading under both drained and undrained conditions and different behaviors were encountered. Under undrained conditions, the extracted caissons carried a layer of clay on the outside and a full plug inside. Under drained conditions, the caissons came out clean. Undrained capacities were several times the drained ones. Cyclic loading tests were performed with a biaxial loading device that allowed independent control of the vertical and horizontal motions. The tests showed that the application of cyclic loads smaller than the long term capacity led to no accumulated movement, whereas application of larger peak loads led to accumulated displacements that depended on the peak load intensity and the number of cycles.

INTRODUCTION

Foundations for offshore structures have undergone a steady evolution since the first offshore structure was constructed in 1947 (McClelland, 1974; Lee and Light, 1992). The first offshore structures were in shallow water and foundation types were a natural extension of those used on land. Because the soil tended to be soft, the foundations were mainly piles driven with hammers operating above the water surface. As structures migrated to deeper waters and became larger, the piles also became larger and special hammers, including ones operated under water, were developed. Driven piles were used for the Cognac platform, in 1025 ft (312.5 m) of water. The piles were 7 ft (2.15 m) in diameter, 625 ft (190 m) long, penetrating about 450 ft (137 m),

---

[1] Suction Caissons Design Specialist, Delmar Systems, Inc., 15915 Katy Freeway #640, Houston, Texas 77094

[2] L. P. Gilvin Professor of Civil Engineering. The University of Texas at Austin, ECJ 9.227, Austin, Texas 78712

with each pile weighing 450 tons (409.1 metric tons) (Lee and Light, 1992). Similar pile foundations were used with the Bullwinkel platform in 1353 ft (407 m) of water (Digre *et al.*, 1989). The Cognac and Bullwinkel were steel jackets that rested directly on the sea floor. As the oil industry ventured into greater water depths, the change was made to floating platforms restrained by vertical tethers to a foundation. "Tension Leg Platforms" (TLP) could be used in considerably greater water depths. The Auger platform was a TLP installed in 2860 ft (872 m) of water. It is moored with sixteen 72-in. (1830-mm)-diameter by 427-ft (130-m)-long steel piles (Ulbricht *et al.*, 1994). Other floating structures are under consideration for water depths exceeding 3000 ft (910 m).

Analyses and data from field tests on suction caissons were reported by Albert *et al.* (1987). They found large capacities for such foundations. When Shell Oil designed the Auger TLP, they used driven piles, even though suction caissons should have been more feasible (Peterson *et al.*, 1994), because the pile technology was better supported by experience.

When the Offshore Technology Research Center was set up in 1988, the goal was to look ahead at where the offshore industry needs would be one or two decades into the future and to consider problems in water depths of 6000 to 10,000 ft (2000 to 3000 m). Interest developed in a diversity of areas, *e.g.*, use of floating structures shaped more like a spar than a floating airport, in use of composite materials, and in foundations other than driven piles.

Driven piles were not the only types of foundations used offshore. In the stiffer soils of the North Sea, the choice was basically to use shallow foundations. Large concrete structures (gravity platforms) were developed with the foundation resting directly on the sea floor. Analyses were rather similar to those used onshore (bearing capacity, settlement) except that the structures were subjected to significant cyclic lateral loading, and there was the danger of scour under the foundations. The structures were provided with "skirts" to increase resistance to scour and sliding (Tjelta *et al.*, 1990). The skirts were walls around the outside of the foundations, extended to a sufficient, but shallow, depth. To ensure that the skirts would penetrate fully, suction was applied at the bottom of the structure during installation.

Once skirts were used successfully, it was a natural extension to use them as the foundation itself. Thus, in place of piles, the foundation consisted of a number of tubes with closed tops that were placed on the sea floor and installed by pumping out the entrapped water. They had the general appearance of inverted buckets, with untapered sides, so the Norwegians tended to call them "bucket foundations" (Tjelta, 1994). As shallow foundations, they had penetrations not significantly exceeding the diameter. To develop the needed capacity, they generally had large diameters, *e.g.*, of the order of 30 to 45 ft (10 to 15 m). They were designed for compressive loading from steel jacket structures. To avoid subjecting the bucket foundations to a tensile loading during storms, they were ballasted with iron ore. The Norwegians developed ingenious methods of analysis for shallow foundations subjected to substantial axial loads and overturning moments (Aas and Andersen, 1992; Andersen and Lauritzsen, 1988; Andersen et al., 1992).

In the deep Gulf of Mexico (GoM), soil is often normally consolidated, or under-consolidated, highly plastic clay to depths of hundreds of feet, often with buried sand layers (Bryant and Simmons, 1992). Use of bucket foundations in such an environment would require loading of the softest and most compressible layers of the soil profile, a procedure that seemed inherently undesirable. The logical evolution was to use foundations shaped more like driven piles, but to install them using suction instead of hammers. We have chosen to call these foundations "suction caissons".

Suction caissons can be used in a variety of ways. For example, at one extreme, a steel template can be constructed onshore with the suction caissons placed at the extremities. The whole unit would be floated out to the site, the caissons flooded, and the template sunk to the bottom. Individual pumps on each of the caissons could be controlled to ensure that the template would be level as the caissons are installed. The floating structure would then be attached to the template. At the other extreme, a single large suction caisson might be installed, the drilling rig installed on the caisson on the bottom of the ocean, and the well drilled through the center of the caisson. A subsea oil production structure could be used and the oil pumped through subsea pipelines to a shallower depth for transmission to the surface, perhaps using an existing structure. A variety of other schemes can be considered.

Suction caissons may also be used as mooring anchors in taut or semi-taut systems, in which close to horizontal loading is applied through chains and cables and transferred to a floating structure that may be several miles away. The suction caisson has the advantage over drag embedment anchors that you can install the caisson precisely where needed and do not have to drag an anchor through a zone that may already have other cables in it.

From a design point of view, the problems with use of suction caissons involves the lack of an experience base. There are serious uncertainties involved with pile foundations as well, but there are thousands of load tests on land, and there are more than 3880 pile-supported offshore structures under the jurisdiction of the Minerals Management Service (Dannenberger, 1993), so designers have a degree of confidence in them. Offshore testing tends to be expensive, so development of an experience base tends to begin with model testing (Clukey and Morrison, 1993; Whittle et al., 1993; Clukey et al., 1995) and theoretical analyses (Whittle et al., 1993) coupled with a small number of offshore tests, with the offshore tests usually proprietary. This manuscript is concerned with experimental studies of model caissons in clay. A companion effort to develop a sophisticated finite element analysis is currently underway.

LABORATORY TESTS WITH MODEL CAISSONS

The details of our experimental procedures have been published previously (El-Gharbawy, 1998) and will not be repeated. The soil used in these tests was a kaolin with a liquid limit of 56% and a plastic limit of 32%. The effective angle of internal friction of the kaolinite, performed using direct shear in the pressure range of 5 to 35 psf (0.24 to 1.67 kPa), was 28°. Kaolinite slurries were mixed in 55-gallon drums, allowed to consolidate under self-weight, and then consolidated using a downward hydraulic gradient of about eight. Thereafter, the samples were allowed to return to equilibrium. The samples had densities and undrained strengths that varied with depth.

Tests were performed using caissons with an outside diameter of 4 in. (100 mm) and with aspect ratios (ratio of length to diameter), of 2, 4, and 6. The caissons for these tests were plastic with a wall thickness of 1/8 in. (3.2 mm). An additional series of tests was performed with a stainless steel caisson, with a diameter of 2 in. (50 mm) and an aspect ratio of 12, with a 1/16-in. (1.6 mm) wall thickness. Caissons were allowed first to penetrate under their own weight with the cap vented. Then, a suction equal to about 2000 psf (100 kPa) was applied and the caissons were allowed to penetrate until the cap made contact with the soil surface. The 4-in. (100-mm)-diameter caissons penetrated at a rate of about 0.12 in./sec (3 mm/sec) with this suction, and the 2-in. (50-mm)-diameter caisson penetrated at a rate of about 0.57 in./sec. (15 mm). Caissons were then allowed to remain in place until the measured pore water pressures adjacent to the walls of the caisson dissipated. The pore pressures generally dissipated in about 30 minutes but the caissons were generally left overnight prior to testing.

A preliminary series of axial tensile loading tests was performed under both undrained and drained conditions, and then a series of tests was performed using cyclic tensile loading in which the angle between the applied force and the vertical axis of the caisson was varied to simulate actual loading conditions for offshore floating structures.

Static Extraction

Extraction under Undrained Conditions. For the tests with undrained soil, the caissons were withdrawn at a rate of 1 to 2 in./sec (25 to 50 mm/sec). A typical curve of tip elevation and applied load versus time (Fig. 1) shows that failure was catastrophic, *i.e.*, once the peak load was reached at a relatively small movement, the caisson steadily lost capacity until it was pulled out completely. Failure typically occurred in 0.1 to 2 seconds.

All caissons came out with about a 1/8" (3-mm) thickness of clay adhering to the outside and with a full plug inside (Fig. 2). Failure was thus clearly in the soil around the caisson and there was an undrained bearing capacity component at the tip.

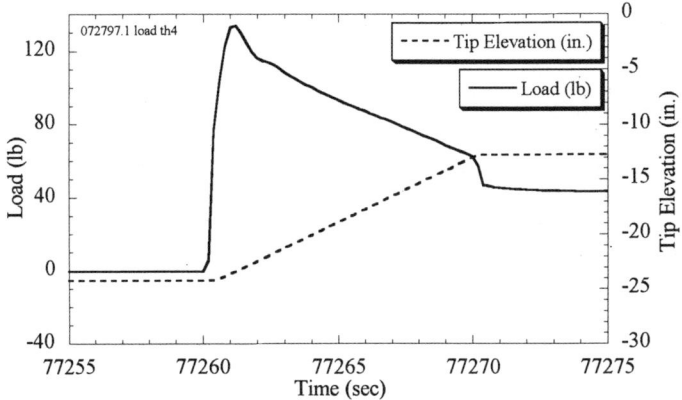

Figure 1: Rapid, Undrained Pullout of Caisson 6.

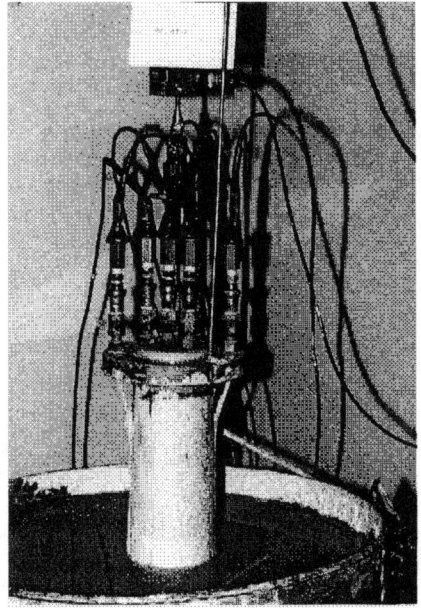

Figure 2: Typical Undrained Pullout

The calculated capacity (Q_c) was:

$$Q_c = W + Q_s + Q_p \tag{1}$$

in which W is the submerged weight of the caisson, Q_s is the side capacity, and Q_p is the point capacity.

It was not feasible to estimate the undrained shearing strength of the clay by testing undisturbed samples because the soil was too soft to be sampled. Instead, tests were performed "in-situ" using a 1" by 4" (25 by 100 mm) vane. The shearing strengths varied nonlinearly with depth (Fig. 3) but the data could be fitted using the equation:

$$c_u = Az + Bz^2 \tag{2}$$

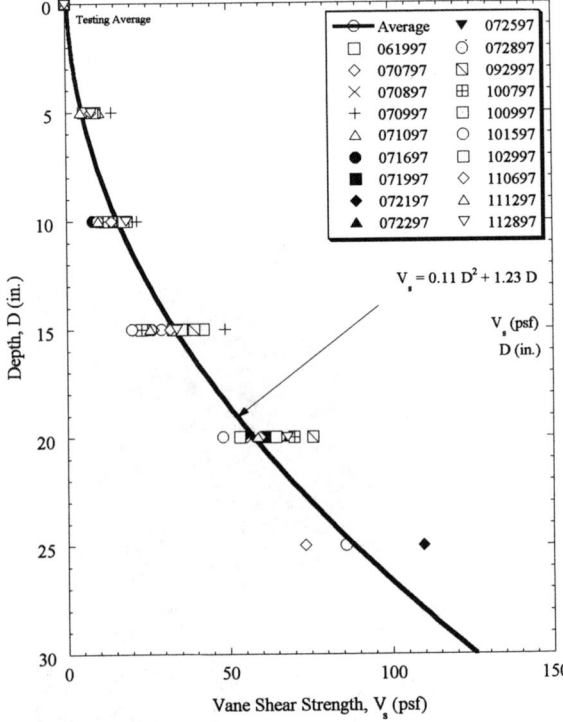

Figure 3: Undrained Soil Shear Strength Profile from Vane Shear Test Data.

The side capacity was thus calculated by integrating the undrained shearing strength over the exterior surface (Eq. 3), with the undrained shearing strength defined using Eq. (2). We used the actual outside diameter of the caisson rather than the slightly larger diameter associated with the apparent failure surface and did not make corrections for viscosity of the soil.

$$Q_s = \int \alpha \ c_u C \ dz \tag{3}$$

The point capacity was calculated using:

$$Q_p = 9 \ c_u \ A_p \tag{4}$$

in which 9 is the bearing capacity factor and A_p is the total area of the tip of the caisson.

The weight of the caisson varied because the weight of numerous cables, connected to the caisson top and used with the load cells and pore pressure transducers, varied depending on how they were suspended for any given test and on the elevation of the caisson. The weights used in the analyses were estimates based on the measured weights in air and the reading of the load cell when the caissons came out of the soil during withdrawal.

The measured and computed capacities are within a scatter band associated with uncertain weights and the scatter in measured shearing strengths. For these tests, the ratios W/Q_c, Q_s/Q_c, and Q_p/Q_c, were 26% to 54%, 10% to 33%, and 36% to 52%, respectively, with the influence of side shear increasing as the penetration depth increased.

Extraction under Drained Conditions. Extraction tests with the soil fully drained were performed using load control and measuring the pore water pressures at various depths on the inside and outside of the caisson. A typical curve of tip elevation and applied load versus time (Fig. 4) shows that the caisson response to the applied load came to equilibrium in approximately one hour, so most of the tests were performed with each load increment maintained for 2.5 to 3.0 hours and the total extraction time being about one day.

When caissons were extracted under drained conditions, they came out as clean (Fig. 5) as if they had just been washed. Failure was thus between the vertical caisson surfaces and the soil with a negligible tip bearing capacity. The capacity was again calculated using Eq. 1, except that Q_p was negligible. The side shear capacity, Q_s, was estimated using:

$$Q_s = \int \sigma_h' \ C \tan (\delta) \ dz \tag{5}$$

in which σ_h' is the horizontal effective stress, C is circumference, δ is the friction angle between the clay and the caisson, and z is depth. The horizontal stress was estimated using:

$$\sigma_h' = \sigma_v' K = K \int \gamma' dz \tag{6}$$

The water contents of the clay were measured at various depths and used to calculate the submerged unit weights. These values could be expressed analytically as:

$$\gamma' = 26.66 + 0.902z - 0.028z^2 \tag{7}$$

in which γ' is in pcf and z in inches.

The usual problem with calculations of this type is the difficulty of estimating values for K, considering that the soil has been disturbed and then reconsolidated. Preliminary calculations using K_o and $\delta = 25°$ led to an underprediction of capacity by up to 29%. A trial value of K=0.75 led to a good comparison of measured and computed capacities (Fig. 6). For that case, the side shear represented 36% to 54% of the total capacity, the rest being caisson weight.

Long Term Capacity. For the tests under drained conditions, there was evidence of long term creep under stresses above about 80% of the measured capacity. No effort was made to study the creep behavior. However, we chose to define the "long term capacity" (LTC) as 80% of the capacity at which the peak load was achieved and to use the LTC as the basis for the cyclic loading tests.

The ratio of the "undrained capacity" to "drained capacity" was 1.9, 2.8, 3.5 and 1.9 for the caissons with aspect ratios of 2, 4. 6, and 12, respectively.

Cyclic Extraction

Cyclic Axial Loading. The model caissons were tested under simulated environmental loading in the form of uniaxial cyclic tension, and combined cyclic axial and lateral loading, with the frequencies of the vertical tensile and lateral loading differing in accord with offshore experience. The monotonic, tensile cyclic loads were applied on top of a static bias load of 60% of the LTC. The peak cyclic load was non-dimensionalized by expressing it as a percentage of the LTC. The cyclic loads were computer-controlled to produce essentially sinusoidal waves. However, the loading system operated at a preset deformation rate, so the frequency of the load was somewhat slower when there were large deformations.

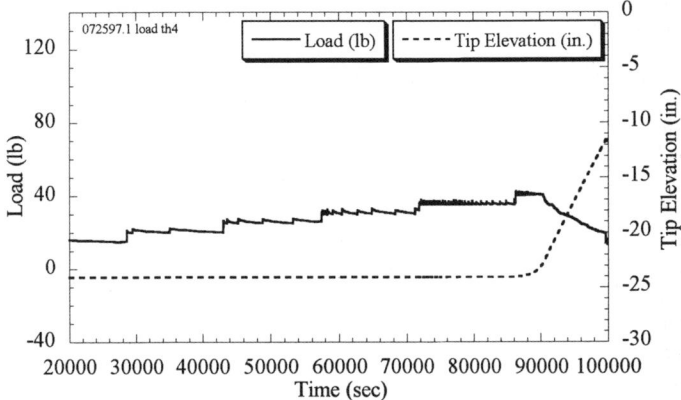

Figure 4: Incremental, Drained Pullout of Caisson 6

Figure 5: Typical Drained Pullout

The displacements for any given number of cycles increased as the peak load increased (Fig. 6). However, application of more than ten thousand cycles of load produced no permanent displacements if the peak load was smaller than the LTC. These results are logical in that application of a single long cycle of loading equal to the LTC would cause failure. The number of cycles of load to reach a given displacement decreased as the load intensity increased (Fig. 7).

Cyclic axial and lateral loading. Only a limited number of tests could be performed with loads that cycled in both the vertical and horizontal direction. The tests had a period for lateral loads that was ten times that for vertical loads (based on data for floating offshore structures) and the resultant force varied from the vertical through angles of 0° to as much as 15°, or -5° to +5°. When normalized relative to the uniaxial vertical cyclic load, the number of cycles to a given displacement can be adjusted for load inclination as shown in Fig. 8.

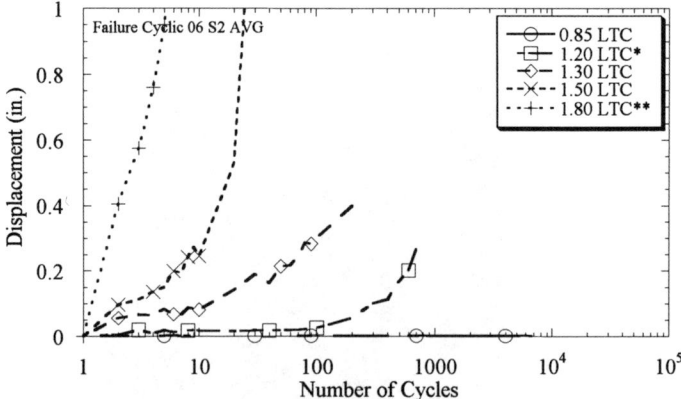

Figure 6: Cyclic Loading Applied Vertically to Caisson 6.
        * Averaged results from 3 tests. ** Averaged results from 2 tests.

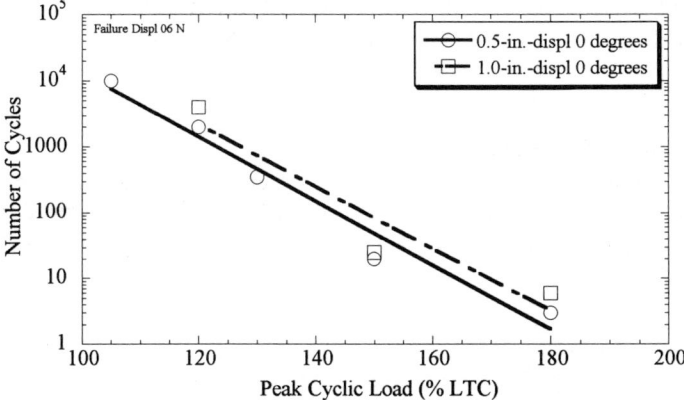

Figure 7: Relationship between Number of Cycles to a Given Displacement and Peak Cyclic Load for Caisson 6 under Vertical Loading.

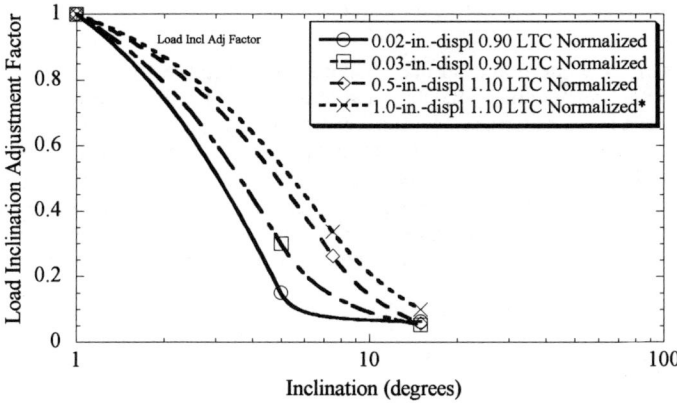

Figure 8: Influence of Inclination of Applied Cyclic Load on the Number of Cycles to a Given Displacement.

PROBABLE CAPACITIES OF OFFSHORE SUCTION CAISSONS

It seems reasonable to try to extend the results of the model tests to examine the possible behavior of full-scale suction caissons in the field.

Soil Shearing Strengths

As a first approximation, we assume that the offshore soil is a normally consolidated clay and has a constant $\lambda=dc_u/dz$. We will use λ_r for the rate of change of remolded shearing strength with respect to depth, and λ_u for the same ratio using undisturbed shearing strengths.

Caisson Installation

The force needed to penetrate the caisson comes from the difference in pressure across the top of the caisson during pumping, which in turn depends on the water depth. Actual penetration lengths will depend largely on the suction pressures that can be offered by the industrial pumps. If the applied suction is Δp, measured relative to gage pressure, the normalized limiting penetration depth is:

$$\left[\frac{L}{D}\right]_{\text{limit}} = \sqrt{\frac{\Delta p}{4\,\alpha\,\lambda_r\,D} + \frac{W}{\pi\,\alpha\,\lambda_r\,D^3}} \tag{8}$$

There remains the question as to the possibility of having the soil plug blow up inside the caisson during suction installation and thus preventing further penetration. If the caisson is examined as a free body, the upward force acting on the plug would be that due to suction, while the weight of the plug, side shear between the plug and the caisson wall, and reverse bearing capacity at the tip would resist the plug blow up. For an applied suction, Δp, a submerged soil unit weight, γ', and a caisson diameter, D, the limiting plug length-to-diameter ratio can be expressed in general terms as the solution of a second order equation (Eq. 9) in which the parameters are defined as follows:

$$\left[\frac{L}{D}\right]_{\text{limit}} = \frac{\sqrt{b^2 - 4ac} - b}{2a} \tag{9}$$

$$a = \frac{\alpha}{2}\lambda_r \tag{10}$$

$$b = \frac{(\gamma'+9\lambda_u)}{4} \tag{11}$$

$$c = \frac{\Delta p}{4D} \tag{12}$$

Extraction with the Soil Undrained

We assume that the soil has reconsolidated after the caisson is installed and, based on the model tests, that failure involves development of the full undrained shearing strength on the outside and an undrained tip bearing capacity failure. The failure load with undrained soil is then estimated using Eq. 1. In turn, the side shear is:

$$Q_s = \int \alpha \lambda_u z C dz = 0.5 \pi D \alpha \lambda_u L^2 \tag{13}$$

and the point capacity is:

$$Q_p = 9 \lambda_u L \pi D^2 / 4 \tag{14}$$

Extraction with the Soil Drained

We will use the same approach as for the model caissons. The side shear on the inside and outside of the caisson is then:

$$Q_s = \pi D L^2 K \gamma' \tan(\delta) \tag{15}$$

The tip bearing capacity for just the thin shell, under drained conditions, is negligible. The capacity is thus from side shear and the dead weight of the caisson. In this analysis, properties that are consistent with the values of λ for the analyses under undrained conditions are used.

CONCLUSIONS

The model tests suggest that rapid (undrained soil) pull out of the caissons leads to failure in the soil surrounding the caisson and an undrained tip bearing capacity failure, with the plug of soil coming out with the caisson. As usual for tension tests on deep foundations, failure is catastrophic in that the capacity diminishes rapidly after the peak load capacity is passed. For the case of slow pull out (drained soil), the caisson came out clean, so failure was between the soil and the caisson on both the inside and outside. Failure was again catastrophic. In the laboratory tests, the capacities under drained conditions were significantly less than under undrained conditions.

Model tests show that application of peak cyclic tensile loads exceeding the static capacity under drained conditions will lead to pull out in a number of cycles that decreases as the peak load intensity increases.

For monotonic loading, capacities were slightly larger for loads applied at an angle than for vertical loads. For cyclic loading, varying the angle of the load resulted in a reduction in capacity.

ACKNOWLEDGMENTS

The work presented herein was funded in part by the Offshore Technology Research Center (NSF Engineering Research Centers Program, grant No. CDR8721512). The authors wish to thank the Offshore Technology Research Center for its support. The authors also wish to thank Undergraduate Research Assistants Jody Kerwin, Neil Martin, Higinio Sanchez, and Pawel Zieminski for their help during this phase of the experimental program.

REFERENCES

Aas, P. M., and K. H. Andersen (1992), "Skirted Foundations of Offshore Structures", *Proceedings*, OSEA Conf., Singapore, Paper 92145, pp. 305-312.

Albert, L. F., R. D. Holtz, and E. Magris (1987), "The Superpile System: A Feasible Alternate Foundation for TLP in Deep Water", *Proceedings*, Offshore Technology Conference, OTC 5392, pp. 307-314.

Andersen, K. H., and R. Lauritzsen (1988), "Bearing Capacity of Foundations with Cyclic Loads", *Jour. of Geotechnical Engineering*, ASCE, Vol. 114, No. GT5, pp. 540-555.

Andersen, K. H., R. Dyvik, and K. Schroder (1992), "Pull-Out Capacity Analyses of Suction Anchors for Tension Leg Platforms", *Proceedings*, Intern. Conf. on Behavior of Offshore Structures, London, Vol. 1, pp. 1311-1322.

Bryant, W. R., and G. R. Simmons (1992), "The Physiography and Engineering Constraints of the Continental Slope in the Northwestern Gulf of Mexico", Civil Engineering in the Oceans V, ASCE, pp. 1036-1050.

Clukey, E. C., and M. J. Morrison (1993), "A Centrifuge Analytical Study to Evaluate Suction Caissons for TLP Applications in the Gulf of Mexico," *Design and Performance of Deep Foundations. Piles and Piers in Soil and Soft Rock*, ASCE Geotechnical Special Publication No. 38, 1993.

Clukey, E. C., M. J. Morrison, J. Gariner, and J. F. Corté (1995), "The Response of Suction Caissons in Normally Consolidated Clays to Cyclic TLP Loading Conditions", *Proceedings*, Offshore Technology Conference, OTC 7796, pp. 909-918.

Dannenberger (1993), oral comments at the Requalification Workshop in New Orleans, December 8th.

Digre, K. A., L. K. Brasted, and P. W. Marshall (1989), "The Design of the Bullwinkle Platform, OTC 6050, pp. 63-80

El-Gharbawy, S. L. The Pullout Capacity of Suction Caisson Foundations for Tension Leg Platforms. *Ph. D. Dissertation*, The University of Texas at Austin, 1998.

Lee, G. C., and L. Light (1992), "Fixed Offshore Platforms - Design and Construction Considerations", notes for the short course *Design of Fixed Offshore Platforms*, Univ. of Texas.

McClelland, B. (1974), Design of Deep Penetration Piles for Ocean Structures, *Jour.*, Geotechnical Engineering Div., ASCE, Vol. 100, No. GR7, pp. 705-748.

McClelland, B., J. A. Focht, Jr., and W. J. Emrich (1969), "Problems in Design and Installation of Offshore Piles", *Jour.*, Soil Mech. and Found. Div., ASCE, Vol. 95, No. SM6, pp. 1491-1513.

Morrison, M. J., and E. C. Clukey (1994), "Behavior of Suction Caissons under Static Uplift Loading", *Proceeding*, Intern. Conf. Centrifuge 94, Singapore, pp. 823-828, Leung, Lee, and Tan Eds., Balkema, Rotterdam.

Peterson, W. H., R. W. Patterson, J. D. Smith, E. B. Denison, D. W. Allen, A. G. Ekvall, E. H. Phifer,, and Y. S. Li (1994), "Auger TLP Well Systems," *Proceeding*, Offshore Technology Conference, Vol. 2, OTC 7617.

Potyondy, J. G. (1961), "Skin Friction between Various Soils and Construction Materials", *Geotechnique*, Vol. 11, No. 4, pp. 339-353.

Poulos, H. G. (1982), "Influence of Cyclic Loading on Axial Pile Response", *Proceeding, Sec. Intern. Conf. on Numerical Methods in Offshore Piling*, pp. 419-440.

Støve, O. J., S. Bysveen, and H. P. Christopherson (1992), "New Foundation System for the Snorre Development," *Proceedings*, Offshore Technology Conference, Vol. 2, OTC 6882.

Tjelta, T. I. (1994), "Geotechnical Aspects of Bucket Foundations Replacing Piles for the Europipe 16/11E Jacket", *Proceeding*, Offshore Technology Conference, OTC 7397, pp. 73-82.

Tjelta, T. I., P. Aas, J. Hermstad, and E. Andenaes (1990), "The Skirt Piled Gullfaks C Platform Installation", *Proceeding*, Offshore Technology Conference, OTC 6473, p, 453-462.

Whittle, A. J., J. T. Germaine, F. D. Cauble, and M. Geer (1993), "Behavior of Suction Caisson Foundations", Progress Report Submitted to the M.I.T. Sea Grant Program and the Joint Industry Consortium.

EVALUATION OF PILE LOAD TESTS IN SOFT COHESIVE LOUISIANA SOILS

Hani Titi[1], Associate Member, ASCE, Murad Abu-Farsakh[1], Associate Member, ASCE, and Mehmet Tumay[2], Fellow, ASCE

ABSTRACT

This paper presents the results of a study conducted on 34 square precast, prestressed concrete piles driven into soft cohesive Louisiana soils. The main objective of this study is to evaluate and compare the ultimate axial bearing capacity of the investigated piles obtained using pile load test interpretation methods and those obtained using the static analysis. These methods are the slope and tangent method, the offset limit method, the α-method, and the β-method.

The evaluation of the measured ultimate pile capacities obtained from the slope and tangent and the offset limit methods showed a small variation between the results obtained by these methods where the coefficient of variation (COV) varies between 0 and 18 percent with an average of 3 percent. On contrary, larger variation was noticed between the results of the ultimate capacities predicted using the static analysis methods. The coefficient of variation for the ultimate bearing capacities obtained by the α-method and the β-method ranges from 3 to 52 percent with an average of 21 percent indicating more discrepancies in predicting the ultimate pile capacity.

INTRODUCTION

Piles are important structural members that carry super-structural loads deep in the ground. Based on the nature of support provided by the surrounding soil, piles may be classified as end-bearing piles and friction piles. While end-bearing piles transfer most of their loads to an end-bearing stratum, friction piles resist a significant portion of their loads via the skin friction developed along the surface of the pile. The behavior of friction piles mainly depends on the interaction between the surrounding soil and the pile shaft.

---

[1] Research Associate, Louisiana Transportation Research Center, Baton Rouge, LA 70808.
[2] Associate Dean for Research and Professor of Civil and Environmental Engineering, College of Engineering, Louisiana State University, Baton Rouge, LA 70803.

The safety and stability of pile-supported structures depend on the behavior of the piles under working loads. Current design of friction piles is generally based on empirical formulas and depends to a large degree on the personal experience and judgment of the engineer. Because of many uncertainties associated with pile foundation analysis and design, pile load tests are usually conducted after pile installation. These tests are expensive and time consuming. However, they are important in assisting design engineers to evaluate the response of piles under real loading conditions. Even though pile load tests are not a substitute for the engineering analysis of the pile behavior, they are important for the verification of the design/analysis of deep foundations. A general design method for pile foundations is not available, but rather solutions are available for given subsurface conditions, pile characteristics, and specific driving/installation methodologies. Therefore, the availability of well-documented case histories is essential for providing the design engineers with real-world examples of bearing capacities of piles. The availability of a wide spectrum of data on deep foundation tests provides a realistic reference for future jobs under similar pile and soil conditions. It will also help in achieving an economical design of piles.

This paper presents an evaluation of 34 load tests conducted on piles driven into soft cohesive Louisiana soils (mainly clay and silty clay deposits). These pile load tests were obtained from the Louisiana Department of Transportation and Development (LA DOTD) files. The investigated piles are 35.6, 40.6, and 45.7 cm (14, 16, and 18 inches respectively) square precast, prestressed concrete (PPC). The ultimate axial bearing capacity (Q_u) of each pile was determined using the Butler-Hoy (1977) and Davisson (1972) pile load test interpretation methods. Then, the static analysis using α-method and β-method was conducted to estimate the ultimate axial bearing capacity of these piles. Ultimate pile capacities predicted using α-method and β-method were compared with those obtained from load test interpretation methods. A critical evaluation of the capability of different methods to predict the ultimate pile capacity was conducted with respect to the ultimate capacity obtained by Butler-Hoy (1977) method, which is the primary load test interpretation method used by LA DOTD.

BACKGROUND

Soil deposits in southern Louisiana are generally characterized as soft and compressible in their nature. Subsurface exploration investigations conducted for LA DOTD projects indicate that a large percentage of soils encountered are saturated soft clay and silty clay deposits with occasional lenses of silt and sand. Due to soil conditions and the presence of lakes, bayous, and waterways encountered in the alignment of highways in Louisiana, it generally becomes necessary for LA DOTD to consider pile foundations for transportation infrastructure.

The practice of LA DOTD is to carry out load tests on deep foundations (piles) for important soil related projects. The purpose of conducting the pile load test is to evaluate the ultimate bearing capacity of the pile and use it to verify the design of the piles in that project. Observation of the actual pile behavior in the field will also foster the confidence in the design of piles. Two methods are used by LA DOTD for interpretation of the compressive static pile load test, namely: the slope and tangent method (Butler and Hoy 1977) - primary method - and the offset limit method (Davisson 1972) -secondary method. Design/analysis of piles is performed using the α-method (Tomlinson 1980). However, the experience and judgment of the engineer and the result of the pile load test play an important role in the design/analysis process. When a pile load test is conducted for a project, the factor of safety for the design of piles is usually taken as 2. This factor of safety is increased to 3 in cases where no pile load test is carried out.

PILE LOAD TESTS

A pile load test database was developed (Tumay and Titi 1998) to expand the Federal Highway Administration's (FHWA) data on deep foundation load tests. The database provides information and data regarding the site, subsurface exploration, soil identification, testing, and deep foundation load tests. The database comprises a wide variety of piles with respect to material, size, cross section, and shape. These include precast, prestressed concrete, H-steel, timber, and step taper piles installed in cohesive and cohesionless soils.

This study was conducted on 34 square PPC piles of different lengths driven into cohesive soils at different locations in Louisiana. The square PPC driven pile is the most commonly used type for LA DOTD projects. The investigated pile load tests are part of the load test database developed by Tumay and Titi (1998). Load test reports for these piles were carefully reviewed for the availability of proper documentation on site location, subsurface exploration and soil testing, and deep foundation data that include installation and testing. Then this data was compiled and analyzed.

Interpretation of Pile Load Tests

For each pile, the ultimate bearing capacity was determined from the load-settlement curve using the Butler-Hoy (1977) and Davisson (1972) methods. The Butler-Hoy (1977) method is the primary interpretation method used by LA DOTD followed by the Davisson (1972) method. Figure 1 depicts a load-settlement curve for a 35.6-cm square PCC pile. This figure also illustrates how to estimate the ultimate pile capacity (Q_u) using Butler-Hoy (1977) and Davisson (1972) methods. In the Butler-Hoy method (also called slope and tangent method), the ultimate pile capacity is obtained from the intersection of the initial slope of the load-settlement curve and a tangent to the load-settlement curve with a slope of 7 mm/kN (1 in./40 kips). In the Davisson (1972) method, the ultimate pile capacity is obtained from the intersection of a line (with slope of EA/L drawn from an initial settlement point of 4+120/b mm) and the load-settlement curve as illustrated in Figure 1. The detailed results of the analyses conducted on the 35.6, 40.6 and 45.7 cm square PPC piles are presented in Tables 1, 2, and 3, respectively.

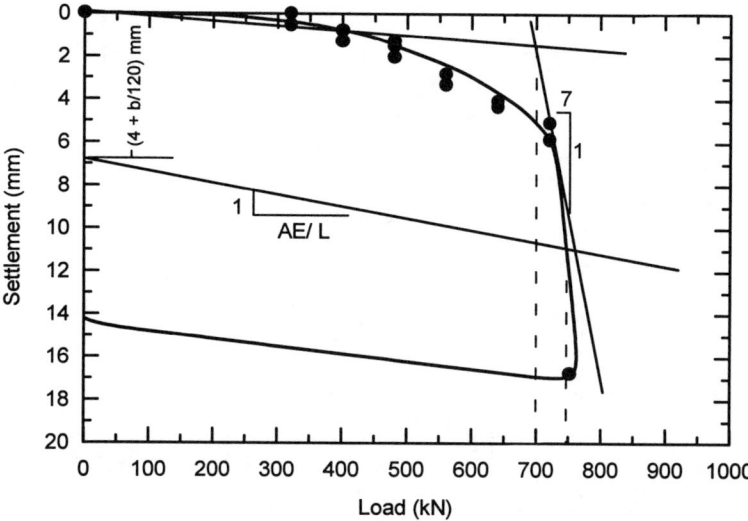

Figure 1: Estimation of the ultimate bearing capacity from pile load test using Butler-Hoy (1977) and Davisson (1972) methods.

Static Analysis

The ultimate pile capacity (Q_u) consists of two components: the pile tip capacity (Q_t) and the shaft frictional capacity (Q_s). Different methods of analysis are used to estimate the ultimate pile capacity based on static and dynamic approaches. Static analysis is based on insitu and/or laboratory evaluation of soil properties. In this study, two methods were selected for the static analysis of the investigated piles: (a) the α-method (total stress approach), in which the skin friction is correlated to the undrained shear strength of the undisturbed soil, where the factor α accounting for the soil disturbance caused by pile installation; and (b) the β-method (effective stress approach), in which the skin friction is correlated to the initial effective overburden stress. The results of the static analysis conducted on the 35.6, 40.6 and 45.7 cm square PPC piles are also presented in Tables 1, 2, and 3, respectively.

EVALUATION OF THE RESULTS

Pile load tests were analyzed in order to evaluate the capability of the load test interpretation methods - Butler-Hoy (1977) and Davisson (1972) – and the static analysis methods - α-method and β-method - to predict the ultimate capacity of piles driven in cohesive Louisiana soils. A comparison of the ultimate bearing capacities determined by these four methods is presented in Figure 2. The ultimate

Table 1: Summary of the analyses conducted on the 35.6 cm (14") square precast prestressed concrete piles.

Project Name and Location	Pile ID	Gross Pile Length (m)	Embedded Pile Length (m)	Predominant Soil Type	Ultimate Capacity from Static Pile Load Test Q_u (kN)		Predicted Ultimate Capacity from Static Analysis, Q_u (kN)					
					Butler-Hoy	Davisson	α - method			β - method		
							Q_s	Q_t	Q_u	Q_s	Q_t	Q_u
Bayou Boeuf Bridge (East Approach), Assumption Parish	TP 5	25.9	24.1	Clay and silty clay	778	703	419	38	457	918	70	988
Southern Pacific railroad Bridge (Lake Cheles), U.S 117, Calcasieu Parish	TP 1	21.0	17.2	Clay	1388	1334	730	154	884	996	571	1566
I-49/U.S 71 Interchange, Rapides Parish	TP 6	21.3	17.7	Clay and silty clay	730	792	576	61	637	558	178	737
Swafford Creek Bridge, Route LA 101, Rapides Parish	TP 1	16.8	14.0	Clay and silty clay	1192	1219	701	157	858	803	500	1304
Bayou Fatma Bridge, Jefferson Parish	TP 1	24.4	20.7	Clay and silty clay	1001	890	419	238	657	493	632	1125
Bayou Boeuf Bridge (East Approach), Assumption Parish	TP 2	24.4	21.6	Clay and silty clay	836	867	431	61	493	473	223	697
Bayou Boeuf Bridge (West Approach), St. Mary Parish	TP1	29.0	27.1	Clay and silty clay	1001	970	724	34	758	1082	74	1157
Lee Drive Extension (Burbank Dr. -Highland Rd.), East Baton Rouge Parish	TP 1	21.3	17.5	Clay and silty clay	698	738	571	17	588	786	36	821
Intracoastal Waterway Bridge (Northeast Approach), Terrebonne and Lafourche Parishes	TP 3	37.2	34.0	Clay	1036	1076	803	40	843	1158	61	1220
Elbow Bayou Bridge, Highland Road Relocation (Ben Hur Rd. - Gardere Lane), East Baton Rouge Parish	TP 1	19.8	16.3	Silty clay	954	970	503	100	603	573	235	808
Bayou Boeuf Bridge (Main Span), Route LA 3052, St. Mary Parish	TP 5	25.9	24.4	Clay	1032	1081	740	43	783	1010	72	1082

Q_u: Ultimate bearing capacity of the pile
Q_s: Ultimate shaft capacity
Q_t: Ultimate tip capacity

Table 2: Summary of the analyses conducted on the 40.6 cm (16") square precast prestressed concrete piles.

Project Name and Location	Pile ID	Gross Pile Length (m)	Embedded Pile Length (m)	Predominant Soil Type	Ultimate Capacity from Static Pile Load Test Q_u (kN)		Predicted Ultimate Capacity from Static Analysis, Q_u (kN)					
							α - method			β - method		
					Buttler-Hoy	Davisson	Q_s	Q_t	Q_u	Q_s	Q_t	Q_u
LA 3188 Interchange, St. John the Baptist Parish	TP 1	21.3	21.2	Clay	1148	1179	532	83	615	560	283	844
Bayou Ramos Bridge, St. Mary Parish	TP 7	25.9	23.5	Clay	952	988	459	59	518	697	240	937
Texas Gas Pipeline Bridge, St. Mary Parish	TP 1	25.9	21.2	Clay and silty clay	881	925	758	65	823	581	205	786
Bayou Liberty Bridge and Approaches, St. Tammany Parish	TP 3	21.3	17.1	Clay and silty clay	943	996	590	80	669	527	314	841
Bayou Liberty Bridge and Approaches - Route LA 433, St. Tammany Parish	TP 1	18.3	14.7	Clay	1566	1601	634	157	791	472	547	1018
Bayou Petite Anse Bridge, Iberia Parish	TP 2	21.3	15.5	Silty clay	934	956	605	78	683	485	305	790
Bayou Des Cannes Bridges, Acadia, Evangeline and St. Landry Parishes	TP 1	22.9	13.1	Silty clay	1041	1059	611	94	705	538	270	807
White Castle - Plaquemine Highway - Route LA 1 (Bayou LA Butte), Iberville Parish	TP 1	22.9	22.3	Silty clay	569	645	594	34	627	435	70	505
Bayou Lacassine Bridge (LA U.S. 90), Jefferson Davis Parish	TP 1	18.3	11.2	Silty clay	1023	1068	452	235	687	516	498	1014
West Grand Marias Ditch (LA U.S. 90), Jefferson Davis Parish	TP2	19.8	12.2	Silty clay	1036	1054	505	120	625	551	350	901
Rocky Canal Bridge, Route LA 1, Ascension Parish	TP1	26.8	26.5	Clay and silty clay	899	952	846	48	895	876	82	958
Bridge over Tributary to Bayou Barbue, Route LA I 49, Natchitoches Parish	TP1	15.2	11.5	Clay	636	818	390	36	425	396	71	467
Washington-Meeker Highway, Route LA I 49, Rapides Parish	TP3	21.3	18.0	Clay and silty clay	925	988	898	94	992	711	230	941
Washington-Meeker Highway, Route LA I 49, Rapides Parish	TP4	18.9	15.7	Clay and silty clay	790	843	589	100	689	594	225	819
Sunshine Bridge-Gramercy Highway (North Section), Route LA 3125, St James Parish	TP1	15.2	10.6	Silty clay	1036	1072	572	223	794	726	504	1231

Q_u: Ultimate bearing capacity of the pile
Q_s: Ultimate shaft capacity
Q_t: Ultimate tip capacity

Table 3: Summary of the analyses conducted on the 45.7 cm (18") square precast prestressed concrete piles.

Project Name and Location	Pile ID	Gross Pile Length (m)	Embedded Pile Length (m)	Predominant Soil Type	Ultimate Capacity from Static Pile Load Test Q_u (kN)		Predicted Ultimate Capacity from Static Analysis, Q_u (kN)					
							α - method			β - method		
					Buttler-Hoy	Davisson	Q_s	Q_t	Q_u	Q_s	Q_t	Q_u
Hooper Road Blackwater Bayou Bridge, East Baton Rouge Parish	TP 2	24.4	20.1	Silty clay	1512	1526	1089	270	1359	1416	504	1920
LA 29 Over Bayou Cocodrie , St. Landry and Evangeline Parishes	TP 2	27.4	24.2	Silty clay	867	890	899	68	967	1055	432	1487
LA U.S. 167 and LA 103 Route I-49, St. Landry Parish	TP 8	21.3	18.5	Clay and silty clay	1290	1334	814	62	876	907	322	1230
Bayou Teche Bridge at Route LA 92 (Extension), St. Martin Parish	TP 1	25.9	23.2	Clay and silty clay	934	979	754	99	853	1051	442	1493
Bayou Des Glaises Diversion Channel Avoyelles Parish	TP1	24.4	24.1	Clay and silty clay	770	774	741	144	885	718	389	1107
Hayes-Lacassine Highway, Route LA101 Calcasieu, and Jefferson Davis Parishes	TP1	13.7	10.5	Clay	905	912	649	225	874	323	518	841
Natalbany River Bridge, Route LA 1040, Tangipahoa and Livingston Parishes	TP1	25.9	18.8	Clay and silty clay	1597	1601	1125	290	1515	715	1054	1769
Belle Place Bridge over Bayou Teche, Iberia Parish	TP1	32.0	21.7	Silty Clay	556	572	512	63	574	322	172	494

Q_u: Ultimate bearing capacity of the pile
Q_s: Ultimate shaft capacity
Q_t: Ultimate tip capacity

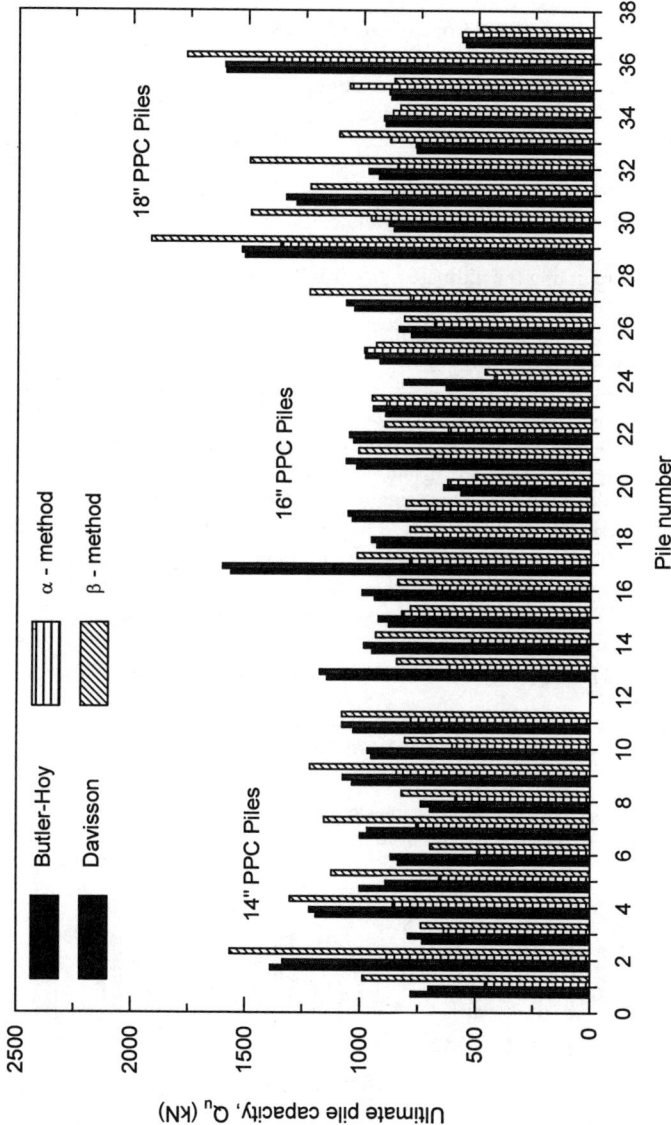

Figure: 2 Ultimate pile capacities of the investigated piles as predicted by the load test interpretation methods and the static analysis methods.

pile capacities (Q_u) estimated from the Butler-Hoy (1977) method ranges from 556 to 1597 kN, while Q_u obtained from the Davisson (1972) method varies between 572 and 1601 kN. On the other hand, Q_u predicted from α-method and β-method ranges from 425 to 1415 kN and from 467 to 1920 kN, respectively. Examination of this figure indicates that the ultimate pile capacities (Q_u) obtained from the interpretation of load tests have small variation, while larger variation is noticed among the ultimate pile capacities predicted using the static analysis methods. This is illustrated in Figure 3 where the coefficient of variation (COV) for the load test interpretation methods varies between 0 and 18 percent with an average of 3 percent. This indicates low variation for the estimated ultimate capacities using these two methods. The COV for the results of the static analysis methods ranges between 3 and 52 percent with an average of 21 percent indicating more discrepancies in predicting the ultimate pile capacities using these methods.

The soil profiles where the investigated piles were driven are mainly successions of clay and silty clay layers with occasionally pockets of silt and sand. These piles are considered friction piles where most of their capacities are derived from the shaft resistance. Figure 4 shows the relationship between the predicted percentage of skin friction versus the pile penetration. Using the α-method, the average ratio of the predicted skin friction is about 86 percent. For β-method, the average ratio of the predicted skin friction is about 70 percent. This generally indicates that these are friction piles. The use of β-method predicted a larger percentage of end-bearing capacity compared to α-method. This is due to the fact that the β-method uses the effective stress at the pile tip to determine the end-bearing capacity. This value usually gives a higher end-bearing capacity than the undrained shear strength of undisturbed soil as in the case of soft cohesive soils, especially for long piles.

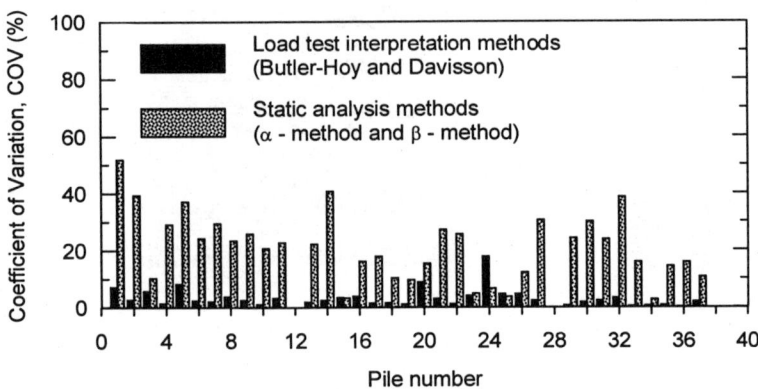

Figure 3: Coefficient of variation among the different methods used in the analyses of the investigated piles.

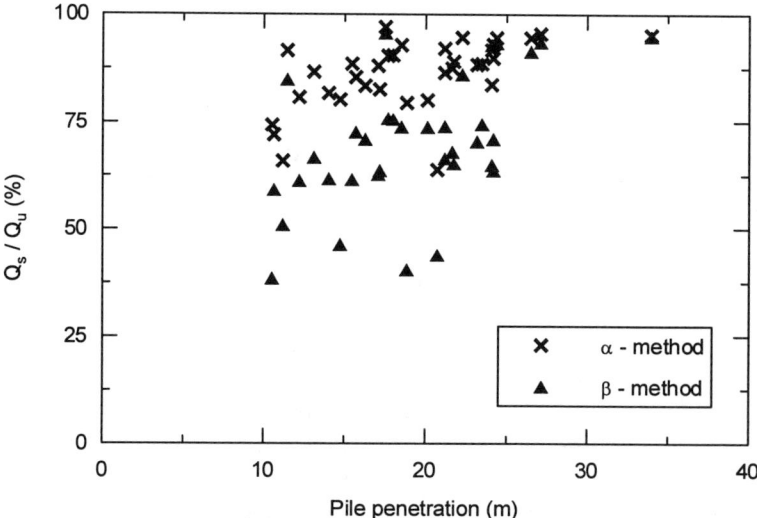

Figure 4: Skin frictional capacity predicted by the static analysis methods versus pile penetration.

Since the Butler-Hoy (1977) method is the primary method used by LA DOTD to estimate the ultimate pile capacity from the load test, it will be used to evaluate the ultimate bearing capacity obtained by the other methods. Figure 5 presents the comparisons of the ultimate bearing capacities estimated by Butler-Hoy (1977) method and those predicted by Davisson (1972) method. In Figures 6a and 6b, a comparison is presented between the ultimate bearing capacities estimated by Butler-Hoy (1977) method and those determined by the α-method, and β-method, respectively.

The ultimate pile capacities estimated by Davisson (1972) method are in compliance with those determined by Butler-Hoy (1977) method. The following equation, obtained from the regression analysis, shows the relationship between these capacities:

$$(Q_u)_{Davisson} = 1.02 \ (Q_u)_{Butler\text{-}Hoy} \qquad\qquad (1)$$

The predicted ultimate pile capacities $(Q_u)_{\alpha\text{-method}}$ and $(Q_u)_{\beta\text{-method}}$ are plotted against the measured pile capacities $(Q_u)_{Butler\text{-}Hoy}$ in Figures 6a and 6b, respectively. The results are widely scattered in which the α-method underpredicts the ultimate capacity of the investigated piles, while the β-method overpredicts the ultimate capacities of those piles.

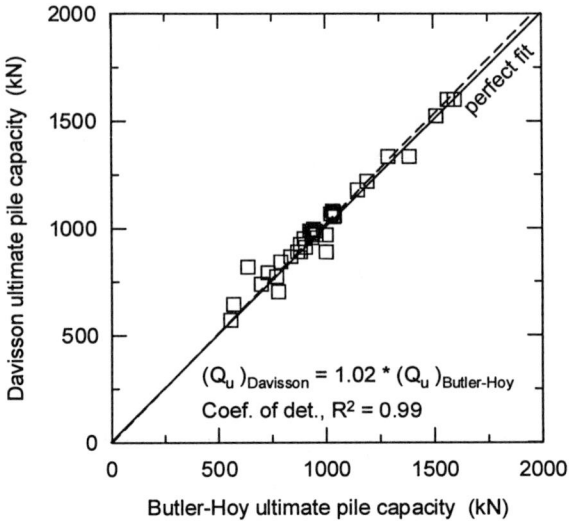

Figure 5: Ultimate pile capacity estimated using Butler-Hoy (1977) method versus Davisson (1972).

CONCLUSIONS

Pile load tests were analyzed in this study in order to evaluate and compare the capability of different methods for predicting the ultimate bearing capacity of piles installed in soft Louisiana soils. These include the Butler-Hoy and Davisson interpretation methods, and α and β static analysis methods. Based on this study, the following conclusions can be drawn out:

Low variation is found between the estimated ultimate pile capacities obtained by Butler-Hoy and Davisson methods with the COV<18 percent. However, large variation is found between static analysis α and β methods with the COV ranges between 3 and 52 percent, indicating more discrepancies in predicting the ultimate pile capacity.

The ultimate pile capacities estimated by Davisson method are in compliance with those determined by Butler-Hoy method. While the static analysis methods predicted widely scattered pile capacities with the α-method tends to underpredict the ultimate capacity, while the β-method tends to overpredict the ultimate capacities of those piles.

The use of β-method predicts a larger end-bearing capacity compared to the α-method. This is due to the fact that the β-method uses the effective stresses at the

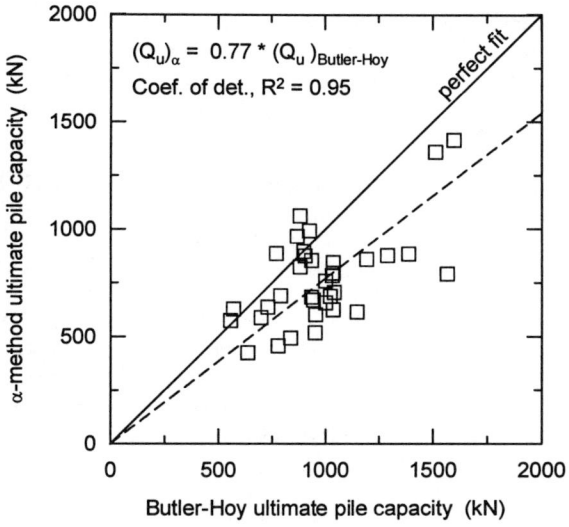

(a) Butler-Hoy method versus α-method

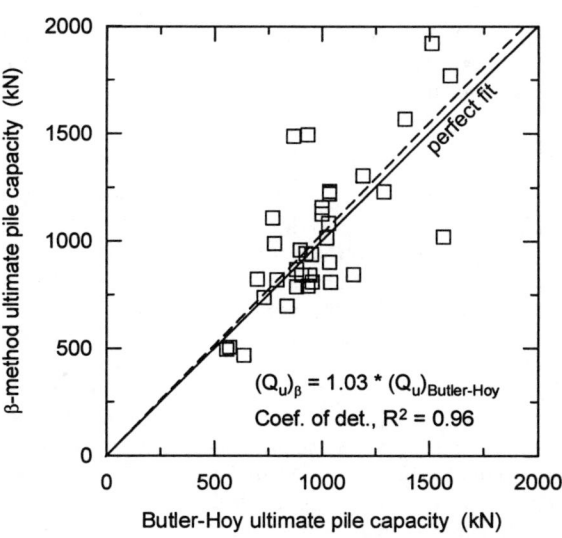

(b) Butler-Hoy method versus β-method

Figure 6: Ultimate pile capacity estimated using Butler-Hoy (1977) method versus α-method and β-method.

pile tip to determine the end-bearing capacity. This value gives usually higher capacity than the undrained shear strength of undisturbed soil as in the case of soft cohesive soils especially for long piles.

ACKNOWLEDGMENTS

Part of this research was financially supported by The U.S. Department of Transportation, Federal Highway Administration Grant No. DTFH61-96-X-00002 through the project supported "Including All Deep Foundation Load Test Data and Soil and Site Investigations for the FHWA Database." The encouragement and support of Albert DiMillio, Carl Ealy, and Raghu Satyanarayana, Turner-Fairbank Highway Research Center, FHWA, are gratefully acknowledged. The assistance of J. B. Esnard and Ed Tavera, LA DOTD Pavement and Geotechnical Design Section, in getting pile load reports from department files is appreciated. Mrs. Elizabeth Hood, Department of Civil and Environmental Engineering – LSU, invested a great effort in the development of the pile load test database.

DISCLAIMER

The contents of this paper reflect the views of the authors who are responsible for the facts and the accuracy of the data presented herein. The contents do not necessarily reflect the views or policies of the Louisiana Department of Transportation and Development, the Louisiana Transportation Research Center, or the Federal Highway Administration. This paper does not constitute a standard, specification, or regulation.

REFERENCES

Butler, H. D., and Hoy, H. E. (1977)."The Texas Quick-Load Method for Foundation Load Testing- User's Manual," Report No. FHWA-IP-77-8.
Davisson, M. T., (1972)."High Capacity Piles," Proceedings of Lecture Series on Innovation in Foundation Construction, American Society of Civil Engineers, ASCE, Illinois Section, Chicago, pp81-112.
Tomlinson, M. J., (1980). *Foundation Design and Construction*, 4th Edition, Pitman Advanced Publishing Program.
Tumay, M. T., and Titi, H. H. (1998)."Including All Deep Foundation Load Test Data and Soil and Site Investigations for the FHWA Database," Final Report, Submitted to the U.S. Department of Transportation/Federal Highway Administration, Grant No. DTFH61-96-X-00002, Louisiana Transportation Research Center, Baton Rouge, Louisiana, 65 pp. + CD-ROM.
Zenon, G. K., Schnore, A. R., Carlo, T. A., and Baily, B. F. (1992) "Static Testing of Deep Foundations," Report No. FHWA-SA-91-042.

An Experiment with Statnamic Lateral Loading of a Drilled Shaft

Dan A. Brown[1]

Abstract

A lateral loading test of a full scale drilled shaft foundation is described, for which the loading is applied using the statnamic device. This device produces a rapid lateral loading with a duration of around 0.1 seconds. In addition to measurements of load and displacement at the top of the shaft, the test shaft was instrumented with strain gauges and accelerometers. A simple method of interpretation of the statnamic loading is presented which provides a means of accounting for inertial and viscous damping effects as well as nonlinear soil resistance. The results of this test are compared similar drilled shafts at the site which were loaded with conventional static methods. All of the testing described was performed at the Auburn University National Geotechnical Experimentation Site (NGES) at Spring Villa, Alabama in a silty soil.

Introduction

In recent years the statnamic loading device has seen increased use as a means of conducting load tests on deep foundations, primarily in the axial direction. The statnamic operates on the principal of Newton's second law, F=ma; a large mass is launched off the foundation element with an acceleration of 20g's or so, thus producing a thrust against the foundation

Figure 1 Statnamic Lateral Test

[1]Gottlieb Assoc. Professor of Civil Engineering, Auburn University, AL 36849

which is 20 times the weight of the mass. The force to launch the mass is applied using combustion within a pressure chamber of a pelletized smokeless powder which burns at a very predictable rate.

This paper describes the use of a statnamic device to apply lateral loading to a deep foundation element, as illustrated on Figure 1. The mass is attached to a flat bottomed sled which slides away as the gas pressure builds and is seen in this figure just after the mass has separated from the piston and the exhaust gasses are venting. The statnamic loading produces a force which builds smoothly to a maximum and decays over a period of about 0.1 sec. This loading period represents a high rate of loading compared to conventional static testing, but is considerably slower than one which might be modeled using wave propagation techniques. Thus the test contains inertial and damping forces characteristic of a dynamic loading event, but has a load duration which is much longer than the typical natural period of the shaft. Such dynamic response may provide insight into the soil response which might be observed during large amplitude dynamic lateral loading events such as vessel impact or seismic loading.

The objective of this paper is to describe the results of one such lateral statnamic test on a drilled shaft foundation. This test shaft was carefully instrumented to provide measurement data important to understanding the performance of the foundation during such a loading event. The test was performed at the Auburn University NGES site, a location which has well documented soil material properties and conventional static lateral load tests which provide a basis for comparison. A total of four lateral statnamic tests were performed, and the response at the top of the shafts were similar; the test described in this paper represents the most well-instrumented shaft.

Soil Conditions

The soils at the test site are composed of a residual soil of decomposed gneiss and quartzite, typical of the Piedmont region of the southeastern U.S. (Vinson and Brown, 1998). These soils are classified as predominantly clayey silts and silty clays (ML-CL) with occasional sand lenses. The weathering is more advanced within the upper 4 meters (which are of most relevance to the lateral test) and these soils tend to be have higher clay content than those at greater depth. Other than the near-surface weathering, the soils within the upper 15 m at the site are relatively uniform on a large scale; on a small scale these soils exhibit significant spatial variability. Groundwater at the time of testing was at a depth of approximately 4 m. Some average soil properties of the soils within the upper 4 meters are provided on Table 1.

Based upon 20 CIUC triaxial tests performed on samples from the 4 to 15 m depth range, these soils have an effective cohesion of 17 kPa and effective friction angle of 31°. The upper 4 m have somewhat higher clay content and are likely to exhibit somewhat higher cohesion (although effective stress strength parameters have not directly been measured for shallow soils).

Property	Average Value	Range
water content	34%	30 to 40%
plasticity index	8%	2 to 20%
std penetration test	11 blows/30cm (b/f)	8 to 16 b/f
cpt tip resistance, q_c	2.5 MPa	2 to 3 MPa
cpt friction ratio	6%	5 to 8%
pressuremeter modulus	8 MPa	6 to 11 MPa
dilatometer modulus	15 MPa	10 to 25 MPa
shear wave velocity	200 m/s	180 to 240 m/s
S_u from UU triaxial	90 kPa	40 to 150 kPa

Table 1 Soil Properties for 0 to 4 m Depth

Test Shaft

All of the test shafts were constructed as drilled and cast-in-place concrete shafts to a diameter of 0.92 m and depth of 11.5 m. The statnamic test shaft and most of the static load test shafts were constructed using a drilling slurry. Reinforcement consisted of 12 #11 longitudinal bars (3.5 cm diameter) with 7.6 cm (3 inches) of cover. The concrete had a compressive strength of approximately 31 MPa (4500 psi). The lateral load was applied at approximately 30 cm above the ground surface in all tests.

Instrumentation during the statnamic test included strain measurements on both the compression and tension sides of the shaft at each of 6 elevations. Strain measurements were obtained using electrical resistance type strain gauges mounted on # 4 sister bars which were tied into the rebar cage. Each sister bar had a pair of T-rosette gauges connected to form a full bridge circuit on each bar, with a gauge factor of 2.6. These gauges were observed to be very stable, with a resolution of less than ½ microstrain.

Below-grade instrumentation also included accelerometers at 3 elevations below the surface (plus one atop the shaft). These accelerometers were placed within inclinometer casings by attaching the gauge to an aluminum mount with spring-loaded wheels designed to fit into the grooves in the casing. In this manner the gauges could be placed at any desired elevation and retrieved after completion of the test. For the relatively low accelerations below grade during the statnamic lateral load test (measured to be less than 5 g's), these accelerometer mounts worked quite well. Numerical integration of the accelerometer measurements provide a measure of displacements at depth. For the shafts used in the static lateral tests, the inclinometer casings were used to measure displacement below grade with a conventional inclinometer device.

Measurements at the top of the shaft during the statnamic loading include load, displacement, and acceleration. Load is measured with a calibrated load cell which is placed between the statnamic piston and a hemispherical bearing on the shaft. Displacement is measured directly using a laser target that senses a remote laser beam; the laser is positioned on the ground surface a sufficient distance from the test shaft so that any ground motion arrivals at the laser itself will not occur within the duration of the test measurements (about 1/4 sec.). An accelerometer on the top of the shaft provides a direct measure of acceleration and, through numerical integration, a redundant measure of displacement of the top of the shaft.

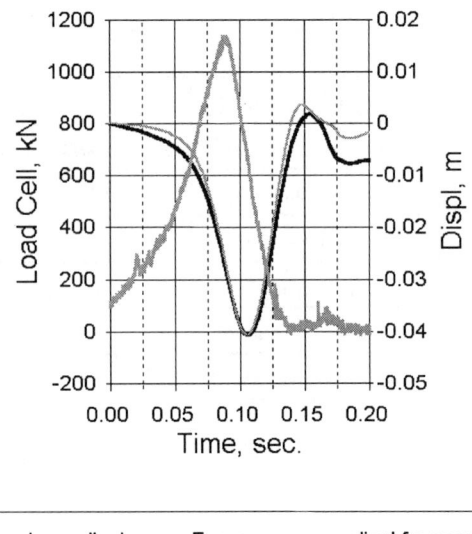

Figure 2 Statnamic Force and Displacement

All instrumentation was monitored with a Megadac® data acquisition system which was programmed to record all data channels at approximately 1 msec intervals.

Test Results

The statnamic loading produced a load and displacement time history as illustrated on Figure 2. The duration of the load pulse is slightly larger than 0.1 seconds with a maximum applied force of 1.1 MN and a maximum displacement of approximately 40 mm. The displacement data suggest that the shaft rebounded with little permanent set and with very little observable oscillation after the initial loading (i.e., the vibration damped out very rapidly).

Measurements of maximum and minimum strain

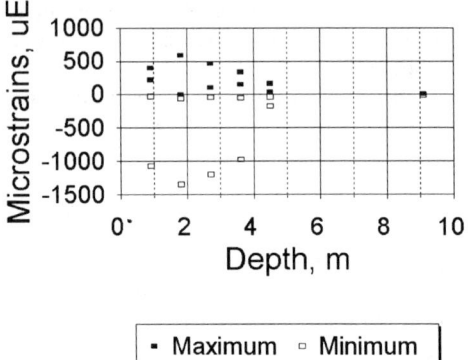

Figure 3 Strain Gauge Peak Values

values at each gauge location are presented on Figure 3; the upper and lower bounds of this figure represent the maxim compressive and tensile strains measured during the test. The maximum bending moment is clearly indicated at a depth of around 2 m. The maximum tensile strains at this level (nearly 1500 microstrains) exceed the maximum compressive strains by a factor of almost 3, suggesting that the shaft was quite severely cracked by this loading. Bending strains below a depth of 4 m are very small.

The downhole accelerometer measurements were integrated to obtain displacement, and these are presented on Figure 4. The peak

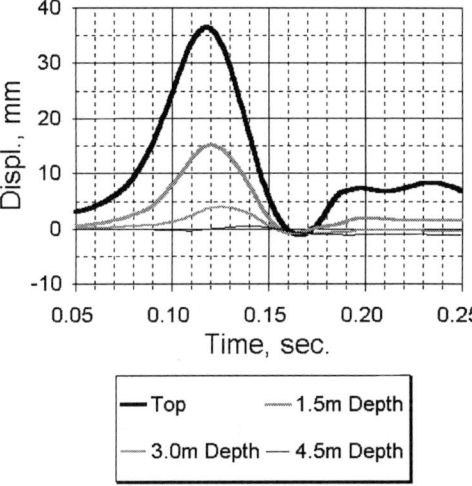

Figure 4 Displacement from Downhole Accel's.

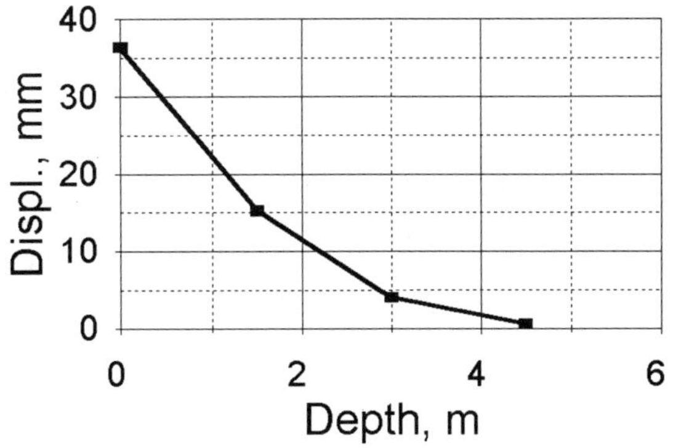

data are seen to exhibit very small phase shifts, thus revealing that wave propagation effects during the statnamic loading are very small. The loading produces a displacement pattern more analogous to that of a rapid static test due to the low frequency relative to the natural frequency of the shaft. The peak values of displacements as a function of depth below grade are presented on Figure 5. These data indicate that the displacements were concentrated within the upper 3 m of the shaft,

Figure 5 Peak Displ. vs Depth During Statnamic Test

and thus the soil resistance within that region would be expected to dominate the behavior.

Analysis and Interpretation

A method of evaluating the results of the statnamic lateral test consists of a simple single degree of freedom dynamic model of the following form:

$$F_{statnamic} = F_{inertia} + F_{damping} + F_{static}$$

where,

$F_{statnamic}$ = measured force on the statnamic load cell
$F_{inertia}$ = inertial resistance from effective mass of the foundation
$F_{damping}$ = effective viscous damping resistance
F_{static} = effective static soil resistance

The inertial resistance is roughly that of a cylinder rotating about its base, with a diameter equal to that of the shaft and a height taken as approximately 3 m, based on the observed displacement pattern. For such a cylinder of radius r, height h, and mass m, the mass moment of inertia about the base, I_y is:

$$I_y = m (r^2/4 + h^2/3) = 14.91 \text{ kN-s}^2/\text{m for 24 kN/m}^3 \text{ concrete}$$

The rotational acceleration of such a cylinder in relation to a displacement x at the top would be ẍ/h and thus summing moments about the base,

$$(F_{inertia})h = (I_y)(\ddot{x}/h)$$

therefore, $F_{inertia} = (I_y/h^2)\ddot{x} = m_e\ddot{x}$

where m_e may be thought of as the effective mass for the foundation. For this test, m_e would be calculated to be 1.66 kN-s^2-m. In reality, there is undoubtedly a mass of soil accompanying the shaft as the passive wedge of soil is displaced, so this mass might rationally be increased by a factor of 2 or so.

The damping resistance is presumed to be represented by a viscous damper in which the force $F_{damping}$ is proportional to velocity, ẋ, by a constant, c (which is in units of force-sec./length). In order to relate this more meaningfully to a system damping parameter, the damping constant is expressed as a percent of the critical damping, c_c, by

$$D = c/c_c = c/[2(k_s m_e)^{1/2}]; \text{ where } k_i = \text{initial static stiffness}$$

thus, $F_{damping} = c \dot{x} = D[2(k_i m_e)^{1/2}]$

The static resistance is modeled as a function of displacement, x, using a spring with stiffness k_s. Because the soil response for lateral loading at large strains is known to be highly nonlinear, this spring is modeled as a nonlinear stiffness which decreases as a function of displacement. For the analyses in this paper, the stiffness has been presumed to decay exponentially from an initial value k_i at x=0 to a value k=α k_i at a displacement of 5% of the shaft diameter (or 0.1r = 0.046 m in this case). Thus the static resistance is modeled using two constants, k_i and α according to:

$$F_{static} = \exp\{Ln\ k_i + Ln[\alpha(.1r)]\}$$

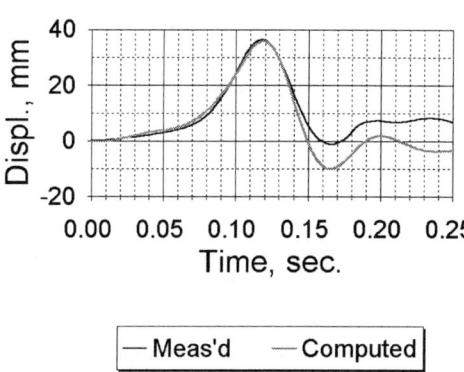

— Meas'd   — Computed

Figure 6 Measured & Model Displ. time history

A simple computational model of the statnamic loading is thus established, involving the effective mass, m_e, damping ratio, D, and stiffness constants k_i and α. With this model, using the applied statnamic force as input, a simple spreadsheet computation of displacement vs time can be computed and compared with the measured displacements. Once a suitable match is obtained, the model may be used to estimate the nonlinear static response or the foundation response to other types of dynamic loadings. For this test, the match shown on Figure 6 was obtained using $m_e = 3.26$, D = 33%, $k_i = 75$ MN/m and $\alpha = 0.25$. The match is observed to be most sensitive to static stiffness. Note that in Figure 6 the measured displacement response indicates some permanent set which is not present in the model; the model does not include a provision for nonlinear unloading and permanent set, so this deviation is not considered particularly important. The unloading stiffness for this simple model was taken as the stiffness at the maximum displacement, so the hysteresis in the unloading response is not simulated.

With the derived computational model, the individual force components contributing to resistance can be estimated as illustrated on Figure 7. Some differing

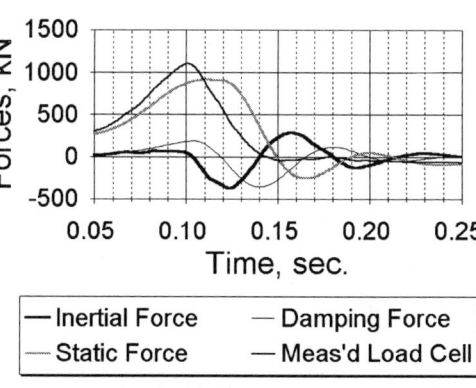

— Inertial Force   — Damping Force
— Static Force   — Meas'd Load Cell

Figure 7 Computed Static and Dynamic Forces

matches were developed using the data presented in Figure 6 in order to evaluate sensitivity to the input parameters. A fairly reasonable match can be achieved with slightly different values of damping resistance and effective mass, but in no case did this result in substantial differences in the derived static forces. The derived forces are presented as a function of displacement (in a manner similar to a static loading test) on

Figure 8. This figure graphically illustrates the nonlinearity in the static component of the resistance, and the relatively small contributions of inertia and damping.

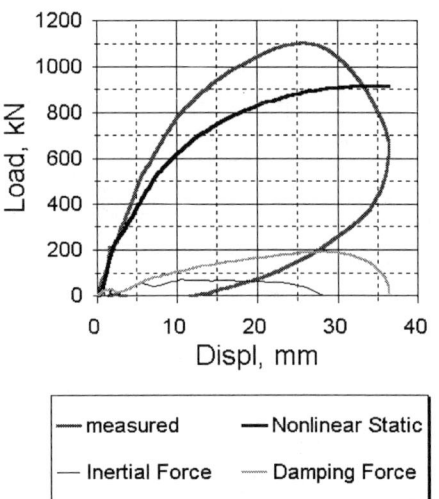

Comparison with Conventional Static Tests

Five shafts which were similar to the statnamic test shaft were subjected to static lateral loading tests. These shafts exhibited a significant variability, illustrating the sensitivity of laterally loaded deep foundations to variations in near-surface soil characteristics. The load - displacement response for the five static tests are shown on Figure 9, along

Figure 8 Derived Force - Displacement Response

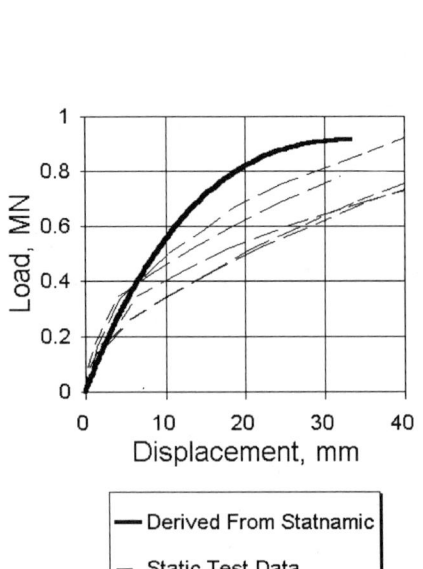

Figure 9 Static & Derived Load vs Displ

with the derived nonlinear static component of force from the statnamic test shaft.

These data reveal that, besides the inherent variability in the lateral stiffness at the site, there appears to be a significant rate effect for this foundation even after accounting for inertia and damping with this simple model. The initial stiffness at small displacements match fairly well, and it appears that the forces achieving large displacements trend toward a consistent relationship. Within the 10 to 30 mm displacements (1 to 3% of the shaft diameter), the stiffness derived from statnamic testing is greater. A large portion of this difference may be attributed to rate effects in the reinforced concrete as this is the region in which cracking and nonlinear bending response (EI) in the shafts occur. According to data presented by Tedesco, McDougal and Ross (1999), the tensile strength of the concrete at strain rates typical of those during the statnamic test is on the order of 2 to 2.5 times higher than for static testing. This increased stiffness (EI) in the test shaft may account for much of the difference between the statnamic and static loading data presented on Figure 9. Rate effects in soil shearing resistance may also contribute. These data also suggest that for transient loadings in which rate effects might be important, designers using simple static p-y models for design may be conservative as there may be additional resistance unaccounted for in a model based on simple static load tests.

Future Research

Additional research with statnamic lateral testing is planned, with several key elements identified:

1)   repeated lateral statnamic testing with several progressively increasing load levels may help to construct a nonlinear load vs displacement relationship with greater reliability in the derived forces across a range of displacement amplitudes,

2)   lateral statnamic testing on linear elastic steel piles is planned, and these tests will eliminate the nonlinearity inherent in reinforced concrete members in bending and allow a more focused attention on soil nonlinearity,

3)   the lateral statnamic test data will be analyzed using dynamic p-y curves in which the variation of soil resistance with depth can be accounted for and the additional damping resistance in the soil response "calibrated" with large scale and large amplitude dynamic test results,

4)   repeated lateral statnamic testing at constant load levels so as to obtain data which can be used to calibrate the gapping around the piles and hysteresis with dynamic p-y curves

5)   testing of groups of steel piles in which dynamic pile-soil-pile interaction is a factor.

Acknowledgements

The author wishes to acknowledge the contributions to this research of the Morris-Shea Bridge Company, the Alabama DOT, Applied Foundation Testing, Inc., the Federal Highway Administration, Professor Gray Mullins of the University of South Florida, and the Auburn University Highway Research Center. The author wishes to express a special appreciation to Professor Lymon Reese for his many years of help and guidance and interest in lateral load testing and in who's honor this paper is presented.

References

Tedesco, J.W., W.G. McDougal, and C.A. Ross, 1999. <u>Structural Dynamics, Theory and Applications</u>, Addison Wesley Longman, Inc., Menlo Park, Calif., 869 p.

Vinson, J.L. and D.A. Brown, 1997. "Site Characterization of the Spring Villa Geotechnical Test Site and a Comparison of Strength and Stiffness Parameters for a Piedmont Residual Soil," Highway Research Center Report No. IR-97-04, Nov., 1997, 385pp.

Subject Index

Page number refers to the first page of paper

Author Index

Page number refers to the first page of paper